W9-BCC-282

Statistical Techniques in Simulation

(IN TWO PARTS)

PART II

STATISTICS

Textbooks and Monographs

A SERIES EDITED BY

D. B. OWEN, Coordinating Editor

Department of Statistics
Southern Methodist University
Dallas, Texas

PETER LEWIS
Naval Postgraduate School
Monterey, California

PAUL D. MINTON
Virginia Commonwealth University
Richmond, Virginia

JOHN W. PRATT
Harvard University
Boston, Massachusetts

VOLUME 1: The Generalized Jackknife Statistic
H. L. Gray and W. R. Schucany

VOLUME 2: Multivariate Analysis
Anant M. Kshirsagar

VOLUME 3: Statistics and Society
Walter T. Federer

VOLUME 4: Multivariate Analysis: A Selected and Abstracted Bibliography, 1957-1972
Kocherlakota Subrahmaniam and Kathleen Subrahmaniam

VOLUME 5: Design of Experiments: A Realistic Approach
Virgil L. Anderson and Robert A. McLean

VOLUME 6: Statistical and Mathematical Aspects of Pollution Problems
edited by John W. Pratt

VOLUME 7: Introduction to Probability and Statistics (in two parts). Part I: Probability. Part II: Statistics
Narayan C. Giri

VOLUME 8: Statistical Theory of the Analysis of Experimental Designs
J. Ogawa

VOLUME 9: Statistical Techniques in Simulation (in two parts)
Jack P. C. Kleijnen

VOLUME 10: Data Quality Control and Editing
Joseph I. Naus

OTHER VOLUMES IN PREPARATION

Statistical Techniques in Simulation

(IN TWO PARTS)

JACK P. C. KLEIJNEN
Katholieke Hogeschool
Tilburg, The Netherlands

PART II

MARCEL DEKKER, INC. New York

COPYRIGHT © 1975 by MARCEL DEKKER, INC. ALL RIGHTS RESERVED.

Neither this book nor any part may be reproduced or transmitted in any form or by any means, electronic or mechanical, including photocopying, microfilming, and recording, or by any information storage and retrieval system, without permission in writing from the publisher.

MARCEL DEKKER, INC.
270 Madison Avenue, New York, New York 10016

LIBRARY OF CONGRESS CATALOG CARD NUMBER: 74-79920
ISBN: 0-8247-6243-6

Current printing (last digit):
10 9 8 7 6 5 4 3 2 1

PRINTED IN THE UNITED STATES OF AMERICA

MATH.-SCI.
1561115
QA
279
.K55
v.2

To Mai and Wilma

CONTENTS OF PART II

CONTENTS OF PART I

PREFACE

This book discusses the statistical design and analysis of simulation and Monte Carlo experiments, with emphasis on digital simulation with models of management and economic systems. Simulation is always a sampling experiment whenever the model contains one or more stochastic variables (although it is a special type of sampling experiment since simulation is performed on an abstract model instead of a real-life object). In any statistical experiment a careful design and analysis is desirable. The analysis should extract as much information from the experiment as possible and, moreover, should reveal the limitations of conclusions based on sampling data. The design should ensure that the experiment contains as much information as possible.

Though simulation is a sampling experiment the technique is usually not applied by statisticians but by engineers, social scientists, etc. (since simulation also requires model building). A thorough statistical design and analysis has been lacking in most simulation studies. It is our hope that this book will make the users of simulation more aware of the statistical aspects of simulation. The purpose of our book is to give the reader a working knowledge of the statistical techniques involved. Because so many techniques are relevant we had to restrict attention to a selected number of topics (but Chapter II gives an overall picture of the statistical aspects of simulation). We have further tried to make all chapters independent of one another, at least to a high degree. In this way the practitioner interested in a specific problem (e.g., runlength determination) may read only the relevant part of the book. While we aim at working knowledge of certain techniques, references are given for further study of related topics. (Each chapter ends with a long list of references; many chapters have a literature section which recommends particular references as a guide for additional study. We would appreciate very much if the reader informed us of any relevant literature omitted by us. We especially welcome references to publications that have recently become available.) To read this book only a "basic" knowledge of mathematical probability and statistics is required. We conjecture that most management scientists and economists do have knowledge of dependence of events, t-tests, simple regression analysis, etc.

Before proceeding to a brief survey of the book we point out that each chapter begins with a section that gives a more detailed summary. Chapter I describes systems, models, and Monte Carlo and simulation techniques. It further discusses random number generation and sampling of stochastic variables. Its purpose is not to present new ideas but to define basic concepts. It also contains appendices with simple applications of the Monte Carlo and simulation methods. Chapter II gives a survey of the statistical aspects of simulation. It further shows how the following chapters fit into the total picture. Chapter III presents variance reduction techniques that may have wide applicability in complicated simulation experiments: stratification, selective sampling, control variates, importance sampling, antithetic variates, and common random numbers. This chapter results in extensions, limitations, and corrections of existing techniques. It also contains an algorithm for the optimal allocation of computer time when applying antithetic and common random numbers. Chapter IV presents experimental designs relevant in simulation. A detailed discussion is given of general factorial experiments (and their analysis of variance and regression analysis), 2^{k-p} designs, and random, supersaturated, and group-screening designs. It further contains a bibliography on response surface methodology. Chapter V covers the effect of the sample size on the reliability of statements based on a sample. It consists of three independent parts. Part A discusses the evaluation of a single population or the comparison of only two populations (or systems). The sample size may be fixed or may be so selected that a predetermined reliability results. Part B covers multiple comparison procedures, i.e., the sample sizes are fixed and k (≥ 2) systems are compared with one another. It also discusses various types of error rates (e.g., the experimentwise rate), selection of a subset containing the best population, and efficiency and robustness of the procedures in simulation. Part C gives multiple ranking procedures, i.e., how many observations should be taken from each of the k (≥ 2) systems in order to select the best system. (The "indifference zone" approach is followed.) Again the efficiency and robustness of these procedures are discussed. Finally, Chapter VI presents a case study demonstrating how various techniques of the preceding chapters can be applied. The case study concerns a Monte Carlo experiment with Bechhofer and Blumenthal's multiple ranking procedure, and investigates the robustness of this procedure. All chapters contain a number of simple exercises and references to the literature for additional exercises. (Note further that stochastic variables are underlined in this book to distinguish them from deterministic quantities.)

Jack P. C. Kleijnen

ACKNOWLEDGMENTS

It is a pleasure to thank the many persons who made this publication possible. Since I began working as a research associate at the Katholieke Hogeschool in Tilburg (Tilburg School of Economics and Business Administration) Professor M. Euwe and later on Professor G. Nielen have given me ample time to engage in the research of the statistical aspects of simulation.

The Netherlands Organization for the Advancement of Pure Research (Z.W.O.) and the Board of the Katholieke Hogeschool enabled me to do research at the University of California at Los Angeles during the academic year 1967-1968. The Katholieke Hogeschool and Professor T. Naylor made it possible for me to visit Duke University during the summer of 1969.

My discussions with Professor H. Lombaers (Technische Hogeschool Delft) resulted in many adaptations of the original Chapter I. The following persons gave their useful comments on parts of the original papers on which this book is based: Professor J. Kriens, Professor D. Neeleman, Mr. H. Tilborghs, Mr. A. van Reeken, Dr. C. Weddepohl (all at the Katholieke Hogeschool), Professor R. Nelson and Professor J. MacQueen (University of California), Professor D. Graham and Professor T. Naylor (Duke University), Professor T. Wonnacott (Western Ontario University), Professor K. Gabriel (Hebrew University), Professor E. Dudewicz (University of Rochester), and Professor P.A.W. Lewis (Naval Postgraduate School). It stands to reason that any errors remain my sole responsibility.

The camera-ready copy was typed by Miss Rosemarie Stampfel of Redwood City, California. The proofreading was done while the author was a Postdoctoral Fellow at the IBM Research Laboratory in San Jose, California.

Statistical Techniques in Simulation

(IN TWO PARTS)

PART II

Chapter IV

THE DESIGN AND ANALYSIS OF EXPERIMENTS

IV.1. INTRODUCTION AND SUMMARY

In this chapter we shall present some experimental designs and their analyses in detail. Of course there are standard textbooks on the design and analysis of experiments like Cochran and Cox (1957) and Davies (1963). However, whereas these two books were revised in 1957 and 1956, respectively, and therefore do not cover recent developments, this chapter is primarily based on articles published after 1960. More recent textbooks are, e.g. John (1971), Mendenhall (1968) and Peng (1967). But these textbooks and other books all have the disadvantage that they do not concentrate on the special kind of experiments that is of interest to us, namely simulation experiments. Therefore such books contain large sections on randomization, blocking, etc., i.e., techniques needed because of incomplete control of the experimental conditions in the usual industrial and agricultural experiments. Owing to space limitations they cannot cover in detail experimental designs relevant for simulation. In this chapter we have eliminated all topics of experimental design that are of little use in simulation experiments.

We just remarked that there are no textbooks tailored to the need of the researcher performing simulation experiments. Therefore we might be tempted to direct the researcher to the articles in the professional journals. Unfortunately a vast literature on experimental design exists; Herzberg and Cox (1969), e.g., list 800 articles published since 1958. Hence, the time it would take the individual researcher to select pertinent publications may very well be prohibitive. We have tried to make such a selection for him.[1)] We selected relevant publications, ordered them and present them here in a simple and uniform way.

The factors in an experiment may be quantitative or qualitative. Many qualitative factors can be quantified, for instance, intelligence is quantified by I.Q., queuing rules by a procedure described by Miller (1968), and distribution type by variation of the parameters of a family of distributions. Nevertheless, as Nauta (1967, p. 75) claims, most industrial simulations involve at least one qualitative factor, e.g., a policy rule. Therefore we shall only briefly discuss so-called response surface methodology which can be applied if all factors are quantitative (but at the end of this chapter we do give a bibliography on this methodology). We shall concentrate on designs that can be used when either all factors are qualitative, all are quantitative, or some are qualitative and some are quantitative.

First we present the general model with interactions that is used in factorial designs. Analysis of variance (or briefly ANOVA) is applied to analyze the results of the factorial experiment. The relationship between ANOVA and regression analysis is shown. Randomization and blocking in simulation are discussed. The ANOVA assumptions, transformations, and coding are examined. The next section covers special factorial designs, namely designs where all factors have only two values. The model for these "2^k designs" is given, together with the analysis of the observations. The following section shows how only a fraction of the full 2^k design can be executed while all important information is still obtained. We demonstrate how a particular pattern of confounding of effects can be realized. We give designs for models with only main effects, designs for estimating main effects even while interactions exist, and designs for estimating both main effects and two-factor interactions (so-called designs of resolution III, IV, and V, respectively). In the next section it is shown how an independent estimate of the experimental error variance σ^2 is obtained by partial duplication of the design. A procedure is presented for re-estimating effects using the additional information of the duplicated design. Instead of duplicating observations we can pool the sums of squares of certain effects. Both procedures can be combined to test the fit of the model. If the model is found not to fit we may proceed to a higher-order model. It is shown that the designs presented in this chapter are easily augmented to higher-order designs (so-called sequentialization of designs). In the next section designs are given for finding the few important factors among many conceivably important factors; so-called screening. The interpretation of fractional factorials if some factors are found not to be important, is discussed. Also designs with randomly selected factor combinations and their analysis are presented. Systematic (i.e., nonrandom) designs with fewer observations than effects are given; so-called supersaturated designs. Then we present several variants of fractional factorials where factors are grouped together in order to reduce the number of factors and observations; the assumptions of these group-screening designs are investigated and found not to be restrictive. The four types of screening-designs are compared with each other. The chapter is concluded with a brief discussion of decision theory and multiple responses. Literature for these two aspects and several more aspects is given.

IV.2. GENERAL FACTORIAL DESIGNS AND THEIR ANALYSIS

We want to consider experiments where the influence of more than one factor is investigated. The traditional method was to change only one factor at a time, the so-called ceteris-paribus method. An alternative method is provided by a factorial design, i.e., all levels of a factor are combined with all levels of each other factor. Fisher (1966, p. 97, 100) showed that factorial experiments are more efficient since they yield more reliable estimates of the (main) effects of the factors, and moreover they give estimates of the interactions among the factors. (We shall give exact definitions of main effects and interactions below.[2]) Examples illustrating the differences between the two methods can be found in Fisher (1966, pp.95-101) and Hicks (1966, pp.75-77). Recently Webb derived some special one-factor-at-a-time designs, but he admits that "these designs do not provide a statistical proof of the reality of effects;" compare Webb (1968b, p. 549). We point out that in the social sciences factorial designs have not been used extensively since it is difficult in these disciplines to combine all factor levels with each other, keeping all other circumstances (i.e., factors not studied in the experiment) constant. In the simulation of social systems, however, we have all factors under control and can combine their levels as we like.

Above we defined a <u>factorial experiment</u> as a design in which all levels of a factor are combined with all levels of each other factor. The various levels of a factor may correspond with qualitative differences (like different queuing disciplines) or quantitative differences (like different number of service stations). If factor f $(f = 1, \ldots, k)$ has L_f levels, then the total number of combinations of levels is[3]

$$L_1 L_2 \cdots L_k = \prod_{f=1}^{k} L_f \tag{1}$$

Hence if the number of levels is the same for each factor, say L, then the total number of levels is L^k. The expression in the left-hand side of Eq. (1) is also used to denote the type of factorial design. For instance, if one factor is at two levels and two factors at three levels this is denoted as a 2×3^2 design.

Consider an example of a factorial experiment with two factors: one factor at two levels, the other factor at three levels; there are two observations per combination of levels. The experimental results are shown in Table 1. From Table 1 it follows that $\underline{y}_{ijg}$ denotes observation g $(g = 1, 2)$ in cell i, j; in this cell factor A is at level i $(i = 1, 2, 3)$ and factor B is at level j $(j = 1, 2)$. Denote the expected value of $\underline{y}_{ijg}$ by η_{ij}. In the design of experiments it is assumed that the following <u>model</u> holds:

$$\underline{y}_{ijg} = \eta_{ij} + \underline{e}_{ijg} \quad (i = 1, \ldots, I)\ (j = 1, \ldots, J)\ (g = 1,2,\ldots) \tag{2}$$

TABLE 1

An Example of a Factorial Design

Factor A	Factor B: Level 1	Factor B: Level 2
Level 1	$\underline{y}_{111}$	$\underline{y}_{121}$
	$\underline{y}_{112}$	$\underline{y}_{122}$
Level 2	$\underline{y}_{211}$	$\underline{y}_{221}$
	$\underline{y}_{212}$	$\underline{y}_{222}$
Level 3	$\underline{y}_{311}$	$\underline{y}_{321}$
	$\underline{y}_{312}$	$\underline{y}_{322}$

where $\underline{e}_{ijg}$, the experimental <u>error</u> (or noise or disturbance), is assumed to be normally and independently distributed with zero-mean and constant variance σ^2; in shorthand notation

$$\underline{e}_{ijk}:\text{NID}(0, \sigma^2) \tag{3}$$

The general or <u>grand mean</u> μ is defined as

$$\mu = \frac{\Sigma_i^I \Sigma_j^J \eta_{ij}}{IJ} = \eta_{..} \tag{4}$$

where a dot means that we have averaged over all values of the corresponding index. If we average the response where A is at level i over all levels of B, then we obtain A_i, i.e.,

$$A_i = \frac{\Sigma_{j=1}^J \eta_{ij}}{J} = \eta_{i.} \tag{5}$$

Then α_i^A, the <u>main effect</u> of factor A at level i, is defined as the excess of this average over the general mean, i.e.,

$$\alpha_i^A = A_i - \mu = \eta_{i.} - \eta_{..} \tag{6}$$

From Eqs. (4) through (6) it follows that the average main effect is zero since

$$\sum_{i=1}^{I} \alpha_i^A = \sum_i \sum_j \eta_{ij}/J - \sum_i \mu = I\mu - I\mu = 0 \tag{7}$$

The main effect of factor B at level j is defined as

$$\alpha_j^B = B_j - \mu = \sum_i \eta_{ij}/I - \mu = \eta_{.j} - \eta_{..} \tag{8}$$

so

$$\sum_j \alpha_j^B = 0 \tag{9}$$

If we assume that no interaction exists then the model equation used in the design of experiments is

$$E(\underline{y}_{ijg}) = \eta_{ij} = \mu + \alpha_i^A + \alpha_j^B \tag{10}$$

From Eq. (10) it follows that, for instance,

$$\eta_{i1} - \eta_{i2} = \alpha_1^B - \alpha_2^B \tag{11}$$

holds for <u>all</u> levels i of factor A. This corresponds with a picture with <u>parallel</u> response curves as in Fig. 1. However, when interaction between factor A and B does exist then a change in factor A

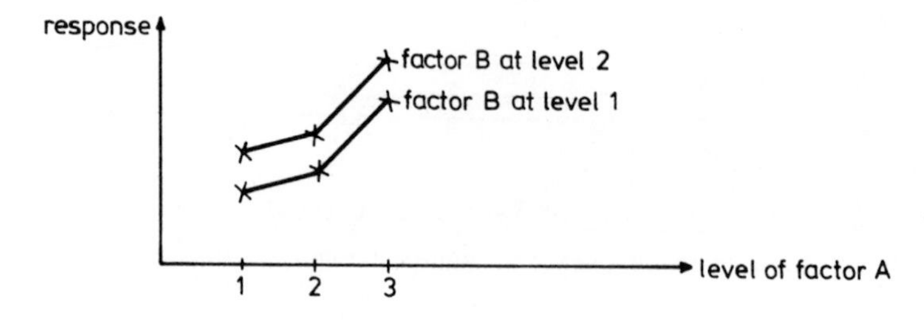

Fig. 1. Response curves if there is no interaction.

yields a change in the response variable at one level of factor B different from that at another level of factor B. The interaction when factor A is at level i and B is at level j, is defined as

$$\alpha_{ij}^{AB} = \eta_{ij} - A_i - B_j + \mu = \eta_{ij} - \eta_{i.} - \eta_{.j} + \eta_{..} \tag{12}$$

Note that as in Eqs. (7) and (9) we have $\alpha_{.j}^{AB} = \alpha_{i.}^{AB} = 0$. The upper index denotes the factors that are interacting and the lower index shows the particular levels for which the effect is defined. The general model used in a factorial design with two factors is[4)]

$$E(\underline{y}_{ijg}) = \mu + \alpha_i^A + \alpha_j^B + \alpha_{ij}^{AB} \tag{13}$$

and this replaces the more specific model given in Eq. (10).

If there are more than two factors, then we can distinguish interactions between two factors, interactions among three factors, etc. The model equation for a factorial experiment involving three factors is

$$\underline{y}_{ijkg} = \eta_{ijk} + \underline{e}_{ijkg} \tag{14}$$

with

$$\eta_{ijk} = \mu + \alpha_i^A + \alpha_j^B + \alpha_k^C + \alpha_{ij}^{AB} + \alpha_{ik}^{AC} + \alpha_{jk}^{BC} + \alpha_{ijk}^{ABC} \tag{15}$$

and

$$\underline{e}_{ijkg} : \mathrm{NID}(0, \sigma^2) \tag{16}$$

In Eq. (15) we distinguish the grand mean

$$\mu = (IJK)^{-1} \sum_i \sum_j \sum_k \eta_{ijk} = \eta_{...} \tag{17}$$

the main effects

$$\alpha_i^A = (JK)^{-1} \sum_j \sum_k \eta_{ijk} - \mu = \eta_{i..} - \eta_{...} \tag{18}$$

$$\alpha_j^B = (IK)^{-1} \sum_i \sum_k \eta_{ijk} - \mu = \eta_{.j.} - \eta_{...} \tag{19}$$

$$\alpha_k^C = (IJ)^{-1} \sum_i \sum_j \eta_{ijk} - \mu = \eta_{..k} - \eta_{...} \tag{20}$$

the two-factor interactions

$$\alpha_{ij}^{AB} = \eta_{ij.} - \eta_{i..} - \eta_{.j.} + \eta_{...} \tag{21}$$

$$\alpha_{ik}^{AC} = \eta_{i.k} - \eta_{i..} - \eta_{..k} + \eta_{...} \tag{22}$$

$$\alpha_{jk}^{BC} = \eta_{.jk} - \eta_{.j.} - \eta_{..k} + \eta_{...} \tag{23}$$

and the three-factor interaction

$$\alpha_{ijk}^{ABC} = \eta_{ijk} - \eta_{ij.} - \eta_{i.k} - \eta_{.jk} + \eta_{i..} + \eta_{.j.} + \eta_{..k} - \eta_{...} \tag{24}$$

More about the rationale of the definitions of the interactions in a three-factor experiment can be found in Scheffé (1964, pp. 119-121). Notice that each effect when averaged over one of its subscripts is zero, i.e.

$$\alpha_.^A = \alpha_.^B = \alpha_.^C = 0 \tag{25}$$

$$\alpha_{i.}^{AB} = \alpha_{.j}^{AB} = \alpha_{i.}^{AC} = \alpha_{.k}^{AC} = \alpha_{j.}^{BC} = \alpha_{.k}^{BC} = 0 \tag{26}$$

$$\alpha_{.jk}^{ABC} = \alpha_{i.k}^{ABC} = \alpha_{ij.}^{ABC} = 0 \tag{27}$$

For the general case of k <u>factors</u> the model involves one grand mean and

$$C_1^k = \frac{k!}{1!\,(k-1)!} = k \qquad \text{main effects}$$

$$C_2^k = \frac{k!}{2!\,(k-2)!} = \frac{k(k-1)}{2} \qquad \text{two-factor interactions}$$

$$\vdots \qquad\qquad\qquad \vdots$$

$$C_q^k = \frac{k!}{q!\,(k-q)!} \qquad \text{q-factor interactions}$$

$$\vdots \qquad\qquad\qquad \vdots$$

$$C_k^k = \frac{k!}{k!\,0!} = 1 \qquad \text{k-factor interactions} \tag{28}$$

The formal definition of the q-factor interaction is given in Scheffé (1964, p. 124).

The observations obtained in a factorial experiment can be used to test the significance of the main effects and interactions. This is done by the well-known analysis of variance. The simplest case where ANOVA is applied is an experiment with a single factor at J levels and I observations per level; compare Table 2.

TABLE 2

One-Factor-Experiment with I Observations per Level

		Factor level	
		1 ... j ... J	
Observation	1	$\underline{y}_{11} \cdots \underline{y}_{1j} \cdots \underline{y}_{1J}$	
	$\vdots$		
	i	$\underline{y}_{i1} \cdots \underline{y}_{ij} \cdots \underline{y}_{iJ}$	
	$\vdots$		
	I	$\underline{y}_{I1} \cdots \underline{y}_{Ij} \cdots \underline{y}_{Ij}$	
Average		$y_{.1} \cdots \underline{y}_{.j} \cdots \underline{y}_{.J}$	$\underline{y}_{..}$

In Appendix IV.1 we derive the following well-known result:

$$\sum_i \sum_j (\underline{y}_{ij} - \underline{y}_{..})^2 = \sum_i \sum_j (\underline{y}_{.j} - \underline{y}_{..})^2 + \sum_i \sum_j (\underline{y}_{ij} - \underline{y}_{.j})^2 \qquad (29)$$

The partitioning of the sum of squares of the left-hand side in Eq. (29) into two parts is the basic idea of ANOVA. (If there are several factors then there will be more than two sum-terms in the right-hand side.) The analysis is an analysis of variances because, as we show in Appendix IV.1, each sum-term in the right-hand side (or in its generalized equivalent for more than one factor) leads to an independent estimator of the error variance σ^2, provided the factor has no influence (or for more than one factor, provided the main effects and interactions are zero). To obtain these estimates of σ^2 we divide the sums of squares by their appropriate degrees of freedom and this yields the mean squares in Table 3. If the factor does influence the response, then the expected value of the "within levels" mean square remains equal to σ^2, whereas the "between levels" mean square can be shown to have expectation

TABLE 3

Analysis of Variance for One Factor with I Observations per Level

Source	Sum of squares $\underline{SS}$	Degrees of freedom df	Mean Square $\underline{MS}$
Between levels	$\sum_i \sum_j (\underline{y}_{.j} - \underline{y}_{..})^2$	J - 1	$\underline{S}S$/df
Within levels	$\sum_i \sum_j (\underline{y}_{ij} - \underline{y}_{.j})^2$	J(I - 1)	$\underline{S}S$/df
	...	...	
Total about grand average	$\sum_i \sum_j (\underline{y}_{ij} - \underline{y}_{..})^2$	IJ - 1	$\underline{S}S$/df

$$E\left[\sum_i \sum_j (\underline{y}_{.j} - \underline{y}_{..})^2/(J - 1)\right] = \sigma^2 + (J - 1)^{-1} I \sum_j \alpha_j^2 \tag{30}$$

where α_j denotes the main effect of the factor at level j. The statistic to test the infleunce of the factor should now be obvious. Under the hypothesis of no factor influence the ratio of the two mean squares

$$\underline{F}_{J-1,J(I-1)} = \frac{\underline{MS} \text{ between levels}}{\underline{MS} \text{ within levels}} \tag{31}$$

is the ratio of two independent estimators of σ^2, i.e., the ratio of two independent χ^2-distributed stochastic variables with degrees of freedom $(J - 1)$ and $J(I - 1)$, respectively, and this ratio has the F-distribution which is tabulated. If the factor does have influence, then Eq. (30) shows that the numerator in relation (31) increases and this makes the F-statistic significant. So high values of F lead to rejection of the hypothesis of no factor influence.

ANOVA can also be used for <u>more than one factor</u>. In that case the total sum of squares

$$\sum_i \sum_j \cdots \sum_k (\underline{y}_{ij\ldots k} - \underline{y}_{\ldots})^2 \tag{32}$$

can be partitioned into several independent sums of squares. Dividing these sums of squares by their corresponding degrees of freedom gives mean squares that are unbiased estimators of the quantities in relation (33), where $C_A, C_B, \ldots, C_{ABC}$ denote known positive constants. [In Eq. (30), e.g. $C_B = (J - 1)^{-1} I$ and all other constants vanish.]

$$\left.\begin{array}{l} \sigma^2 \\ \sigma^2 + C_A \sum_i (\alpha_i^A)^2 \\ \sigma^2 + C_B \sum_j (\alpha_j^B)^2 \\ \vdots \end{array}\right\} \text{main effects}$$

$$\left.\begin{array}{l} \sigma^2 + c_{AB} \sum_i \sum_j (\alpha_{ij}^{AB})^2 \\ \vdots \end{array}\right\} \text{two-factor interactions}$$

$$\sigma^2 + c_{ABC} \sum_i \sum_j \sum_k (\alpha_{ijk}^{ABC})^2$$

$$\vdots$$

$$\text{and so on} \tag{33}$$

To test whether a particular effect (main effect or interaction) is significant, we simply calculate the ratio of the corresponding mean square and the mean square for pure error and compare this ratio with the significance level found in the F-table. To test if both A <u>and</u> B have no main effect we sum or "pool" their sums of squares. Since the individual sums of squares are independent their sum is still χ^2-distributed with the number of degrees of freedom being equal to the sum of the individual degrees of freedom. Hence, we divide the pooled sums of squares by their pooled degrees of freedom, and this yields a mean square. This mean square divided by the mean square for error is F-distributed, if at least the hypothesis of no A and B main effects holds. Details of these calculations can be found in, e.g., Hicks (1966) and Scheffé (1964). Computer programs for ANOVA of factorial experiments are given, e.g. by Beaton (1969), Bock (1963), Fowlkes (1969), and Peng (1967, pp. 219-230). In the following sections we shall give computation formulas for the special factorials that are of interest to us.

ANOVA has been applied in a few <u>simulation</u> and <u>Monte Carlo</u> studies. For example, in their simulation experiments with the Samuelson-Hicks model of the National Economy and with the model of a firm Naylor et al. (1967b and 1968) applied ANOVA to an experiment with one factor at five levels. Jensen (1967) gives an elaborated example of the ANOVA for a 2^4 experiment with a simulated accounting system. Kaczka and Kirk (1967) used a 2^5 design in their simulated organizational model. Sasser et al. (1970) describe a ANOVA for a $5^2 \times 2$ experiment with a multi-item inventory simulation.[5)] Emshoff and Sisson (1971, pp. 211-214) discuss a 2^2 experiment with a maintenance model. Sasser (1969) and Schink and Chiu (1966) used ANOVA in a 3^2 Monte Carlo experiment on the performance of several regression analysis techniques. Note that Balderston and Hoggatt (1962, pp. 119-120) applied nonparametric ANOVA to their simulated market processes.

If a factor is found to be significant then the levels of the factor can be ranked in accordance with the average response obtained for the corresponding level. Confidence intervals can be determined for the differences among the responses for different levels. A survey of these so-called <u>multiple comparison</u> procedures will be given in the next chapter.

The models used in factorial designs can also be expressed as the linear models familiar from <u>regression analysis</u>. Since social scientists are much more familiar with regression analysis than ANOVA, we shall next discuss the relations between both techniques in some detail. As an example we consider the two-factor model for the design in Table 1. For simplicity of presentation we shall assume that there are no interactions and no replications. From the model in Eqs. (2) and (10) it follows that

$$\begin{aligned} \underline{y}_{11} &= \mu + \alpha_1^A + \alpha_1^B + \underline{e}_{11} \\ \underline{y}_{12} &= \mu + \alpha_1^A + \alpha_2^B + \underline{e}_{12} \\ &\vdots \\ \underline{y}_{32} &= \mu + \alpha_3^A + \alpha_2^B + \underline{e}_{32} \end{aligned} \tag{34}$$

or in matrix notation[6)]

$$\vec{\underline{y}} = \vec{X}\vec{\beta} + \vec{\underline{e}} \tag{35}$$

where $\vec{\underline{y}}$ is the vector of observations

$$\vec{\underline{y}}\,' = (\underline{y}_{11}, \underline{y}_{12}, \cdots, \underline{y}_{32}) \tag{36}$$

$\vec{X}$ is the matrix of explanatory or independent ("dummy") variables

$$X = \left[\begin{array}{c|ccc|cc} 1 & 1 & 0 & 0 & 1 & 0 \\ 1 & 1 & 0 & 0 & 0 & 1 \\ 1 & 0 & 1 & 0 & 1 & 0 \\ 1 & 0 & 1 & 0 & 0 & 1 \\ 1 & 0 & 0 & 1 & 1 & 0 \\ 1 & 0 & 0 & 1 & 0 & 1 \end{array}\right] \qquad (37)$$

the dashed lines partitioning $\vec{X}$ such that column 1 corresponds with μ, columns 2 through 4 with α_i^A $(i = 1, 2, 3)$ and columns 5 and 6 with α_j^B $(j = 1, 2)$; $\vec{\beta}$ is the vector of effects or parameters

$$\vec{\beta}' = (\mu, \alpha_1^A, \alpha_2^A, \alpha_3^A, \alpha_1^B, \alpha_2^B) \qquad (38)$$

and $\vec{\underline{e}}$ is the vector of "errors"

$$\vec{\underline{e}}' = (\underline{e}_{11}, \underline{e}_{12}, \ldots, \underline{e}_{32}) \qquad (39)$$

$\underline{e}_{ij}$ satisfying Eq. (3). The matrix $\vec{X}$ is not of full rank since, e.g. summing the columns 2 through 4 or the columns 5 and 6 yields column 1. It can be shown that the rank of $\vec{X}$ is four. From Eqs. (7) and (9) two side conditions follow, namely

$$\alpha_1^A + \alpha_2^A + \alpha_3^A = 0 \qquad (40)$$

and

$$\alpha_1^B + \alpha_2^B = 0 \qquad (41)$$

These two restrictions and the so-called normal equations

$$\vec{X}' \vec{y} = \vec{X}' \vec{X} \vec{\beta} \qquad (42)$$

following from Eq. (35), yield unique least squares estimators; see e.g. John (1971, pp. 26-28). It is a well-known result in regression analysis that these least squares estimators are also maximum likelihood estimators and minimum variance linear unbiased estimators if Eq. (3) above holds; compare Johnston (1963, pp. 108-116).

It can be found in, e.g. Peng (1967, pp. 175-179) and also in Scheffé (1964, pp. 98-102) that working out Eqs. (40) through (42) yields the following estimators, which are well-known from ANOVA (and intuitively acceptable).

$$\hat{\underline{\mu}} = \underline{y}_{..} \tag{43}$$

$$\hat{\underline{\alpha}}_i^A = \underline{y}_i - \underline{y}_{..} \qquad (i = 1, 2, 3) \tag{44}$$

$$\hat{\underline{\alpha}}_j^B = \underline{y}_{.j} - \underline{y}_{..} \qquad (j = 1, 2) \tag{45}$$

We can further test whether certain parameters in $\vec{\beta}$ are zero. Testing formulas are given in, e.g. Johnston (1963, pp. 115-126), Peng (1967, pp. 173-175) and Scheffé (1964, pp. 25-45). Working out these formulas for the particular models in factorial designs yields the familiar ANOVA sums of squares tables like Table 3 above; compare, e.g. Peng (1967, pp. 178-179). For a comparison of ANOVA and regression analysis we also refer to Seeger (1966, pp. 6-13) and Smith (1969).

Summarizing, the models for factorial designs are special cases of the general linear regression model. The vector of parameters $\vec{\beta}$ consists of the grand mean, the main effects, and the interactions; the matrix of independent variables $\vec{X}$ shows only ones and zeros. (In Section IV.3 we shall see a slightly different formulation, where $\vec{X}$ has only plus and minus ones.) The experiment is designed, i.e., $\vec{X}$ is selected, such that the estimators have certain desirable properties to which we shall return later on.

ANOVA is suitable for both qualitative and quantitative factors. If all factors are quantitative we can use ANOVA to test if a factor has effect, without specifying (in the form of a regression curve) how the response reacts to the factor over the whole experimenthal region. If we want to estimate the response for some point in the experimental region then a regression curve is more useful; compare Fig. 2(a) and (b).

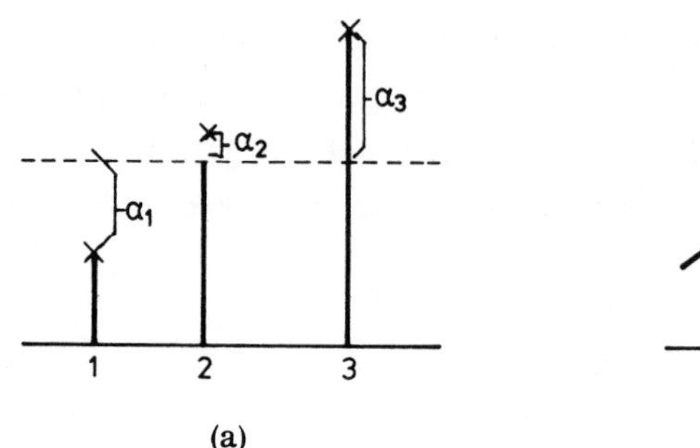

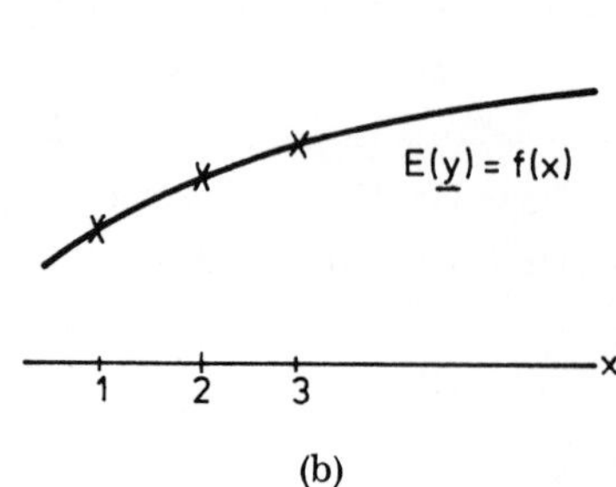

Fig. 2. (a) ANOVA-representation; (b) regression-representation.

For the reader who is more familiar with ANOVA, we remark that various problems treated in the literature on factorial experiments will not be covered here. For instance, we consider only fixed effect models as in the following sections the levels of the factors are deliberately fixed and not randomly sampled. Further, randomization and blocks are introduced in the literature because the experimental conditions cannot be fully controlled (compare different fertility of soil in agricultural experiments; material from different batches in industrial experiments; time trends). In simulation, however, all conditions are under control; pure experimental error is generated by using random numbers. Or as Naylor et al. (1967a, p. 323) state, "the role which uncontrolled and unobserved factors play in the real world is played in a computer simulation model by the random character of exogenous variables." Overholt (1968, p. 22) points out that there may be a special type of error in simulation experiments. For the simulation model may fail to contain relevant factors and then the model is not realistic. This specification error should be detected in the validation phase of the modeling process.

Experimental error leads us to the three basic assumptions of ANOVA. As we specified in Eq. (3), in the ANOVA of factorial experiments (and in the design of such experiments) it is assumed that the experimental errors are normally and independently distributed with a constant variance. In simulation the experimental errors can be made independent by using different sequences of random numbers in each run. In general, however, the errors need not be normally distributed with common variance. Therefore we shall consider the effects of nonnormality and heterogeneity of variance in ANOVA. Scheffé (1964, pp. 345, 350, 358) states that, if the number of degrees of freedom is very large, then nonnormality has not much effect on the power of the F-test. Inequality of the variances does not effect the power much in the case of equal number of observations per "cell" (equal to the combination of levels), but has a serious effect with unequal cell numbers. For a small number of degrees of freedom Scheffé suggests that nonnormality still does not effect the F-test, especially with equal cell numbers. In the case of unequal variances an equal number of observations should be used unless we are certain that some cells show much larger variances in which case we take more observations for those cells. Later on an investigation of the robustness of the F-test was done by Donaldson (1966). He investigated the effect of large deviations from the assumptions of normality and equal experimental error variances, limiting his investigation to the case of a single-factor experiment with an equal number of observations per factor level. In the cases he investigated, he found very acceptable effects on the α- and β-error of the F-test. Nevertheless, in his simulation study Jensen (1966, pp. 235-236) did test for normality (applying a Kolmogorov-Smirnov test) and homogeneity of variances (applying Bartlett's test).[7] Deviations from the ANOVA assumptions may be detected by the analysis of the estimated residuals $\underline{u} = \underline{y} - \hat{\underline{y}}$ ($\hat{\underline{y}}$ being estimated response); see Anscombe and Tukey (1963). Because of the normality assumption in ANOVA we may apply nonparametric or distribution-free ANOVA. An advanced exposé on such nonparametric analysis (including ANOVA and other techniques, possibly extended to multiresponse situations) is given by Puri and Sen (1971, for instance, pp. 103, 221, 266-277, 286-300, 331-337). However, as Donaldson (1966, p. 44) points out, the F-test in parametric ANOVA is very insensitive to nonnormality, while heterogeneous variances disturb both parametric and nonparametric ANOVA.

We have just seen that the F-test is not very sensitive to nonnormality and heterogeneity of the variances. Nevertheless, we may try to force the data of the experiment to be (more) normally distributed with common variance. A special opportunity exists in the simulation of nonterminating systems (defined in Section II.8). With such systems we can continue the simulation until the run for a particular factor combination has reached a fixed variance. Another, general technique for obtaining a constant variance consists of the transformation of the original observations. The derivation of the appropriate transformation can be found in Mendenhall (1968, pp. 206-208), Peng (1967, pp. 183-185) or Scheffé (1964, pp. 364-368). They show that if the original variable $\underline{y}$ has mean μ and variance σ^2 such that σ^2 can be expressed as a function of μ, say

$$\sigma^2 = f(\mu) \tag{46}$$

then the new variable $\underline{y}^*$ has an approximately constant variance if

$$y^* = \int \frac{1}{\sqrt{f(y)}} \, dy \tag{47}$$

Application of Eq. (46) and (47) yields Table 4.

We observe that the last case in Table 4, $f(\mu) = c^2\mu^2$, is valid if the response $\underline{y}$ is the estimated variance of some variable $\underline{x}$, i.e.

$$\underline{y} = \underline{s}_x^2 \tag{48}$$

Scheffé (1964, p. 83) proves that

$$\mu = E(\underline{y}) = E(\underline{s}_x^2) = \sigma_x^2 \tag{49}$$

and

TABLE 4

Transformation for Constant Variance,
if $var(\underline{y}) = f(\mu)$ with $\mu \equiv E(\underline{y})$

$var(\underline{y}) = f(\mu)$	y^*	Approximate $var(\underline{y}^*)$
μ (Poisson)	$(y)^{1/2}$	1/4
μ $(1-\mu/n)$ (Binomial)	arc $\sin(y/n)^{1/2}$ (in radians)	1/(4n)
μ $(1-\mu)/n$ (Binomial)	arc $\sin(y)^{1/2}$	1/(4n)
$c^2\mu^2$ (c a constant)	ln y	c^2

$$\sigma^2 = var(\underline{y}) = var(\underline{s}_x^2) = \sigma_x^4\left(\frac{2}{n-1} + \frac{\gamma_2}{n}\right) = \mu^2\left(\frac{2}{n-1} + \frac{\gamma_2}{n}\right) = \mu^2 c^2 \quad (50)$$

where γ_2 measures the kurtosis and reduces to zero if $\underline{x}$ is normally distributed. Additional references for transformations are Andrews (1971), Dolby (1963), and Draper and Hunter (1969). Transformations can also be used to obtain more normally distributed variables. Since nonnormality is less serious than heterogeneity of variances, most transformations are used to realize a constant variance. (Normality and common variance may require conflicting transformations as Scheffé (1964, p. 367) observes.) Naylor et al. (1968) applied the logarithmic transformation for the analysis of $\underline{s}_x^2$ where $\underline{x}$ is simulated national income. This log transformation reduced both the nonnormality and the heterogeneity of variance. A serious problem may be that the transformation changes the <u>interpretation of the response</u>, e.g. what is the arc sin of national income? More examples are given by Scheffé (1964, p. 366). No problems arise in an experiment with a single factor, as in the above-mentioned study by Naylor et al. For, if $E(\underline{y})$ does not vary with the levels of that factor then $E[g(\underline{y})]$ does not either.

Transformations lead us to the area of <u>coding</u> of the explanatory variables or factors. If the factors are qualitative then we may assign numbers to the levels of these factors just as a mnemonic device. However, if the factors are quantitative then the original variables are usually coded such that "normalized" variables result, i.e. variables with average value 0 and ranging from -1 to +1. As we shall see in the next sections such coding simplifies the calculation of the parameters since only diagonal matrices are to be inverted; rounding errors are reduced. As an example of the effects of coding we consider a regression equation, Eq. (51). (As we shall see in Section IV.3, Eq. (51) may represent an ANOVA model with main effects β_i and two-factor interactions β_{ij}.)

$$\underline{y} = \beta_0 + \sum_{i=1}^{k} \beta_i x_i + \sum_{i=1}^{k-1} \sum_{j=i+1}^{k} \beta_{ij} x_i x_j + \underline{e} \tag{51}$$

where x_i is a normalized variable obtained by coding the original variable z_i as in (52).

$$x_i = a_i z_i + b_i \tag{52}$$

From (51) and (52) it follows that

$$\underline{y} = \gamma_0 + \sum_{1}^{k} \gamma_i z_i + \sum_{i=1}^{k-1} \sum_{j=i+1}^{k} \gamma_{ij} z_i z_j + \underline{e} \tag{53}$$

where

$$\gamma_0 = \beta_0 + \sum_{1}^{k} b_i \beta_i + \sum \sum b_i b_j \beta_{ij} \tag{54}$$

$$\gamma_1 = a_i \beta_i + \sum_{j \neq i} a_i b_j \beta_{ij} \tag{55}$$

$$\gamma_{ij} = a_i a_j \beta_{ij} \tag{56}$$

Hence, if, e.g. the standardized variables show no interactions, i.e. $\beta_{ij} = 0$, then x_i having no main effect (i.e, $\beta_i = 0$) implies that the original variable has no main effect (i.e. $\gamma_i = 0$). (If there are interactions then the influence of a particular factor varies with the levels of the other factors and is not measured by the main effect alone.) Notice that both Eqs. (51) and (53) yield the same response change if a particular factor is changed. For example, it follows from Eq. (53) that for $z_j = 0 \quad (j \neq i)$

$$\frac{\partial y}{\partial z_i} = \gamma_i \tag{57}$$

and from Eqs. (51) and (52) it follows that for $z_j = 0$, or equivalently $x_j = b_j$

$$\frac{\partial y}{\partial z_i} = \frac{\partial y}{\partial x_i} \frac{\partial x_i}{\partial z_i} = (\beta_i + \sum_{j \neq i} \beta_{ij} b_j)\, a_i = \gamma_i \tag{58}$$

More about coding can be found in Mendenhall (1968, pp. 221-229, 251-257). Note that the regression parameters are marginal responses; also see Mihram (1972, pp. 359-360).

IV.3. FACTORIAL DESIGNS WITH ALL FACTORS AT TWO LEVELS

A special class of factorial designs consists of experiments where each factor is at only two levels. Hence, if there are k factors, then the total number of combinations is 2^k. These experiments are called 2^k factorial experiments. In the experiment one or more factors may be qualitative, e.g. one factor may be queuing discipline, the two levels of the factor being the "first in, first out" and the "first in, last out" rule. For a quantitative factor the levels will correspond with the extreme values that the factor can assume in the experiment.

As an example we shall consider an experiment with two factors, A and B, without replications, as in Table 5. From Eqs. (2) and (13) it follows that the model is

TABLE 5

A 2^2 Experiment Without Replication

	Factor B	
Factor A	Level 1	Level 2
Level 1	$\underline{y}_{11}$	$\underline{y}_{12}$
Level 2	$\underline{y}_{21}$	$\underline{y}_{22}$

$$\begin{aligned}
\underline{y}_{11} &= \mu + \alpha_1^A + \alpha_1^B + \alpha_{11}^{AB} + \underline{e}_{11} \\
\underline{y}_{12} &= \mu + \alpha_1^A + \alpha_2^B + \alpha_{12}^{AB} + \underline{e}_{12} \\
\underline{y}_{21} &= \mu + \alpha_2^A + \alpha_1^B + \alpha_{21}^{AB} + \underline{e}_{21} \\
\underline{y}_{22} &= \mu + \alpha_2^A + \alpha_2^B + \alpha_{22}^{AB} + \underline{e}_{22}
\end{aligned} \tag{59}$$

which can also be represented in matrix notation analogous to Eqs. (35) thorugh (39). The least squares estimators of the parameters in Eq. (59) can be determined taking into account the following side conditions which follow from Eq. (7), (9), and (12).

$$\alpha_1^A = -\alpha_2^A \tag{60}$$

$$\alpha_1^B = -\alpha_2^B \tag{61}$$

$$\alpha_{12}^{AB} = -\alpha_{11}^{AB} \tag{62}$$

$$\alpha_{21}^{AB} = -\alpha_{11}^{AB} \tag{63}$$

$$\alpha_{22}^{AB} = -\alpha_{21}^{AB} \qquad (= \alpha_{11}^{AB}) \tag{64}$$

We can also substitute Eqs. (60) through (64) directly into Eqs. (59) which then yields

$$\underline{y}_{11} = \mu - \alpha_2^A - \alpha_2^B + \alpha_{11}^{AB} + \underline{e}_{11}$$

$$\underline{y}_{12} = \mu - \alpha_2^A + \alpha_2^B - \alpha_{11}^{AB} + \underline{e}_{12}$$

$$\underline{y}_{21} = \mu + \alpha_2^A - \alpha_2^B - \alpha_{11}^{AB} + \underline{e}_{21}$$

$$\underline{y}_{22} = \mu + \alpha_2^A + \alpha_2^B + \alpha_{11}^{AB} + \underline{e}_{22} \qquad (65)$$

In matrix notation, Eqs. (65) become

$$\vec{\underline{y}} = \vec{X}\,\vec{\beta} + \vec{\underline{e}} \qquad (66)$$

with

$$\vec{\underline{y}}' = (y_{11}, y_{12}, y_{21}, y_{22}) \qquad (67)$$

$$\vec{X} = \begin{bmatrix} +1 & -1 & -1 & +1 \\ +1 & -1 & +1 & -1 \\ +1 & +1 & -1 & -1 \\ +1 & +1 & +1 & +1 \end{bmatrix} \qquad (68)$$

$$\vec{\beta}' = (\mu, \alpha_2^A, \alpha_2^B, \alpha_{11}^{AB}) \qquad (69)$$

$$\vec{\underline{e}}' = (\underline{e}_{11}, \underline{e}_{12}, \underline{e}_{21}, \underline{e}_{22}) \qquad (70)$$

Notice that all columns of $\vec{X}$ are orthogonal, i.e.,

$$\vec{x}_i'\,\vec{x}_j = 0 \qquad (i \neq j) \qquad (71)$$

where $\vec{x}_i$ and $\vec{x}_j$ may be any two columns of $\vec{X}$. Obviously $\vec{X}$ is nonsingular and therefore the least squares estimators of $\vec{\beta}$ are given by

$$\vec{\underline{b}} = (\vec{X}'\,\vec{X})^{-1}\,\vec{X}'\,\underline{y} \qquad (72)$$

Because of Eq. (71) and

$$\vec{x}_i' \vec{x}_i = N \tag{73}$$

N denoting the number of observations (here $N = 4$), it follows that

$$(\vec{X}' \vec{X}) = N\vec{I} \tag{74}$$

Further, element h of $\vec{X}' \vec{\underline{y}}$ is given by relation (75). (h runs from 1 to H, H being the total number of parameters; in this example $H = 4$.)

$$\sum_{g=1}^{N} x_{gh}\underline{y}_g \tag{75}$$

where x_{gh} is the gth element of $\vec{x}_h$. Substitution of Eqs. (74) and (75) into Eq. (72) yields

$$\underline{b}_h = \frac{1}{N} \sum_{g=1}^{N} x_{gh}\underline{y}_g \tag{76}$$

Hence,

$$\underline{b}_1 = \underline{\hat{\mu}} = \frac{1}{4} (\underline{y}_{11} + \underline{y}_{12} + \underline{y}_{21} + \underline{y}_{22}) = \underline{y}_{..} \tag{77}$$

$$\underline{b}_2 = \underline{\hat{\alpha}}_2^A = \frac{1}{4} (-\underline{y}_{11} - \underline{y}_{12} + \underline{y}_{21} + \underline{y}_{22})$$

$$= \frac{1}{4} [-(y_{11} + y_{12} + y_{21} + y_{22}) + 2(y_{21} + y_{22})]$$

$$= -\underline{y}_{..} + y_{2.} \tag{78}$$

Comparing Eqs. (78) with the definition of α_2^A

$$\alpha_2^A = \eta_{2.} - \eta_{..} \tag{79}$$

shows that the estimator of the effect is analogous to the effect itself. In the same way we can derive that the least squares estimators of the main effect α_2^B and the interaction α_{11}^{AB} are the sample analogous of the definitions of these effects that follow from Eqs. (8) and (12).

It is important to observe that in the matrix of explanatory variables $\vec{X}$, column 1 corresponds with the grand mean μ and consists of only +1's. Columns 2 and 3 correspond with the main effects α_2^A and α_2^B of factor A and B, respectively; element g (g = 1, ..., N) of the column is -1 if the corresponding factor is at its "low" level and is +1 if the factor is at its "high" level in the gth observation. (For a qualitative factor "low" and "high" are only mnemonic symbols for the two levels of that factor.) Column 4 of $\vec{X}$ corresponds with the interaction α_{11}^{AB}; the gth element of this column can be obtained by multiplying the gth element of the columns 2 and 3 with each other! Hence we can write the regression model in Eq. (66) as in Eq. (80).

$$\underline{y}_g = \beta_0 + \sum_{s=1}^{2} d_{gs}\beta_s + (d_{g1}d_{g2})\beta_{12} + \underline{e}_g \quad (g = 1, \ldots, N) \quad (80)$$

where $d_{gs} = -1$ if factor s is at its low level in observation g, and $d_{gs} = +1$ if that factor is at its high level in observation g; β_0 denotes the grand mean μ, β_s is a main effect of factor s (i.e. $\beta_1 = \alpha_2^A = -\alpha_1^A$ and $\beta_2 = \alpha_2^B = -\alpha_1^B$) and β_{12} denotes an interaction between factors 1 and 2 ($\beta_{12} = \alpha_{11}^{AB} = -\alpha_{12}^{AB} = -\alpha_{21}^{AB} = \alpha_{22}^{AB}$). Observe that Eq. (80) is a regression equation consisting of a second degree polynomial except for the "pure quadratic" terms $(d_{g1}^2\beta_{11})$ and $(d_{g2}^2\beta_{22})$ that are missing in the ANOVA model.

It is customary to define the effects in a design where all factors are at only two levels, slightly different from the above general definitions. So "the" main effect of factor A is defined as the average response at the high level of factor A minus the average response at the low level of A, i.e.

$$\alpha^A = \eta_{2.} - \eta_{1.} \quad (81)$$

Hence

$$\begin{aligned}\alpha^A &= (\eta_{2.} - \eta_{..}) - (\eta_{1.} - \eta_{..}) \\ &= \alpha_2^A - \alpha_1^A = \alpha_2^A + \alpha_2^A = 2\alpha_2^A \\ &= -2\alpha_1^A \end{aligned} \tag{82}$$

So the new definition of the main effect of a factor in a 2^k design gives twice the value of the general definition (and possibly a change in sign but this sign is arbitrary since it depends on what we call the high and low level of the factor). The interaction between A and B is redefined as follows. If factor B is at its high level then the effect of factor A is

$$\eta_{22} - \eta_{12} \tag{83}$$

and if B is at its low level then the effect of A is

$$\eta_{21} - \eta_{11} \tag{84}$$

Hence if Eqs. (83) and (84) differ then there is an interaction. The interaction is defined as the "average" difference between Eqs. (83) and (84), i.e.

$$\alpha^{AB} = \frac{1}{2}[(\eta_{22} - \eta_{12}) - (\eta_{21} - \eta_{11})] \tag{85}$$

From Eq. (12) it follows that

$$\begin{aligned} &2(\eta_{11} - \eta_{1.} - \eta_{.1} + \eta_{..}) \\ &\quad = 2\left[\eta_{11} - \frac{(\eta_{11} + \eta_{12})}{2} - \frac{(\eta_{11} + \eta_{21})}{2} + \frac{(\eta_{11} + \eta_{12} + \eta_{21} + \eta_{22})}{4}\right] \\ &\quad = \frac{1}{2}\eta_{11} - \frac{1}{2}\eta_{12} - \frac{1}{2}\eta_{21} + \frac{1}{2}\eta_{22} \\ &\quad = \alpha^{AB} \end{aligned} \tag{86}$$

So Eq. (86) together with Eqs. (62) through (64) show that the new definition of "the" interaction gives twice the value of the old definitions of α_{11}^{AB}, α_{12}^{AB}, α_{21}^{AB}, and α_{22}^{AB} with a possible change of sign.

After the above example with two factors we shall consider the general case of k factors at two levels. Instead of the 2×2 layout of Table 5 we use a representation as shown in Table 6 for $k = 3$. In this table the low value of a factor is denoted by -1 and its high value by +1. For k factors the column of factor s ($s = 1, \ldots, k$) contains $2^{(s-1)}$ times -1, $2^{(s-1)}$ times +1, $2^{(s-1)}$ times -1, etc. The response is denoted as follows. If factor A is at its high level in a combination then we use the letter a, if factor A is at its low level in a combination, this is indicated by the absence of the letter a. The same holds for the other factors. The combination where all factors are at their low level is indicated by the number 1. The combination of lower case letters is usually also employed, not to indicate the response at a particular factor combination but the combination itself.

TABLE 6

A 2^3 Experiment (Without Replication)

	Factor			
Combination	A	B	C	Response
1	-1	-1	-1	1
2	+1	-1	-1	a
3	-1	+1	-1	b
4	+1	+1	-1	ab
5	-1	-1	+1	c
6	+1	-1	+1	ac
7	-1	+1	+1	bc
8	+1	+1	+1	abc

Analogous to Eq. (81) the new definition of the main effect of factor A in a 2^k design is

$$\alpha^A = \eta_{2....} - \eta_{1....} \tag{87}$$

where the number of dots is (k-1). It is easy to show that the new definition of the main effect gives values twice as large as the original definition. However, if a factor has no effect, both definitions result in a main effect with the value zero. The new definition defines the main effect of A as the difference between the mean response at the high and low levels of A. Therefore, in order to obtain an estimate of α^A we subtract the average response when factor A is at its low level (while the other factors assume low and high levels) from the average response when factor A is at its high level. Let us denote the total number of runs in the experiment by N. Then N is 2^k in this factorial. From the way a 2^k factorial is constructed, it follows that in one half of the combinations factor A is at its high level, and in the other half it is at its low level; compare Table 6 where we have alternate plus and minus signs for factor A. So

$$\underline{\hat{\alpha}}^A = \frac{\Sigma_i \underline{y}_i}{N/2} - \frac{\Sigma_j \underline{y}_j}{N/2} \tag{88}$$

the index i indicating responses of combinations where factor A is at its high level; j indicates responses of combinations with A at its low level. Hence Eq. (88) is equal to

$$\begin{aligned} \underline{\hat{\alpha}}^A &= \frac{2}{N} \left\{ \sum_i (+1)\, \underline{y}_i + \sum_j (-1)\, \underline{y}_j \right\} \\ &= \frac{2}{N} \sum_{g=1}^{N} x_{g1} \underline{y}_g \end{aligned} \tag{89}$$

where x_{g1} is the gth element in the column of factor 1 (i.e., factor A). In general the estimator of the main effect of factor s is given by

$$\hat{\underline{\alpha}}^s = \frac{2}{N} \sum_{g=1}^{N} x_{gs}\underline{y}_g \qquad (s = 1, \ldots, k) \tag{90}$$

where x_{gs} is -1 if factor s is at its low level in factor combination g, and x_{gs} is +1 if factor s is at its high level in combination g. Analogous to Eqs. (59) through (79) it can be shown that the estimator in Eq. (90) is the least squares estimator of the main effect α^s of factor s. In Appendix IV.2 an example is given of the estimation of a two-factor interaction in a 2^3 experiment. It can be proven that the least squares estimator of the interaction between the factors $j, m, \ldots,$ and r is

$$\hat{\underline{\alpha}}^{j,m,\ldots,r} = \frac{2}{N} \sum_{g=1}^{N} (x_{gj}\, x_{gm} \ldots x_{gr})\underline{y}_g \tag{91}$$

The least squares estimator of the grand mean is

$$\hat{\underline{\mu}} = \bar{\underline{y}} = \frac{1}{N} \sum_{g=1}^{N} x_{g0}\underline{y}_g \tag{92}$$

where

$$x_{g0} = 1 \qquad (g = 1, \ldots, N) \tag{93}$$

In a 2^k factorial experiment there are 2^k combinations or "experimental points." Each experimental point can be represented by a point in the k-dimensional space having coordinates $(\pm 1, \pm 1, \ldots, \pm 1)$. If we denote the number of combinations or "runs" in the experiment by N, we can define the so-called design matrix $\vec{D}$:

$$\vec{D} = \{d_{ij}\}, \qquad i = 1, 2, \ldots, N;\ j = 1, 2, \ldots, k \tag{94}$$

where d_{ij} is -1 if factor j is at its low level in combination i, etc. Adding a column of +1's to $\vec{D}$ (i.e. the identity column $\vec{I}$) and multiplying all pairs of columns, triplets, etc. in $\vec{D}$ we obtain $\vec{X}$, the "matrix of independent variables." For $k = 3$, $\vec{D}$ and $\vec{X}$ are given in Table 7 where for convenience only the plus and minus signs without the number 1 are given.

TABLE 7

Design Matrix and Matrix of Independent Variables for a 2^3 Experiment

Design Matrix $\vec{D}$			Matrix of independent variables $\vec{X}$							
$\vec{1}$	$\vec{2}$	$\vec{3}$	$\vec{I}$	$\vec{1}$	$\vec{2}$	$\vec{3}$	$\vec{12}$	$\vec{13}$	$\vec{23}$	$\vec{123}$
-	-	-	+	-	-	-	+	+	+	-
+	-	-	+	+	-	-	-	-	+	+
-	+	-	+	-	+	-	-	+	-	+
+	+	-	+	+	+	-	+	-	-	-
-	-	+	+	-	-	+	+	-	-	+
+	-	+	+	+	-	+	-	+	-	-
-	+	+	+	-	+	+	-	-	+	-
+	+	+	+	+	+	+	+	+	+	+

The grand mean, the main effects, and all interactions can be estimated by multiplying the corresponding column of $\vec{X}$ with the observations $\underline{y}$ as in Eqs. (92), (90), and (91), respectively. Obviously the regression equation for the ANOVA model with k factors at two levels is given by Eq. (95), the generalization of Eq. (80).

$$
\begin{aligned}
\underline{y}_i &= \sum_{j=1}^{J} x_{ij}\gamma_j + \underline{e}_i \\
&= \beta_0 + \sum_{s=1}^{k} d_{is}\beta_s + \sum_{s=1}^{k-1} \sum_{z=s+1}^{k} (d_{is}d_{iz})\beta_{sz} \\
&\quad + \sum_{s=1}^{k-2} \sum_{z=s+1}^{k-1} \sum_{v=z+1}^{k} (d_{is}d_{iz}d_{iv})\beta_{szv} \\
&\quad + \cdots + (d_{i1}d_{i2} \cdots d_{ik})\beta_{12\ldots k} + \underline{e}_i
\end{aligned}
\tag{95}
$$

where x_{ij} and d_{is} are the elements of $\vec{X}$ and $\vec{D}$, respectively;

$J = 2^k$ is the number of regression parameters γ_j; these parameters γ_j denote the grand mean β_0, the main effects β_s, the two-factor interactions β_{sz}, ..., the k-factor interaction $\beta_{12\ldots k}$, provided we use the general definitions for these effects given in Section IV.2. If we use the special definitions for 2^k experiments given in this section then γ_j denotes half of the main effects and interactions.

Having obtained the estimators of the effects, we shall next derive their <u>variances</u> and <u>covariances</u>. We still assume that the experimental errors are independent and have a common variance σ^2. We shall use the formula for the covariance matrix $\vec{\Omega}_\gamma$ of the least squares estimators of the γ_j in (95). It is well-known, see e.g. Johnston (1963, p. 110), that

$$\vec{\Omega}_\gamma = \sigma^2(\vec{X}'\,\vec{X})^{-1} \tag{96}$$

where $\vec{X}$ denotes the matrix of independent variables. From the construction of the factorial experiment it follows that the columns $\vec{x}_i$ of $\vec{X}$ satisfy

$$\begin{aligned} \vec{x}_i'\,\vec{x}_j &= 0 \quad \text{if } i \neq j \\ &= N \quad \text{if } i = j \end{aligned} \tag{97}$$

Hence

$$\vec{\Omega}_\gamma = \frac{\sigma^2}{N}\vec{I} \tag{98}$$

Now we have to remember that the main effects and interactions (but not the grand mean) are usually defined as twice the corresponding γ_j. Consequently the variances of the estimated main effects and interactions are $4\sigma^2/N$; the variance of the estimated grand mean remains σ^2/N and all covariances remain zero.

Next we shall see how we can <u>test</u> the hypothesis that a particular effect (a main effect or interaction) is not important. The number of degrees of freedom for any effect is one, since there are two levels per factor; compare the general formulas for the degrees of freedom in, e.g. Scheffé (1964, p. 125). From Hicks (1966, pp. 102, 106-107) it follows that $\underline{SS}_j$, the sum of squares of effect j ($j = 1, 2, \ldots, 2^k-1$), can be calculated applying the general formulas or applying "Yates's method" for 2^k designs. This yields

$$\underline{SS}_j = r \cdot 2^{k-2} \cdot (\hat{\underline{effect}}_j)^2 \qquad (j = 1, \ldots, 2^k-1) \tag{99}$$

where r is the number of replications per experimental point ($r \geq 1$). If $r > 1$ then the error sum of squares which goes into the denominator of the F-statistic is given by Eq. (100).

$$\underline{SS}_e = \underline{SS}_{total} - \sum_{j=1}^{2^k-1} \underline{SS}_j \tag{100}$$

with

$$\underline{SS}_{total} = \sum_{i=1}^{N} \underline{y}_i^2 - \frac{(\Sigma_1^N \underline{y}_i)^2}{N} \qquad (N = 2^k \cdot r) \tag{101}$$

The number of degrees of freedom for $\underline{SS}_e$ is

$$2^k(r - 1) \tag{102}$$

i.e., $(r - 1)$ degrees of freedom per experimental point. In case of no duplication (i.e., $r = 1$), Eq. (102) results in zero degrees of freedom. Therefore, if there is no prior independent estimate of σ^2 then some interactions may be assumed zero and be pooled to give an estimate of σ^2. For instance, if the effects corresponding with $\underline{SS}_1, \underline{SS}_2, \ldots, \underline{SS}_m$ are supposed to be zero, then a pooled estimator of σ^2 is

$$\frac{\underline{SS}_1 + \underline{SS}_2 + \cdots + \underline{SS}_m}{m} \qquad (m \geq 1) \tag{103}$$

In Section IV.6 we shall return to the problem of obtaining an independent estimator for the experimental error variance σ^2. Note that to test a single effect with one degree of freedom, we can also use the t-statistic (cf. $\underline{F}_{1,v} = \underline{t}_v^2$). To test several effects jointly we use the F-statistic. An example will be given in Chapter VI, Eqs. (89) and (92).

We want to return briefly to the advantages of a factorial design as compared with the one-factor-at-a-time method. Suppose we take N observations. To measure the main effect of the first factor we take $N/2$ observations at its low level and $N/2$ observations at its high level. In a factorial design we distribute the $N/2$ observations at the low level of factor 1 evenly between the high and low levels of the remaining $(k-1)$ factors; the same for the $N/2$ observations at the high level of factor 1. The precision of our estimate of the main effect of the first factor will not change, but the factorial design makes it possible to estimate the effects of all the other factors at the same time.

Equation (28) showed that there are C_q^k q-factor interactions $(q = 2, \ldots, k)$ and C_1^k main effects; together (2^k-1) effects. So the $(r \cdot 2^k)$ observations can be used to estimate these (2^k-1) effects plus the grand mean, all together 2^k estimates; also compare Hicks (1966, p. 107). As k increases the number of effects and observations needed for the estimation of these effects increases sharply as is demonstrated in Table 8. However, high values of k result in interactions of very high order. These high-order interactions are often assumed to be negligible, because of a priori considerations, previous experiments, general experience, or other grounds. An example of those "other grounds" will be seen in Section IV.5, where all factors are quantitative and the higher-order interactions correspond with high-order terms in the regression polynomial; these high-order terms are assumed to be zero, because a low-order polynomial is supposed to give an adequate regression equation. If some effects are assumed to be zero we do not take observations at all 2^k experimental points; a fraction of those 2^k points will be sufficient. These fractional factorials form the subject of the next section.

TABLE 8

The Number of Factors (k) and the Number of Observations (2^k)

k	1	2	3	4	5	6	7	8
2^k	2	4	8	16	32	64	128	256

IV.4. FUNDAMENTALS OF FRACTIONAL FACTORIAL DESIGNS WITH ALL FACTORS AT TWO LEVELS

In the previous section we saw that if the number of factors increases, the number of observations increases sharply, even if all factors are at only two levels. We also remarked that we may assume that certain effects are zero so that less than 2^k observations are needed. The problem we shall consider in this section is, which observations to take. To make this choice we need to know what the consequences are of dropping observations. We shall start with an example.

TABLE 9
The 2^3 Factorial

	Matrix of independent variables $\vec{X}$								
Run	$\vec{I}$	$\vec{1}$	$\vec{2}$	$\vec{3}$	$\vec{12}$	$\vec{13}$	$\vec{23}$	$\vec{123}$	$\vec{E}(\vec{\underline{y}})$
1	+	-	-	-	+	+	+	-	1
2	+	+	-	-	-	-	+	+	a
3	+	-	+	-	-	+	-	+	b
4	+	+	+	-	+	-	-	-	ab
5	+	-	-	+	+	-	-	+	c
6	+	+	-	+	-	+	-	-	ac
7	+	-	+	+	-	-	+	-	bc
8	+	+	+	+	+	+	+	+	abc

Suppose there are three factors as shown in Table 9. Suppose we do not make all eight runs, but only those four runs for which

$$x_1 x_2 x_3 = +1 \tag{104}$$

Deleting the other four runs gives Table 10. The remaining rows, shown in Table 10, tell us how to calculate the estimates of the effects from this incomplete factorial. For instance, column $\vec{1}$ results in

$$\underline{\hat{\alpha}}^{A} = \frac{2}{N}(\underline{y}_2 - \underline{y}_3 - \underline{y}_5 + \underline{y}_8) \tag{105}$$

where the number of runs is now

$$N = 4 \tag{106}$$

Formula (105) is obviously correct, as factor 1 is at its high level in the runs 2 and 8 and at its low level in the runs 3 and 5. Hence, the effect of factor A is measured by

$$\frac{a + abc}{2} - \frac{b + c}{2} = \frac{1}{2}(a - b - c + abc) \tag{107}$$

Let us next consider column $\vec{23}$, resulting in

$$\underline{\hat{\alpha}}^{BC} = \frac{2}{N}(\underline{y}_2 - \underline{y}_3 - \underline{y}_5 + \underline{y}_8) \tag{108}$$

TABLE 10

An Incomplete Factorial $(x_1 x_2 x_3 = +1)$

Run	$\vec{I}$	$\vec{1}$	$\vec{2}$	$\vec{3}$	$\vec{12}$	$\vec{13}$	$\vec{23}$	$\vec{123}$	$E(\vec{\underline{y}})$
2	+	+	-	-	-	-	+	+	a
3	+	-	+	-	-	+	-	+	b
5	+	-	-	+	+	-	-	+	c
8	+	+	+	+	+	+	+	+	abc

Equation (108) is correct as can be seen in the following. Analogous to Eqs. (83) through (85) the interaction between B and C is the average difference between the effects of B when C is at its high and low level, respectively. The effect of B at high C can be measured by abc - c; the effect of B at low C by b - a. Half of the difference between those effects is

$$\frac{(abc - c) - (b - a)}{2} = \frac{1}{2}(a - b - c + abc) \tag{109}$$

Next let us compare Eqs. (108) with (105). Then we see that we obtain the same values for $\hat{\underline{\alpha}}^{A}$ and $\hat{\underline{\alpha}}^{BC}$. Or, to put it in another way, using the last column of Table 10 gives

$$E(\underline{y}_2 - \underline{y}_3 - \underline{y}_5 + \underline{y}_8) = a - b - c + abc \tag{110}$$

The right-hand side of Eq. (110) can be written as in Eq. (111), where Eq. (106) still holds.

$$\begin{aligned} a - b - c + abc \\ &= \frac{2}{N}(-1 + a - b + ab - c + ac - bc + abc) \\ &+ \frac{2}{N}(+1 + a - b - ab - c - ac + bc + abc) \end{aligned} \tag{111}$$

Finally, using Eqs. (90) and (91), we can write Eq. (111) as follows.

$$a - b - c + abc = \alpha^{A} + \alpha^{BC} \tag{112}$$

Combining Eq. (112) with Eq. (110) yields

$$E(\underline{y}_2 - \underline{y}_3 - \underline{y}_5 + \underline{y}_8) = \alpha^{A} + \alpha^{BC} \tag{113}$$

So the fractional factorial of this example implies that we obtain the same values for the main effect of factor A and the interaction between the factors B and C; actually the value calculated measures the sum of both effects. The effects are said to be confounded or to be each other's alias. If, however, the BC-interaction is zero, then $\underline{y}_2 - \underline{y}_3 - \underline{y}_5 + \underline{y}_8$ is an unbiased estimator of α^A.

Generalizing, for a 2^k factorial we can obtain a one-half fraction by deleting from the table for the full 2^k factorial those rows that give a plus sign for a particular effect (or alternatively, that give a minus sign). For, remember that each effect has as many plus signs as minus signs in its column. Suppose we choose to take only those rows corresponding with the plus sign for a particular effect, say effect α^{ABC}. This means that in the resulting table for the fractional design there are only plus signs in the column for the effect α^{ABC}. Hence, that effect is confounded with the grand mean. This is indicated by the "defining relation"

$$\vec{I} = \vec{ABC} \tag{114}$$

where $\vec{I}$ is the identity column consisting of only +1's and $\vec{ABC}$ is called the generator of the design. If we had chosen the alternative fraction corresponding with the minus sign of α^{ABC}, we would have gotten

$$\vec{I} = -\vec{ABC} \tag{115}$$

instead of Eq. (114). There is a very simple procedure to determine which effects are confounded when choosing a particular fraction. "Multiply" the letters indicating an effect with the members of the defining relation, where the power of a letter is taken modulo 2, i.e. if the power coefficient is c then c modulo 2 is

$$\begin{aligned} c \pmod 2 &= 0 \quad \text{if } c \text{ is an even number } 0, 2, 4, 6, \ldots \\ c \pmod 2 &= 1 \quad \text{if } c \text{ is an odd number } 1, 3, 5, \ldots \end{aligned} \tag{116}$$

For example, in Table 10 the alias of the main effect A is found from the defining relation for the first fraction

$$\vec{I} = \overrightarrow{ABC} \tag{117}$$

The multiplication rule gives

$$\vec{A}\cdot\vec{I} = \vec{A}\cdot(\overrightarrow{ABC}) \tag{118}$$

or

$$\vec{A} = \overrightarrow{A^2BC} \tag{119}$$

Modulo 2 gives

$$\vec{A} = \overrightarrow{A^0\ BC} = \overrightarrow{BC} \tag{120}$$

which agrees with Eq. (113). In the same way we obtain

$$\vec{B} = \overrightarrow{AC} \tag{121}$$

$$\vec{C} = \overrightarrow{AB} \tag{122}$$

If we had taken the other fraction corresponding with Eq. (115), then Eq. (117) would be replaced by

$$\vec{I} = -\overrightarrow{ABC} \tag{123}$$

or

$$\vec{A} = -\overrightarrow{BC} \tag{124}$$

$$\vec{B} = -\overrightarrow{AC} \tag{125}$$

$$\vec{C} = -\overrightarrow{AB} \tag{126}$$

From this procedure it follows that we shall usually use the highest order interaction as generator, say

$$\overrightarrow{A\ B \cdots K} \equiv \overrightarrow{1\ 2 \cdots k} \tag{127}$$

as this means that the main effects of the factors A, B, ... , K are confounded with interactions among (k-1) factors and it are high-order interactions that are assumed to be zero in fractional designs. Two factor interactions are confounded with interactions among (k-2) factors and therefore, for not too small k, we might be able to obtain unbiased estimators of the two-factor interactions.

Next we shall consider the construction of the design matrix (i.e. the N experimental points in k dimensions) for which the interaction among all k factors is the generator. Write down the design matrix for a full factorial in (k-1) factors. This is a matrix with 2^{k-1} rows and (k-1) columns. Add the column for the interaction among all (k-1) factors, that is the column

$$1\cdot 2\cdot \overrightarrow{\cdots} \cdot (k-1) \tag{128}$$

Identify the kth factor with the column in (128), i.e.

$$\vec{k} = 1\cdot 2\cdot \overrightarrow{\cdots} \cdot (k-1) \tag{129}$$

or

$$\vec{k}\cdot\vec{k} = 1\cdot 2\cdot \overrightarrow{\cdots} \cdot (k-1)\cdot\vec{k} \tag{130}$$

Because multiplying a column by the same column results in a column of only plus one's, we obtain

$$\vec{k}\cdot\vec{k} = \vec{I} \tag{131}$$

Hence, Eq. (130) becomes

$$\vec{I} = 1\cdot 2\cdot \overrightarrow{\cdots} \cdot (k-1)\cdot k \tag{132}$$

So we indeed constructed a design where the k-factor interaction is the generator. The other half of the 2^k factorial can be obtained by flipping the signs in the $1\cdot 2\cdot \overrightarrow{\cdots} \cdot (k-1)$ column, i.e.

$$k = -1\cdot 2\cdot \overrightarrow{\cdots} \cdot (k-1) \tag{133}$$

or

$$\vec{I} = -1\cdot 2\cdot \overrightarrow{\cdots} \cdot (k-1)\cdot k \tag{134}$$

For large k even a one-half fraction, i.e., 2^{k-1} observations, might take too much time and moreover, may not be needed, as many interactions of high order may be assumed to be zero. Therefore we then take smaller fractions of the complete factorial. A $(1/2)^p$ fraction of a full design in k factors is called a 2^{k-p} design.[9] Let us consider the design for seven factors. A full factorial would require $2^7 = 128$ observations. A one-half fraction could be generated from

$$\vec{I} = 1 \cdot 2 \cdot \vec{3} \cdot 4 \cdot 5 \cdot 6 \cdot 7 \tag{135}$$

which implies that main effects are confounded with interactions among six factors, two-factor interactions with five-factor interactions and three-factor interactions with four-factor interactions. If we are interested in the main effects and assume that interactions are nonexistent, then we may take a one-sixteenth fraction, i.e., $p = 4$. Because a 2^{7-4} design consists of $2^{7-4} = 2^3 = 8$ experimental points or runs, we start by writing down the full factorial for three factors.[10] Then we associate the remaining four factors with the four possible interactions among the first three factors. This results in Table 11. So

TABLE 11

Design Matrix for a 2^{7-4} Experiment

$\vec{1}$	$\vec{2}$	$\vec{3}$	$\vec{4} = \vec{12}$	$\vec{5} = \vec{13}$	$\vec{6} = \vec{23}$	$\vec{7} = \vec{123}$
-	-	-	+	+	+	-
+	-	-	-	-	+	+
-	+	-	-	+	-	+
+	+	-	+	-	-	-
-	-	+	+	-	-	+
+	-	+	-	+	-	-
-	+	+	-	-	+	-
+	+	+	+	+	+	+

$$\overrightarrow{4} = \overrightarrow{12}, \quad \overrightarrow{5} = \overrightarrow{13}, \quad \overrightarrow{6} = \overrightarrow{23}, \quad \overrightarrow{7} = \overrightarrow{123} \tag{136}$$

or

$$\overrightarrow{I} = \overrightarrow{124}, \quad \overrightarrow{I} = \overrightarrow{135}, \quad \overrightarrow{I} = \overrightarrow{236}, \quad \overrightarrow{I} = \overrightarrow{1237} \tag{137}$$

Hence, Eq. (137) yields the p (= 4) generators $\overrightarrow{124}$, $\overrightarrow{135}$, $\overrightarrow{236}$, and $\overrightarrow{1237}$. But if $\overrightarrow{I} = \overrightarrow{124}$ and $\overrightarrow{I} = \overrightarrow{135}$, then clearly

$$\overrightarrow{I} \cdot \overrightarrow{I} = \overrightarrow{I} = (\overrightarrow{124})(\overrightarrow{135}) = \overrightarrow{1^2 2345} = \overrightarrow{2345} \tag{138}$$

Consequently, by multiplying the p generators with each other in pairs, triplets, etc., we obtain the 2^p members or "words" of the defining relation. In the 2^{7-4} example we have

$$\overrightarrow{I} = \overrightarrow{124} = \overrightarrow{135} = \overrightarrow{236} = \overrightarrow{1237} \qquad \text{(the generators)}$$

$$= \overrightarrow{2345} = \overrightarrow{1346} = \overrightarrow{347} = \overrightarrow{1256} = \overrightarrow{257} = \overrightarrow{167} \qquad \text{(pairs)}$$

$$= \overrightarrow{456} = \overrightarrow{1457} = \overrightarrow{2467} = \overrightarrow{3567} \qquad \text{(triplets)}$$

$$= \overrightarrow{1234567} \qquad \text{(p-tuple)} \tag{139}$$

To find the aliases of an effect we simply "multiply" the defining relation with that effect. This gives, for instance,

$$\overrightarrow{1} = \overrightarrow{24} = \overrightarrow{35} = \overrightarrow{1236} = \overrightarrow{237} = \overrightarrow{12345} = \overrightarrow{346} = \overrightarrow{1347} = \overrightarrow{256}$$

$$= \overrightarrow{1257} = \overrightarrow{67} = \overrightarrow{1456} = \overrightarrow{457} = \overrightarrow{12467} = \overrightarrow{13567} = \overrightarrow{234567} \tag{140}$$

So in this example the main effect of factor 1 is not confounded with other main effects but only with interactions.

Other fractions then the one in Table 11 can be obtained by flipping signs in one of the colmns $\overrightarrow{4}$, $\overrightarrow{5}$, $\overrightarrow{6}$, or $\overrightarrow{7}$. For instance, flipping signs in column $\overrightarrow{4}$ means

$$\overrightarrow{4} = -\overrightarrow{12} \tag{141}$$

In total sixteen fractions can be constructed from

$$\vec{4} = \pm \vec{12}, \quad \vec{5} = \pm \vec{13}, \quad \vec{6} = \pm \vec{23}, \quad \vec{7} = \pm \vec{123} \tag{142}$$

With each of the sixteen combinations in Eqs. (142) there corresponds a defining relation which yields the aliases. The one particular combination corresponding with taking only positive signs in Eqs. (142) gives the principal generators, principal identifying relation, and principal fraction. The sixteen possible fractions are said to belong to the same "family" of fractions. As Peng (1967, pp. 123-126) observes--see also John (1971, pp. 159-160)--a 2^{k-p} design can be conveniently analyzed by looking at the 2^{k-p} fractional as a full factorial design with only q $(= k-p)$ factors, i.e. by neglecting $(k-p)$ factors temporarily. The particular factors to be neglected are those factors that yield a full 2^q design in the remaining factors. For example, it follows from the way we constructed Table 11 that neglecting the factors 4 through 7 results in a full design in q $(= k-p = 7-4 = 3)$ factors. After having analyzed the design in the q factors, we introduce the aliases. For instance, if the sum of squares for the interaction 23 is significant, we conclude that factor 6 is significant, as compared with Eqs. (136). Peng (1967, pp. 237-247) gives a FORTRAN program for the analysis of 2^{k-p} designs (with the practical restrictions $k \leq 16$ and $2^{k-p} \leq 256$). We observe that in some publications on fractional designs group theory is applied. A simple introduction to the use of group theory in 2^{k-p} designs is given by Peng (1967, pp. 121-123).

IV.5. DESIGNS OF RESOLUTIONS III, IV, and V

Box and Hunter (1961a, p. 319) define the following types of design.

(i) Designs of resolution III: no main effect is confounded with any other main effect, but main effects are confounded with two-factor interactions and two-factor interactions with each other.

(ii) Designs of resolution IV: no main effect is confounded with any other main effect or two-factor interaction, but two-factor interactions are confounded with each other.

(iii) Designs of resolution V: no main effect or two-factor interaction is confounded with any other main effect or two-factor interaction, but two-factor interactions are confounded with three-factor interactions.

In general the resolution of a design is equal to the smallest number of characters in any word of the defining relation. In this section we shall have a closer look at the above types of design.

1. Resolution III Designs

There are resolution III designs available that require only N runs to study (N-1) factors provided N is a multiple of four. For values of N (= k+1), which are not a multiple of four we obtain a resolution III design from the resolution III design for the next value of N, say N' = k'+1, which is a multiple of four and deleting any (k'-k) columns (i.e. factors) from this design. If N is not only a multiple of four but also a power of two, then these resolution III designs are fractional factorial designs, denoted as 2^{k-p}_{III} designs. We shall first consider these 2^{k-p}_{III} designs and then the so-called Plackett and Burman designs where it is only required that N is a multiple of four.

If N must be a power of two then N is 2, 4, 8, 16, 32, etc. Hence k, the number of factors that can be studied in N (= k+1) runs is 1, 3, 7, 15, 31, etc. Consequently the 2^{k-p}_{III} designs for k is 3 through 31 are given by Table 12. We have seen, e.g. in Table 11, how 2^{k-p} designs are constructed by writing down the full factorial in q (= k-p) factors and equating the last p factors to the interactions among the first q factors. Analogous to Eqs. (136) through (140) we can find the alias.

Above we remarked that for values of N which are not a multiple of four we obtain a resolution III design by taking the

TABLE 12

2^{k-p} Designs of Resolution III

N	k	p	Design
4	3	1	2^{3-1}
8	7	4	2^{7-4}
16	15	11	2^{15-11}
32	31	26	2^{31-26}

next value of N, say $N' = k'+1$, being a multiple of four and deleting any $(k'-k)$ columns from the $2_{III}^{k'-p}$ design. These designs are no longer "saturated," i.e. no longer is the number of runs N, equal to the number of parameters $(k+1)$. The consequences are that for these nonsaturated designs we may (i) study more factors in the same number of runs; (ii) estimate some two-factor interactions; (iii) estimate the experimental error variance σ^2. As an example let us consider an experiment with five factors. For five the next multiple of four is eight, i.e. $N' = 8$. With eight runs we could study seven factors [see possibility (i)]. Hence we may drop two factors. If a factor is dropped from a factorial, then the alias structure remains the same except for all words containing the dropped factor; these words vanish. So in the 2_{III}^{7-4} design we had the defining relation given in (139). When dropping e.g. the factors 3 and 5 this relation reduces to

$$\vec{I} = 1\vec{2}4 = 1\vec{6}7 = 2\vec{4}67 \tag{143}$$

As there are eight runs, whereas only five main effects plus the grand mean are estimated, two interactions can also be estimated. From Eqs. (143) it follows that the following six two-factor interactions cannot be estimated since they are confounded with the main effects; 12, 14, 24, 16, 17, and 67. As the number of two-factor

TABLE 13

Plackett-Burman Designs of Resolution III

```
N = 12   ++-+++---+-
N = 20   ++--++++-+-+----++-
N = 24   +++++-+-++--++--+-+----
N = 28
                   +-++++---  -+---+--+  ++-+-++-+
                   ++-+++---  --++--+--  -++++-++-
                   -++++++---  +---+--+-  +-+-++-++
                   ---+-++++  --+-+---+  +-+++-+-+
                   ---++-+++  +----++--  ++--++++-
                   ----+++++  -+-+---+-  -+++-+-++
                   +++---+-+  --+--+-+-  +-++-+++-
                   +++---++-  +--+----+  ++-++--++
                   +++----++  -+--+-+--  -++-+++-+
N = 38   -+-++ +---+ ++++- +++-- +---- +-+-+ +--+-
N = 40   Double design for  N = 20
N = 44   ++--+ -+--+ ++-++ +++-- -+-++ +---- -+--- ++-+- ++-
N = 48   +++++ -++++ --+-+ -+++- -+--+ +-++- --+-+ -++-- --+-- --
N = 52
         +  +-+-+-+-+-  +-+-+-+-+-  +-+-+-+-+-  +-+-+-+-+-  +-+-+-+-+-
            -------------------------------------------------------
         +  -+--------  ++----++++  ++++--++--  ++--++--++  ++++++----
         -  ++-+-+-+-+  +--+-++-+-  +-+--++--+  +--++--++-  +-+-+---+-+
         +  ---+------  ++++----++  --++++--++  ++++--++--  --++++++--
         -  -+++-+-+-+  +-+--+-++-  -++-+--++-  +-+--++--+  -++-+-+--+
         +  -----+----  ++++++----  ++--++++--  --++++--++  ----++++++
         -  -+-+++-+-+  +-+-+--+-+  +--++-+--+  -++-+--++-  -+-++-+-+-
         +  --------+--  --++++++--  --++--++++  ++--++++--  ++----++++
         -  -+-+-+++-+  -++-+-+--+  -++--++-+-  +--++-+--+  +--+-++-+-
         +  ---------+  ----++++++  ++--++--++  --++--++++  ++++----++
         -  -+-+-+-+++  -+-++-+-+-  +--++--++-  -++--++-+-  +-+--+-++-
```

interactions is $k(k-1)/2 = 5 \times 4/2 = 10$, there are four more two-factor interactions to be considered. These interactions are: 26, 27, 46, and 47. Equations (143) give $\vec{26} = \vec{47}$, $\vec{27} = \vec{46}$. So besides the grand mean and the main effects we can study one pair of the following four pairs of two-factor interactions: (26, 27), (26, 46), (47, 27), or (47, 46). Notice that from Eq. (139) it follows that $\vec{3} = \vec{26} = \vec{47}$ and $\vec{5} = \vec{27} = \vec{46}$, i.e. the two-factor interactions that can be studied correspond with the main effects that were dropped. In general we conclude that those $(k'-k)$ factors should be dropped which result in the alias-structure that we find most attractive, e.g. drop the factors 3 and 5 if the two-factor interactions 26 and 27 are most interesting among the two-factor interactions. If the interactions that can be estimated are nonexistent then their sums of squares are unbiased estimators of the pure experimental error variance σ^2 [compare possibility (iii)].

Next we shall consider resolution III designs where k factors are studied in $N = k+1$ runs, and N is a multiple of four but not necessarily a power of two. Such designs have been derived by Plackett and Burman (1946). If N is a power of two, then their designs are the same as the fractional factorials we have just discussed. Placket and Burman (1946, pp. 323-324) tabulated resolution III designs for $N \leq 100$.[11)] We have reproduced their designs in Table 13 for those values of N that are a multiple of four but not a power of two. The use of Table 13 is demonstrated below.

(i) For $N = 12$ Table 13 shows

$$+ + - + + + - - - + -$$

This row is written down as the first column in Table 14. The other columns are obtained by shifting this column one place cyclically. Finally one row of minus signs is added.

(ii) For some values of N (like $N = 28$) these cyclical permutations are performed on the blocks themselves. So if the three (9×9) blocks for $N = 28$ are denoted by A, B, and C, then the design is:

(ii) Designs of <u>resolution IV</u>: no main effect is confounded with any other main effect <u>or</u> two-factor interaction, but two-factor interactions are confounded with each other.

(iii) Designs of <u>resolution V</u>: no main effect or two-factor interaction is confounded with any other main effect or two-factor interaction, but two-factor interactions are confounded with three-factor interactions.

In general the <u>resolution</u> of a design is equal to the smallest number of characters in any word of the defining relation. In this section we shall have a closer look at the above types of design.

1. Resolution III Designs

There are resolution III designs available that require only N runs to study (N-1) factors provided N is a multiple of four. For values of N (= k+1), which are not a multiple of four we obtain a resolution III design from the resolution III design for the next value of N, say N' = k'+1, which is a multiple of four and deleting any (k'-k) columns (i.e. factors) from this design. If N is not only a multiple of four but also a power of two, then these resolution III designs are <u>fractional</u> factorial designs, denoted as 2^{k-p}_{III} designs. We shall first consider these 2^{k-p}_{III} designs and then the so-called <u>Plackett</u> <u>and</u> <u>Burman</u> designs where it is only required that N is a multiple of four.

If N must be a <u>power</u> <u>of</u> <u>two</u> then N is 2, 4, 8, 16, 32, etc. Hence k, the number of factors that can be studied in N (= k+1) runs is 1, 3, 7, 15, 31, etc. Consequently the 2^{k-p}_{III} designs for k is 3 through 31 are given by Table 12. We have seen, e.g. in Table 11, how 2^{k-p} designs are constructed by writing down the full factorial in q (= k-p) factors and equating the last p factors to the interactions among the first q factors. Analogous to Eqs. (136) through (140) we can find the alias.

Above we remarked that for values of N which are <u>not</u> a multiple of four we obtain a resolution III design by taking the

TABLE 12

2^{k-p} Designs of Resolution III

N	k	p	Design
4	3	1	2^{3-1}
8	7	4	2^{7-4}
16	15	11	2^{15-11}
32	31	26	2^{31-26}

next value of N, say $N' = k'+1$, being a multiple of four and deleting any $(k'-k)$ columns from the $2_{III}^{k'-p}$ design. These designs are no longer "saturated," i.e. no longer is the number of runs N, equal to the number of parameters $(k+1)$. The consequences are that for these nonsaturated designs we may (i) study more factors in the same number of runs; (ii) estimate some two-factor interactions; (iii) estimate the experimental error variance σ^2. As an example let us consider an experiment with five factors. For five the next multiple of four is eight, i.e. $N' = 8$. With eight runs we could study seven factors [see possibility (i)]. Hence we may drop two factors. If a factor is dropped from a factorial, then the alias structure remains the same except for all words containing the dropped factor; these words vanish. So in the 2_{III}^{7-4} design we had the defining relation given in (139). When dropping e.g. the factors 3 and 5 this relation reduces to

$$\vec{I} = \vec{124} = \vec{167} = \vec{2467} \tag{143}$$

As there are eight runs, whereas only five main effects plus the grand mean are estimated, two interactions can also be estimated. From Eqs. (143) it follows that the following six two-factor interactions cannot be estimated since they are confounded with the main effects; 12, 14, 24, 16, 17, and 67. As the number of two-factor

```
N = 56  Double design for  N = 28
N = 60  ++-++ +-+-+ --+-- +++-+ +++-- +++++ ----- ++--- -+---
              ++-++ -+-+- --+-
N = 68  ++--+ -+--+ +---+ +++-+ -++++ ++--+ ---+- +++-+ +----
              --+-+ ----+ ++--+ +-+-+ +-
N = 72  +++++ ++-++ +-+-- ++-++ +---+ +-+-+ +-+-- -+++- +--+-
              +--++ +---+ --++- +---+ ----- -
N = 76  + +- +- +- +- +- +- +- +- +- +- +- +- +- +- +- +- +- +- +- +- +- +-
        -----------------------------------------------------------------
        + -+ -- ++ -- -- ++ ++ -- ++ -- -- -- -- ++ ++ ++ -- ++ ++ ++ ++ --
        - ++ -+ +- -+ -+ +- +- -+ +- -+ -+ -+ -+ +- +- +- -+ +- +- +- +- -+

           +- +- +- +- +- +- +- +- +- +- +- +- +- +- +-
           ---------------------------------------------
           ++ ++ ++ -- -- -- -- ++ -- ++ ++ -- -- ++ --
           +- +- +- -+ -+ -+ -+ +- -+ +- +- -+ -+ +- -+
N = 80  +++-+ +--++ ++-+- -+-++ ++++- ++--- -++-- -+-+- +-+-+
              ++--+ +++-- +---- --+-+ +-+-- --++- -+--
N = 84  ++-++ --+-+ +++-- -++-- -+-+- +++++ ++-+- -+++- ++--+
              ---++ -+--- ----+ -+-++ +--++ +---- +-++- -+-
N = 88  Double design for  N = 44
N = 92  This design has not yet been obtained
N = 96  Double design for  N = 48
```

TABLE 13 (Continued)

```
N = 100
  + +-+-+-+-+-+-+- +-+-+-+-+-+-+- +-+-+-+-+-+-+- +-+-+-+-+-+-+-
    -------------------------------------------------------------
  + -+------------ ++--++++--++-- ++----++++--++ ++++++------++
  - ++-+-+-+-+-+-+ +--++-+--++--+ +--+-++-+--++- +-+-+--+-+-++-
  + ---+---------- --++--++++--++ ++++----++++-- ++++++++------
  - -+++-+-+-+-+-+ -++--++-+--++- +-+--+-++-+--+ +-+-+-+--+-+-+
  + -----+-------- ++--++--++++-- --++++----++++ --++++++++----
  - -+-+++-+-+-+-+ +--++--++-+--+ -++-+--+-++-+- -++-+-+-+--+-+
  + -------+------ --++--++--++++ ++--++++----++ ----++++++++--
  - -+-+-+++-+-+-+ -++--++--++-+- +--++-+--+-++- -+-++-+-+-+--+
  + ---------+---- ++--++--++--++ ++++--++++---- ------++++++++
  - -+-+-+-+++-+-+ +--++--++--++- +-+--++-+--+-+ -+-+-++-+-+-+-
  + -----------+-- ++++--++--++-- --++++--++++-- ++------++++++
  - -+-+-+-+-+++-+ +-+--++--++--+ -++-+--++-+--+ +--+-+-++-+-+-
  + -------------+ --++++--++--++ ----++++--++++ ++++------++++
  - -+-+-+-+-+-+++ -++-+--++--++- -+-++-+--++-+- +-+--+-+-++-+-

                   +-+-+-+-+-+-+- +-+-+-+-+-+-+- +-+-+-+-+-+-+-
                   ----------------------------------------------
                   ++++------++++ ++++--++++---- ++--++--++++--
                   +-+--+-+-++-+- +-+--++-+--+-+ +--++--++-+--+
                   ++++++------++ --++++--++++-- --++--++--++++
                   +-+-+--+-+-++- -++-+--++-+--+ -++--++--++-+-
                   ++++++++------ ----++++--++++ ++--++--++--++
                   +-+-+-+-+-+-+- -+-++-+--++-+- +--++--++--++-
                   --++++++++---- ++----++++--++ ++++--++--++--
                   -++-+-+-+--+-+ +--+-++-+--++- +-+--++--++--+
                   ----++++++++-- ++++----++++-- --++++--++--++
                   -+-++-+-+-+--+ +-+--+-++-+--+ -++-+--+--++-
                   ------++++++++ --++++----++++ ++--++++--++--
                   -+-+-++-+-+-+- -++-+--+-++-+- +--++-+--++--+
                   ++------++++++ ++--++++----++ --++--++++--++
                   +--+-+-++-+-+- +--++-+--+-++- -----++-+--++-
```

TABLE 14

Plackett-Burman Design for Eleven Factors

$\vec{1}$	$\vec{2}$	$\vec{3}$	$\vec{4}$	$\vec{5}$	$\vec{6}$	$\vec{7}$	$\vec{8}$	$\vec{9}$	$\vec{10}$	$\vec{11}$
+	-	+	-	-	-	+	+	+	-	+
+	+	-	+	-	-	-	+	+	+	-
-	+	+	-	+	-	-	-	+	+	+
+	-	+	+	-	+	-	-	-	+	+
+	+	-	+	+	-	+	-	-	-	+
+	+	+	-	+	+	-	+	-	-	-
-	+	+	+	-	+	+	-	+	-	-
-	-	+	+	+	-	+	+	-	+	-
-	-	-	+	+	+	-	+	+	-	+
+	-	-	-	+	+	+	-	+	+	-
-	+	-	-	-	+	+	+	-	+	+
-	-	-	-	-	-	-	-	-	-	-

A B C
C A B
B C A
\- - -

where the last row consists of only minus signs. In the cases N = 52, 76, and 100 the last row does not consist of only minus signs but of plus and minus signs as indicated above the blocks. If N = k+1 is not a multiple of four we again select the next value of N that is a multiple of four and drop columns not needed.

It is interesting to see that in 2_{III}^{k-p} designs each two-factor interaction is completely confounded with a particular main effect, i.e. the main effect and the associated interaction have identical columns in the matrix of independent variables $\vec{X}$. However, in a Plackett-Burman design a two-factor interaction is not completely confounded with a particular main effect. For example, in Table 14 we can construct the column for the two-factor interaction 12, and this column is not identical with any one of the columns $\vec{1}$ through $\vec{11}$. Of course $\vec{12}$ can be expressed as a linear combination of the eleven main effect columns and the grand mean column as these twelve columns span the twelve-dimensional space. Hence the estimate of an interaction in a Plackett-Burman design is biased by a linear combination of the main effects and grand mean. The consequence is that if an interaction is large, then its estimate may indicate this, even while bias by some main effects occurs. In a 2_{III}^{k-p} design, however, the interaction is completely confounded with one particular main effect. Tukey (1959b, pp. 170-171) has quantified the degree of "concealment" of interaction effects in Plackett-Burman designs, fractional factorials and "random" designs; the latter designs will be discussed later on. His comparisons show that Plackett-Burman designs are best in this respect.[12)] For other types of resolution III designs we refer to John (1971, p. 172).

Let us review the definition of the resolution III designs. We observe that even if we assume that all interactions among three or more factors are zero, then we still obtain only an estimator of the sum or difference of a main effect and some two-factor interactions.[13)] An unambiguous conclusion is possible only if we also assume that all two-factor interactions are zero. More precise conclusions are possible if we proceed to <u>resolution IV designs</u>. Then main effects are no longer confounded with two-factor interactions; the two-factor interactions are confounded with each other.

2. Resolution IV Designs

In order to understand the construction of resolution IV designs we first study the consequence of <u>adding</u> a new fraction to the original 2^{k-p} fraction. Consider the following example of a 2^{7-4} design. Equation (139) shows that the principal fraction has the defining relation (144) if we omit interactions among four or more factors.

$$\vec{I} = \vec{124} = \vec{135} = \vec{236} = \vec{347} = \vec{257} = \vec{167} = \vec{456} \qquad (144)$$

Hence, main effects are confounded with two-factor interactions as specified by Eqs. (145).

$$\begin{aligned} \vec{1} &= \vec{24} = \vec{35} = \vec{67} \\ \vec{2} &= \vec{14} = \vec{36} = \vec{57} \\ \vec{3} &= \vec{15} = \vec{26} = \vec{47} \text{ etc.} \end{aligned} \tag{145}$$

Define the linear combination $\underline{\ell}_j$ of the observations $\underline{y}_i$ as in Eq. (146).

$$\underline{\ell}_j = \frac{2}{N} \sum_{i=1}^{N} \underline{y}_i d_{ij} \qquad (j = 1, \ldots, k) \tag{146}$$

where d_{ij} is the ith element of the jth column of the design matrix $\vec{D}$. Confounding implies that $\underline{\ell}_j$ estimates the main effect of factor j and plus or minus the effects aliased with this main effect. So in this example it follows from Eqs. (145) that

$$\begin{aligned} E(\underline{\ell}_1) &= E\left(\frac{2}{N} \sum_{i=1}^{N} \underline{y}_i d_{i1}\right) = \alpha^1 + \alpha^{24} + \alpha^{35} + \alpha^{67} \\ E(\underline{\ell}_2) &= E\left(\frac{2}{N} \sum \underline{y}_i d_{i2}\right) = \alpha^2 + \alpha^{14} + \alpha^{36} + \alpha^{57} \\ E(\underline{\ell}_3) &= E\left(\frac{2}{N} \sum \underline{y}_i d_{i3}\right) = \alpha^3 + \alpha^{15} + \alpha^{26} + \alpha^{47} \quad \text{etc.} \end{aligned} \tag{147}$$

Suppose that in an additional one-sixteenth fraction the signs in column $\vec{1}$ have been switched, i.e. the generators of the additional fraction are not given by Eqs. (136) but by Eqs. (148).

$$\vec{4} = -\vec{12}, \quad \vec{5} = -\vec{13}, \quad \vec{6} = \vec{23}, \quad \vec{7} = -\vec{123} \tag{148}$$

so (omitting four-factor interactions and higher) the defining relation equals Eqs. (149) instead of Eqs. (139).

$$\vec{I} = -\vec{124} = -\vec{135} = \vec{236} = \vec{347} = \vec{257} = -\vec{167} = \vec{456} \tag{149}$$

Therefore

$$\vec{1} = -\vec{24} = -\vec{35} = -\vec{67}$$
$$\vec{2} = -\vec{14} = \vec{36} = \vec{57}$$
$$\vec{3} = -\vec{15} = \vec{26} = \vec{47} \quad \text{etc.} \tag{150}$$

or

$$E(\underline{\ell}_1') = \alpha^1 - \alpha^{24} - \alpha^{35} - \alpha^{67}$$
$$E(\underline{\ell}_2') = \alpha^2 - \alpha^{14} + \alpha^{36} + \alpha^{57}$$
$$E(\underline{\ell}_3') = \alpha^3 - \alpha^{15} + \alpha^{26} + \alpha^{47} \quad \text{etc.} \tag{151}$$

Combining linear combinations from both fractions gives, for example,

$$E\left(\frac{\underline{\ell}_1 + \underline{\ell}_1'}{2}\right) = \alpha^1$$
$$E\left(\frac{\underline{\ell}_2 - \underline{\ell}_2'}{2}\right) = \alpha^{14}$$
$$E\left(\frac{\underline{\ell}_3 - \underline{\ell}_3'}{2}\right) = \alpha^{15} \quad \text{etc.} \tag{152}$$

So Eqs. (152) show that the main effect of factor 1 and all two-factor interactions between that factor and the other factors can be estimated without confounding with two-factor interactions. Generalizing, we may add a special fraction switching the signs in the column of one particular factor, say factor f, in a 2^{k-p} design of resolution III (or higher). Then we can estimate the main effect of factor f and the two-factor interactions between f and the other factors, <u>without</u> confounding by two-factor interactions. Such a second fraction is useful if the first fraction hints that some factor is of particular importance.

The <u>generators</u> in the above example where two fractions were combined, can be determined as follows. Equation (137) shows that the generators for the principal fraction are

$$\vec{I}_8 = \vec{124} = \vec{135} = \vec{236} = \vec{1237} \tag{153}$$

where the index 8 indicates that there are eight plus signs in the column $\vec{I}$. Now we run an additional fraction from this family of designs, namely the fraction in which factor 1 is switched. Hence the generators for this fraction are

$$\vec{I}_8 = -\vec{124} = -\vec{135} = \vec{236} = -\vec{1237} \tag{154}$$

As in both fractions

$$\vec{I}_8 = \vec{236} \tag{155}$$

we have already obtained one generator for the combined design, which is a 2^{7-3} design with $p = 3$ generators. This generator is

$$\vec{I}_{16} = \vec{236} \tag{156}$$

Further Eqs. (153) and (154) show that

$$\vec{124}\cdot\vec{135} = (-\vec{124})\cdot(-\vec{135}) = \vec{2345} = \vec{I}_{16} \tag{157}$$

and

$$\vec{124}\cdot\vec{1237} = (-\vec{124})\cdot(-\vec{1237}) = \vec{347} = \vec{I}_{16} \tag{158}$$

So Eqs. (156), (157), and (158) give the three generators (from which the defining relation follows). Notice that

$$(\vec{135})\,(\vec{1237}) = (-\vec{135})\,(-\vec{1237}) = \vec{257} = \vec{I}_{16} \tag{159}$$

is not an independent generator as this word can be obtained from Eqs. (157) and (158):

$$\vec{257} = \vec{2345}\cdot\vec{347} \tag{160}$$

In general the (p-1) generators for a combination of two 2^{k-p} fractions from the same family (and any resolution) are obtained by (i) taking the generators from the original 2^{k-p} fractions, with the same sign; and (ii) multiplying an even number of original generators with unequal signs.

We saw above that if a fraction with linear contrasts $\underline{\ell}_j$ as in Eqs. (147) is combined with another fraction with contrasts $\underline{\ell}_j'$ as in Eqs. (151), then

$$\frac{1}{2}(\underline{\ell}_j + \underline{\ell}_j') \tag{161}$$

and

$$\frac{1}{2}(\underline{\ell}_j - \underline{\ell}_j') \tag{162}$$

show which effects can be estimated from the combined fractions. These effects will have less confounding, as is quite natural since their estimates are based on more observations. There are two particularly useful additional fractions. In one fraction, which we have just studied, the signs in the column of one particular factor are switched so that the effects involving that factor can be better studied; in the other fraction the signs in the columns of all factors are switched. The latter fraction permits the construction of a resolution IV design from a resolution III design. As an example Table 15 gives the 2_{III}^{7-4} design, all signs of the principal fraction in Table 11 being switched.

Neglecting interactions among four or more factors, the defining relation of the design in Table 15 is

$$\vec{I} = -\vec{124} = -\vec{135} = -\vec{236} = -\vec{347} = -\vec{257} = -\vec{167} = -\vec{456} \tag{163}$$

Consequently

$$\vec{1} = -\vec{24} = -\vec{35} = -\vec{67}$$
$$\vec{2} = -\vec{14} = -\vec{36} = -\vec{57} \quad \text{etc.} \tag{164}$$

TABLE 15

Switching all Signs in the Principal Fraction of the 2^{7-4} Experiment

Run	$\vec{1}$	$\vec{2}$	$\vec{3}$	$\vec{4} = -\vec{12}$	$\vec{5} = -\vec{13}$	$\vec{6} = -\vec{23}$	$\vec{7} = \vec{123}$
1	+	+	+	-	-	-	+
2	-	+	+	+	+	-	-
3	+	-	+	+	-	+	-
4	-	-	+	-	+	+	+
5	+	+	-	-	+	+	-
6	-	+	-	+	-	+	+
7	+	-	-	+	+	-	+
8	-	-	-	-	-	-	-

or

$$E(\underline{\ell}_1') = \alpha^1 - \alpha^{24} - \alpha^{35} - \alpha^{67}$$

$$E(\underline{\ell}_2') = \alpha^2 - \alpha^{14} - \alpha^{36} - \alpha^{57} \quad \text{etc.} \tag{165}$$

Together with the principal fraction with $E(\underline{\ell}_j)$ given in Eqs. (147) this yields

$$E\left(\frac{\underline{\ell}_1 + \underline{\ell}_1'}{2}\right) = \alpha^1$$

$$E\left(\frac{\underline{\ell}_2 + \underline{\ell}_2'}{2}\right) = \alpha^2 \quad \text{etc.} \tag{166}$$

Further we obtain groups of three two-factor interactions:

$$E\left(\frac{\underline{\ell}_1 - \underline{\ell}_1'}{2}\right) = \alpha^{24} + \alpha^{35} + \alpha^{67}$$

$$E\left(\frac{\underline{\ell}_2 - \underline{\ell}_2'}{2}\right) = \alpha^{14} + \alpha^{36} + \alpha^{57} \quad \text{etc.} \tag{167}$$

So the duplicated fraction with reversed signs together with the original 2^{7-4}_{III} fraction gives a <u>resolution IV</u> design, namely a 2^{7-3}_{IV} design. If we also switch the identity column I_8 and associate factor 8 with the $\vec{I}$ column we can even obtain a resolution IV design in eight factors. This design is called a <u>fold over</u> design since it is constructed by augmenting the original design with a duplicated design with reversed signs.

The <u>generators</u> for the above combined design can be obtained in the usual manner, as described in Eqs. (153) through (160). So the generators for the first 2^{8-5} fraction are

$$\vec{I} = \vec{8} = \vec{124} = \vec{135} = \vec{236} = \vec{1237} \tag{168}$$

where $\vec{I} = \vec{8}$ follows from the association of factor 8 with the identity column; the remaining generators are the generators we derived before; compare for instance Eqs. (137). The generators for the second 2^{8-5} fraction are

$$\vec{I} = -\vec{8} = -\vec{124} = -\vec{135} = -\vec{236} = \vec{1237} \tag{169}$$

As we switched signs in the second fraction, words consisting of an odd number of factors got a minus sign. From Eqs. (168) and (169) we see immediately that $\vec{1237}$ is a generator of the combined fraction; the remaining three independent generators of the 2^{8-4}_{IV} design are, for instance,

$$\begin{aligned} \vec{8}\cdot\vec{124} &= (-\vec{8})\cdot(-\vec{124}) = \vec{1248} \\ \vec{8}\cdot\vec{135} &= (-\vec{8})\cdot(-\vec{135}) = \vec{1358} \\ \vec{8}\cdot\vec{236} &= (-\vec{8})\cdot(-\vec{236}) = \vec{2368} \end{aligned} \tag{170}$$

The above derivation demonstrates that in general the generators for a <u>fold over</u> design are (i) the generators in the original fraction with an even number of factors (in the example $\vec{1237}$); (ii) the generators in the original fraction with an odd number of factors, augmented with the extra factor associated with the $\vec{I}$-column (in the example $\vec{1248}$, $\vec{1358}$, $\vec{2368}$).

Besides the fold over technique there is a second method for the construction of 2_{IV}^{k-p} designs; compare (128) through (135). Write down the full factorial in $(k-p) = q$ factors and associate the remaining p factors with the interactions among an odd number (3, 5, ...) of those first q factors. For example, the previous 2_{IV}^{8-4} design could be constructed as follows. Write down the 2^4 full factorial. For comparison with the fold over technique, mention those q (= 4) factors: 1, 2, 3, and 8. Associate the remaining p (= 4) factors as in Eqs. (171).

$$
\begin{aligned}
\vec{128} &= \vec{4} \\
\vec{138} &= \vec{5} \\
\vec{238} &= \vec{6} \\
\vec{123} &= \vec{7}
\end{aligned}
\tag{171}
$$

Comparing Eqs. (171) with (170) shows that the designs are the same.

We saw above how we can construct a 2_{IV}^{k-p} design from a $2_{III}^{k'-p}$ $(k = k'+1)$ design by the fold over technique. Box and Wilson (1951, p. 35) proved the following theorem.

> "Suppose we use a particular design for the estimation of the k linear effects (not necessarily an incomplete factorial or even an orthogonal design) with $N \times k$ matrix $\vec{D}_1$; also suppose that $\vec{X}$ is the corresponding $N \times (k+1)$ matrix of independent variables. Then
>
> $$\vec{D}_2 = \begin{bmatrix} \vec{X} \\ -\vec{X} \end{bmatrix}$$
>
> will provide a design matrix for the estimation of the linear effects of $k+1$ factors and, however biased by second-order effects were the estimates of first-order effects using $\vec{D}_1$, the first-order estimates using $\vec{D}_2$ will be completely free of bias due to this cause."

From this theorem it follows that, since a Plackett and Burman design is a resolution III design, its fold over is a resolution IV design. The fold over technique doubles the number of runs in the resolution III designs, but allows one more factor to be studied by association of that factor with the identity-column in the resolution III design. We have seen that resolution III designs exist for studying k factors in only $N = k+1$ runs, if N is a multiple of four. So the number of runs in a resolution IV design for studying k factors is $2k$ if k is a multiple of four. This is also shown in Table 16.

If k is not a multiple of four we can take the resolution IV design that corresponds with the next value of k that is a multiple of four. For example, such a resolution IV design for three factors would consist of eight runs; for five, six or seven factors it would consist of sixteen runs. However, Webb (1968a) has derived nonorthogonal resolution IV designs for k factors in $2k$ runs, where k is not a multiple of four. His designs are also based on the fold over technique. By definition these resolution IV designs allow estimates of the main effects not biased by two-factor interactions, while two-factor interactions are confounded with each other.

TABLE 16

Resolution III and IV Designs of Minimal Size for Studying k Factors in N runs

Resolution III		Resolution IV	
k	N	k	N
3	4	4	8
7	8	8	16
11	12	12	24
15	16	16	32
19	20	20	40
⋮	⋮	⋮	⋮

Since Webb's designs are nonorthogonal the estimators of the main effects are mutually dependent and their variances are larger than if an orthogonal design existed. In Table 17 Webb's designs are summarized. In this table the factor combination in the last column is denoted as in the last column of Table 6. Only the first N/2 runs are given since the next N/2 runs are simply the fold over of the first half of the design. To illustrate the use of Table 17 we have written down the complete matrix of independent variables up to two-factor interactions for k = 3; see Table 18.

TABLE 17

Webb's Nonorthogonal Resolution IV Fold Over Designs

k	N	variance for main effects	runs in half of design
3	6	$1/4\ \sigma^2$	a, b, c
5	10	$1/9\ \sigma^2$	a, b, c, d, e
6	12	$1/10\ \sigma^2$	ab, ac, bc, d, e, f
7	14	$11/100\ \sigma^2$	a, b, c, d, e, f, g

TABLE 18

Webb's Design for Three Factors

Run	I	A	B	C	AB	AC	BC
1	+	+1	-1	-1	-1	-1	+1
2	+	-1	+1	-1	-1	+1	-1
3	+	-1	-1	+1	+1	-1	-1
4	+	-1	+1	+1	-1	-1	+1
5	+	+1	-1	+1	-1	+1	-1
6	+	+1	+1	-1	+1	-1	-1

Notice that in Table 18 any main effect column is orthogonal to any two-factor interaction column and to the grand mean column. No main effect column is orthogonal to any other main effect column and no two-factor interaction column is orthogonal to any other two-factor interaction column or the grand mean column. So not all columns of the main effects, the confounded two-factor interactions and the grand mean are orthogonal, and therefore the estimation of the effects is more complicated than in 2_{IV}^{k-p} designs. We have to express the ANOVA model as the general regression model (95), i.e.

$$
\begin{aligned}
\underline{y}_i &= \sum_{j=1}^{J} \gamma_j x_{ij} + \underline{e}_i \\
&= \beta_0 + \sum_{s=1}^{k} \beta_s d_{is} + \sum_{s=1}^{k-1} \sum_{z=s+1}^{k} \beta_{sz}(d_{is} d_{iz}) + \underline{e}_i
\end{aligned}
$$

$$(i = 1, \ldots, N) \tag{172}$$

where β_0 denotes the grand mean, β_s is half the main effect α^s of factor s, and β_{sz} is half the interaction α^{sz} between the factors s and z together; $J = 1 + k + k(k-1)/2$ parameters. Since there are only N $(= 2k)$ observations we cannot estimate all J parameters. In Appendix IV.3 we derive that the main effects

$$\vec{\beta}_M' = (\beta_1, \beta_2, \ldots, \beta_k) \tag{173}$$

have the least squares estimators

$$\underline{\vec{b}}_M = \frac{1}{2} (\vec{U}'\vec{U})^{-1} (\vec{U}', -\vec{U})\underline{y} \tag{174}$$

$\vec{U}$ corresponding with the main effects and being defined by

$$\vec{U} = (d_{uj}) \qquad u = 1, \ldots, \frac{1}{2} N, \quad j = 1, \ldots, k \tag{175}$$

and

$$\underline{\vec{y}}' = (\underline{y}_1, \underline{y}_2, \ldots, \underline{y}_N) \tag{176}$$

It can be shown that Eq. (174) gives estimators that are not biased by two-factor interactions or the grand mean. Further we derived that the matrix of variances and covariances of $\vec{\underline{b}}_M$ is equal to

$$\vec{\Omega}_M = \frac{1}{2}\,\sigma^2(\vec{U}'\vec{U})^{-1} \tag{177}$$

The two-factor interactions cannot be uniquely determined for there are not enough observations available. Some interactions can be assigned arbitrary values which then determine the value of the remaining interaction estimates. We put (J-N) two-factor interactions equal to zero, i.e. the corresponding vector of parameters, say $\vec{\beta}_3$, is zero and their estimators satisfy

$$\vec{\underline{b}}_3 = \vec{0} \tag{178}$$

The interactions in $\vec{\beta}_3$, assumed zero, should be selected such that the remaining $(N-k-1) = k-1$ interactions together with the grand mean form a matrix of independent variables $\vec{X}_2$ which together with the matrix for the main effects

$$\vec{X}_1 = \begin{bmatrix} U \\ \\ -U \end{bmatrix} \tag{179}$$

yields a <u>nonsingular</u> matrix $[\vec{X}_1, \vec{X}_2]$; as compared with (3.13) in Appendix 3. The grand mean and those interactions that correspond to $\vec{X}_2$ are estimated by

$$\vec{\underline{b}}_2 = (\vec{X}_2'\vec{X}_2)^{-1}\,\vec{X}_2'\vec{\underline{y}} \tag{180}$$

It can easily be shown that $\vec{\underline{b}}_2$ is unbiased if the interactions in $\vec{\beta}_3$ are zero. Further the matrix of variances and covariances for $\vec{\underline{b}}_2$ is given by

$$\vec{\Omega}_2 = \sigma^2(\vec{X}_2'\vec{X}_2)^{-1} \tag{181}$$

The t-test can be used to test a single regression parameter β, the F-test to test several parameters jointly [compare the general theory on testing regression parameters, e.g. Johnston (1963, pp. 115-135)]. Note that first we can estimate the main effects from Eq. (174). Then we can set those two-factor interactions to zero, which correspond with small main effects. This last procedure is limited by the requirement that $\vec{X}_2$ should be chosen such that $[\vec{X}_1, \vec{X}_2]$ is nonsingular.

A further discussion of resolution IV designs is given by John (1971, pp. 173-174), Marjolin (1969), and Srivastava and Anderson (1970); also compare Srivastava and Anderson (1969). If in a resolution IV design all estimates of the linear combinations of two-factor interactions are small, we are inclined to conclude that only main effects are significant. This last conclusion is not foolproof, as there remains the possibility of large two-factor interactions that cancel out exactly in their linear combination. Therefore, if we want to estimate all two-factor interactions separately, or if some linear combination of two-factor interactions estimated in a resolution IV design is large, then we may want to proceed to a resolution V design.

3. Resolution V designs

Resolution V designs give estimates of main effects and two-factor interactions which are not confounded with other main effects or two-factor interactions. This means that all words in the defining relation comprises five or more characters. These better estimates are obtained at the expense of more observations. Resolution V fractionals for two, three, or four factors do not exist. Table 19 shows the fractionals of resolution V and higher as given by Box and Hunter (1961b, p. 450). To illustrate the use of Table 19 let us consider the design for eleven factors. The 2^{11-4} design consists of $2^7 = 128$ runs. So start by writing down the full factorial for seven factors. Associate the remaining four factors as indicated in the table, i.e.

TABLE 19

Fractional Factorials of Resolution V and Higher

Number of factors (k)	Number of runs (N)	Fraction (2^{-p})	Type of design	Remaining factors
5	16	$\frac{1}{2}$	2_V^{5-1}	$\pm$ 5 = 1234
6	32	$\frac{1}{2}$	2_{VI}^{6-1}	$\pm$ 6 = 12345
7	64	$\frac{1}{2}$	2_{VII}^{7-1}	$\pm$ 7 = 123456
8	64	$\frac{1}{4}$	2_V^{8-2}	$\pm$ 7 = 1234
				$\pm$ 8 = 1256
9[a]	128	$\frac{1}{4}$	2_{VI}^{9-2}	$\pm$ 9 = 14578
				$\pm$ 10 = 24678
10	128	$\frac{1}{8}$	2_V^{10-3}	$\pm$ 8 = 1237
				$\pm$ 9 = 2345
				$\pm$ 10 = 1346
11	128	$\frac{1}{16}$	2_V^{11-4}	$\pm$ 8 = 1237
				$\pm$ 9 = 2345
				$\pm$ 10 = 1346
				$\pm$ 11 = 1234567

[a] The nine factors in this design are labeled 1, 2, 4, 5, 6, 7, 8, 9, and 10.

$$\begin{aligned} \pm\ 8 &= 1237 \\ \pm\ 9 &= 2345 \\ \pm\ 10 &= 1346 \\ \pm\ 11 &= 1234567 \end{aligned} \tag{182}$$

From (182) the generators and the identifying relation follow immediately. We remark that Table 19 shows that the number of runs for seven and eight factors are the same. Therefore in an experiment with seven factors it might be worthwhile to look for a possible eighth factor as this does not lead to more runs. On the other hand, however, the design for seven factors is of resolution VII rather than V. Analogous remarks can be made for the designs with nine, ten, or eleven factors.[14)] The reader who wants to construct his own fractionals of resolution V for more than eleven factors is advised to read the exposé by Box and Hunter (1961b, p. 455).

Though the designs of Table 19 are incomplete they may imply considerably more observations than the number of effects to be estimated. Daniel (1956, p. 92) gives Table 20 for resolution V designs, where E measures "efficiency" defined as the number of main effects and two-factor interactions divided by the total number of degrees of freedom. The degrees of freedom are equal to the number of observations less one degree of freedom for the grand mean, i.e.

$$E = \frac{k + k(k-1)/2}{N-1} = \frac{k(k+1)}{2(N-1)} \tag{183}$$

Owing to the inefficiency of the 2_V^{k-p} designs other types of fractionals have been derived, such as the "reduced" designs of Whithwell and Morbey (1961) and the "irregular fractions" or "3.2^{-p} replicates" of Addelman (1961). These special designs of resolution V require less observations than the 2_V^{k-p} factorials. Unfortunately, this advantage is realized by assuming certain (not all) two-factor interactions to be zero or by nonorthogonality (i.e. the estimators of the effects are correlated). We find the fractionals constructed later on by Rechtschaffner (1967) more attractive. His fractions of 2^k factorials are saturated, i.e. there are as many observations as there are parameters to be estimated. The number of parameters is: one general mean, k main effects, and $k(k-1)/2$ two-factor interactions. The construction of Rechtschaffner's designs is very simple.

TABLE 20

Efficiency of 2_V^{k-p} Designs

k	p	2^{-p}	N	E
5	1	$\frac{1}{2}$	16	1.00
6	1	$\frac{1}{2}$	32	0.68
7	1	$\frac{1}{2}$	64	0.44
8	2	$\frac{1}{4}$	64	0.56
9	2	$\frac{1}{4}$	128	0.35
10	3	$\frac{1}{8}$	128	0.43
11	4	$\frac{1}{16}$	128	0.52
12	4	$\frac{1}{16}$	256	0.31
13	5	$\frac{1}{32}$	256	0.36
14	6	$\frac{1}{64}$	256	0.41
15	7	$\frac{1}{128}$	256	0.47

Associate one "generator" with each kind of effect as indicated in Table 21. The remaining factor combinations are obtained by permuting the elements of each generator. This is demonstrated for five factors in Table 22.

The saturated fraction of the 2^5 factorial in Table 22 demonstrates the construction of fractions for other numbers of factors. Actually this design is a special case since for five factor Rechtschaffner's design is exactly the same as the 2_V^{5-1} design. For other numbers of factors, however, the saturated designs give correlated estimators of the parameters, whereas 2_V^{k-p} designs allow orthogonal estimation. Hence, the variance of the response, averaged

TABLE 21

Design Generators for Rechtschaffner's Saturated Fractions of 2^k Factorial Designs

Number	Type	Generator	
I	Grand mean	(-1, ..., -1)	for all k
II	Main effects	(-1, 1, ..., 1)	for all k
III	Two-factor Interactions	(-1, -1, 1)	for $k = 3$
		(1, 1, -1, ..., -1)	for $k > 3$

TABLE 22

Saturated Fraction of 2^5 Design

Run	Generator	x_1	x_2	x_3	x_4	x_5
1	(-1, -1, -1, -1, -1)	-1	-1	-1	-1	-1
2	(-1, 1, 1, 1, 1)	-1	1	1	1	1
3		1	-1	1	1	1
4		1	1	-1	1	1
5		1	1	1	-1	1
6		1	1	1	1	-1
7	(1, 1, -1, -1, -1)	1	1	-1	-1	-1
8		1	-1	1	-1	-1
9		1	-1	-1	1	-1
10		1	-1	-1	-1	1
11		-1	1	1	-1	-1
12		-1	1	-1	1	-1
13		-1	1	-1	-1	1
14		-1	-1	1	1	-1
15		-1	-1	1	-1	1
16		-1	-1	-1	1	1

over the experimental region and taken per observation, is larger for the saturated designs. This is shown in Table 23, taken from Rechtschaffner (1967, p. 573), and illustrates that the saturated designs are less efficient. However, the number of runs in these designs is much smaller, especially for a large number of factors. We therefore suggest that the experimenter who wants to choose between the traditional 2_V^{k-p} designs and Rechtschaffner's saturated designs, takes into account the following.

(a) Saturated designs are attractive if there is little (computer) time available for experimentation, as these designs imply much fewer observations especially for a larger number of factors.

(b) Lower variances of the estimated response and parameters mean more reliable estimates, i.e. lower probability of false conclusions; false conclusions imply losses. Traditional 2_V^{k-p} designs give these more reliable estimates, even if we take into account that this higher reliability is achieved by taking more observations. So when computer time for experimentation is not restricted in a technical sense, we might be willing to take more observations, if we think that the gain in reliability (which is equal to the decrease in possible losses) is worth the increase in computer time costs. The exact solution of the problem of decreasing losses and increasing experimental cost should be found along the lines of decision theory. Unfortunately such a solution is not available and in general we can say that at present decision theory usually leads to analytical formulas which are hard to evaluate numerically.

The experimenter may not want a saturated design if he desires smaller variances and covariances or degrees of freedom remaining for estimating the experimental error σ^2. (Moreover, in real world experiments as opposed to simulation the number of experimental units N may be given by the available machines, etc.) We refer to Srivastava and Chopra (1971) for resolution V designs which are not necessarily saturated. They tabulated designs for four, five, and six factors (and $11 \leq N \leq 28$, $16 \leq N \leq 32$, $22 \leq N \leq 40$, respectively).

TABLE 23. Saturated vs. 2_V^{k-p} Fractional Factorial Designs

		Number of runs N			Efficiency for parameters	
Number of factors (k)	Fraction 2^{-p} in 2_V^{k-p}	2_V^{k-p}	Saturated design	Efficiency for response (E_y)[a]	E_μ[b]	$E_i = E_{ij}$[c]
3	1	8	7	57.1	57.1	57.1
4	1	16	11	72.7	93.5	65.5
5	1/2	16	16	100.0	100.0	100.0
6	1/2	32	22	86.1	82.6	87.0
7	1/2	64	29	62.8	45.1	68.5
8	1/4	64	37	43.7	21.2	53.6
9	1/4	128	46	30.7	10.4	42.6
10	1/8	128	56	22.2	5.6	34.5

a)
$$E_y = \frac{(\text{"averaged" variance of response in } 2_V^{k-p})/N_V}{(\text{"averaged" variance of response in saturated design})/N_S} \times 100 \quad \text{where} \quad N_V = 2^{k-p},$$
$$N_S = 1 + k + k(k-1)/2.$$

b)
$$E_\mu = \frac{(\text{var}(\underline{\hat{\mu}}) \text{ in } 2_V^{k-p})/N_V}{(\text{var}(\underline{\hat{\mu}}) \text{ in saturated design})/N_S} \times 100$$

c) Same as E_μ with μ replaced by the main effect α^i or the interaction α^{ij}.

A special, rather sophisticated use of fractionals is given by Box and Hunter (1961a, pp. 318-319, 349-350). They use "blocks" to estimate the main effects and all interactions of the "major" factors while at the same time estimating possible effects of "minor" factors, i.e. factors that are assumed to show only main effects, but no interactions at all. We have paid no attention to "blocking" as this procedure is not needed in computer simulation experiments. The reader interested in blocking (either for the elimination of heterogeneous influences in real life experiments or for the construction of special designs) is referred to Hicks (1966) for definitions and to Box and Hunter (1961a, pp. 330-333, 345-349; 1961b, pp. 452-458) for actual blocking arrangements of 2^{k-p} designs. A catalogue of 2^{k-p} designs (k ranging from 5 through 16 and p from 1 through 8) together with their blocking arrangement is given in Fractional Factorial Designs for Factors at Two Levels (1957). The reader may also wish to consult Draper and Mitchell (1967; 1968). An extensive discussion of so-called balanced and partially balanced incomplete block (or PBIB) designs is given by John (1971, pp. 219-328). For simulation experiments some authors, e.g. Naylor et al. (1967a, p. 324) suggest considering the random number sequence as a block when simulating several systems with the same random numbers. However, no replication error remains, and observations within a block are dependent, so ANOVA of such a design becomes questionable. Mihram (1972, p. 401) proposes the creation of a block by generating observations that partly use the same random numbers, e.g. only arrivals use the same random numbers but service times do not. Yet ANOVA is still complicated by such an approach. Moreover, from a design point of view it is recommended that all random numbers be the same when comparing several systems; see Chapter III.

If all factors are quantitative a more general model is usually assumed which besides the two-factor interactions also contains pure quadratic terms, i.e.

$$E(\underline{y}_i) = \beta_0 + \sum_{s=1}^{k} \beta_s d_{is} + \sum_{s=1}^{k-1} \sum_{z=s+1}^{k} \beta_{sz}(d_{is}d_{iz}) + \sum_{s=1}^{k} \beta_{ss} d_{is}^2 \quad (184)$$

So it is assumed that the response can be adequately approximated by a second-order polynomial regression equation. It may be that a transformation is first applied so that Eq. (184) will be an adequate approximation; compare Box and Tidwell (1962). We observe that Eq. (184) is only an approximation because we actually know that the computer simulation program is a better model. Nevertheless such an approximation is useful for detecting the important factors and their desirable levels; also compare Naylor and Hunter (1969, p. 24). Note that the approximation can be interpreted as the Taylor series expansion of the true function; see, e.g. Mihram (1972, pp. 359-360). Purely quantitative factors are the domain of response surface methodology which we briefly discussed in Sec. II.6. A multitude of designs has been developed for RSM. Here we give only the basic design used to fit a second-order polynomial such as Eq. (184), namely, the central composite design. In order to be able to estimate all the coefficients in (184) this design combines the fractional (or for low k the full) factorial 2^{k-p} design with a "star" design and center points, i.e. the 2^{k-p} experimental points of the unit cube denoted by $(\pm 1, \pm 1, \ldots, \pm 1)$ are augmented with the $2k$ "axial" points of the star design

$$
\begin{array}{c}
(+a, \quad 0, \ldots, \quad 0) \\
(-a, \quad 0, \ldots, \quad 0) \\
(\ 0, +a, \ldots, \quad 0) \\
(\ 0, -a, \ldots, \quad 0) \\
\ldots \\
(\ 0, \quad 0, \ldots, +a) \\
(\ 0, \quad 0, \ldots, -a)
\end{array}
\tag{185}
$$

and, say, n_0 replicates at the center point $(0, \ldots, 0)$. For $k = 2$ the central composite design is pictured in Fig. 3. The

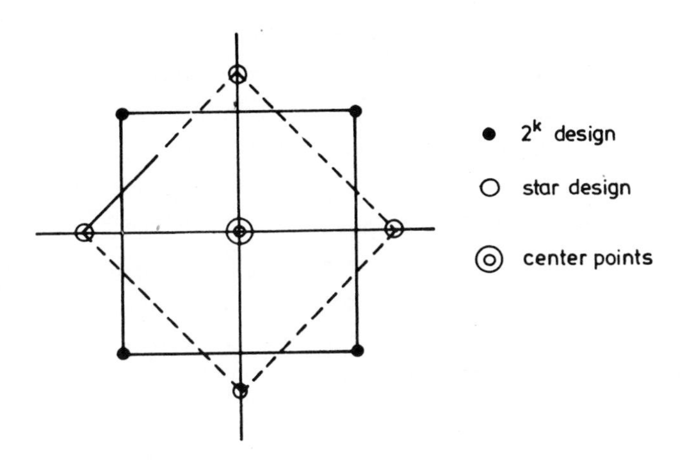

Fig. 3. Central composite design for two factors.

value of a and n_0 can be so chosen that the variance of the estimated response is constant at a fixed distance from the center of the design (so-called "rotatability" of the design) and the bias because of possible third order coefficients in (184) is minimized. For literature on these designs, many other designs and RSM in general we refer to the special bibliography on RSM at the end of this chapter.

Most simulation and Monte Carlo studies do not utilize the results of the theory on experimental design. There are exceptions, e.g. Smith (1968, pp. 20, 24) employed a 2^{3-1}_{III} and a 2^{7-4}_{III} design while Bonini (1967, pp. 85-96) used a 2^{8-2}_{IV} design (arranged in blocks); for a summary see Bonini (1971). Ignall (1972) used RSM in an inventory simulation (together with common random numbers); Boyd (1964, pp. 74-78) applied a rotatable RSM design in a company simulation; also see the bibliography on RSM (Category D, Applications); Mihram (1970) discussed factorial experimentation in the simulation of a military airlift system. Clough et al. (1965, p. 127), Doornbos (1965, p. 209), and McQuie (1969) applied "mixed" designs, i.e. designs in which not all factors have the same number of levels (but, e.g. some have two levels and other ones have three levels). Overholt (1970) discussed a number of experimental design and analysis techniques and applied them to a military simulation experiment. We shall apply a 2^{7-3}_{IV} design in the Monte Carlo experiment reported in Chapter VI. Ignall (1971) and Jacoby and Harrison (1962, p. 134) pointed out that in simulation the number of factors (k) may be reduced by a theoretical analysis of the system prior to experimentation. Nolan and Sovereign (1972) applied linear programming to a global, aggregated model of the system and used simulation to study the most important factors in detailed models of subsystems; this simulation employed a 2^3 design.

IV.6. EXPERIMENTAL ERROR VARIANCE, LACK OF FIT, AND SEQUENTIALIZATION

We have seen that in a 2^k design the $N = 2^k$ observations are used to estimate the grand mean and the (2^k-1) possible effects (main effects and interactions). So no degrees of freedom remain for the estimation of the experimental error variance σ^2. In a 2^{k-p} design with $p \geq 1$ the grand mean, confounded with the (2^p-1) effects in the defining relation, can be estimated. The remaining degrees of freedom are used by the $(2^{k-p}-1)$ estimates of confounded effects, not mentioned in the defining relation. Again no degrees of freedom remain for estimating error. Nevertheless, an estimate of σ^2 is needed to establish confidence intervals for the various effects and to test these effects.

To obtain an estimate of experimental error we may _duplicate_ one or more runs. True duplication can be a problem in real-life experimentation as Dykstra (1959, p. 63) pointed out.[15] However, in simulation independent duplicates are obtained simply by using new sequences of random numbers. Webb (1969, p. 430) states that in simulation experiments "there is no such thing as experimental error in the usual sense." Nevertheless, we think that in simulation we have pure experimental error as opposed to real-life experiments, where systematic factors like time-trends may disturb pure error. If in an experimental design we obtain $J_g \geq 2$ observations for the gth experimental point $(g = 1, 2, \ldots, G \leq N)$, then an unbiased estimator of the experimental error variance at point g is

$$\hat{\sigma}_g^2 = \frac{\Sigma_{j=1}^{J_g} (y_{gj} - \bar{y}_g)^2}{(J_g - 1)} \qquad (g = 1, \ldots, G) \tag{186}$$

where y_{gj} is the jth observation at point g and $\bar{y}_g$ is the average at point g. Obviously the estimator in (186) has (J_g-1) degrees of freedom. In ANOVA the experimental error variance is assumed to be constant over all points. Hence the weighted average (with weights J_g-1) or the pooled estimator of this common variance σ^2 is

$$\hat{\sigma}^2 = \left[\sum_{g=1}^{G} \hat{\sigma}_g^2 (J_g - 1) \right] \left[\sum_{g=1}^{G} (J_g - 1) \right]^{-1} \tag{187}$$

where G points are duplicated. We can write Eq. (187) as a pooled mean square. For define

$$SS_g = \hat{\sigma}_g^2 (J_g - 1) = \sum_{j=1}^{J_g} (y_{gj} - \bar{y}_g)^2 \tag{188}$$

and

$$df_g = (J_g - 1)$$

Then Eq. (187) can be written as

$$\hat{\sigma}^2 = \left(\sum_{g=1}^{G} SS_g \right) \Big/ \left(\sum_{g=1}^{G} df_g \right) \tag{190}$$

The number of degrees of freedom of this pooled estimator is equal to[16)]

$$N_2 = \sum_{g=1}^{G} (J_g - 1) = \sum_{g=1}^{G} J_g - G \tag{191}$$

Let us say that we have $(J_g - 1)$ "duplicates" at point g if we have J_g observations at that point. So the total number of duplicates is N_2. For the estimator of experimental error it does not matter how we assign the N_2 duplicates to the various experimental points in so far as any assignment results in N_2 degrees of freedom. So we could take all N_2 observations at the same point. The additional observations, however, can also be used to <u>re-estimate the effects</u>. Then the allocation of the N_2 observations influences the reliability of the various effect-estimates. Therefore Dykstra (1959) suggests a particular pattern of duplication such that besides an estimate of the experimental error variance more reliable estimates of the effects are obtained. He considers only 2_V^{k-p} designs $(p \geq 0)$ and proposes to make the duplicates form a 2^{k-q} design $(q > p)$. We have summarized his designs in Table 24, where the last column refers to the pages in Dykstra (1959). For the sake of completeness we mention that Patel (1963) gives partial duplicates of the irregular fractional designs of Addelman (1961), these duplicates requiring less observations than Dykstra; Dykstra (1960) also derived duplication patterns for composite designs in RSM. More recently, Dykstra (1971, a and b) discussed the selection of additional experimental points, when these points are not necessarily duplicates of old points but may be new points. An additional point is so selected that the determinant of $\vec{X}'\vec{X}$ is maximized. [Maximization of $|\vec{X}'\vec{X}|$ is a standard criterion for designs that are not necessarily orthogonal; compare Dykstra (1971a, pp. 682-683) and Box and Draper (1971, pp. 732-733) and (1972).] He showed that for each candidate point, say $\vec{x}_0$, the corresponding variance of the estimated response, i.e. $\mathrm{var}(\hat{\underline{y}}|\vec{x}_0)$ should be estimated and the point corresponding with the greatest variance should be selected.[17)] Examples of this procedure are provided by Hebble and Mitchell (1972). Once we have obtained the duplicates (or new experimental points), we have to analyze the augmented design. Of course we can use a general regression analysis program. If we wish to take advantage of the factorial layout then we could apply Dykstra's (1959) special formulas for the designs of

TABLE 24

Some Partial Duplicates of 2_V^{k-p} Factorials

Number of factors (k)	Number of runs in original 2^{k-p} (N_1)	Original generators	Number of additional runs (N_2)	Additional generators	Dykstra (1959) (page numbers)
3	8	...	4	- ABC	64
4	16	...	8	ABCD	70
5	16	- ABCDE	8	AB - CDE	70
6	32	ABCDEF	16	- ABC	70
7	64	- ABCDEFG	16	- ABC - CDE	71
8	64	- ACDFH - BDEGH	16	- ABC - CDE	72
9	128	ABDEGH - ACDEFGJ	16	- ABC - EFG - ADG	72
10	128	- ABEGH - ACFGK ABCDFJ	16	- ADJ - BCF - BDH	73
11	128	- ABEFL - BCFGK ADEFGH BCDEFJ	16	- ABC - CDE - EFG	74

Table 24. We prefer applying the simpler formulas derived by Box (1966) for augmented designs. His formulas are valid for Dykstra's partially duplicated designs but also for nonoverlapping and partially overlapping additional design points; compare Box (1966, p. 186). The procedure runs as follows. The original 2^{k-p} design with $N_1 = 2^{k-p}$ runs gives the independent least squares estimators $\underline{\ell}_1$, $\underline{\ell}_2, \ldots, \underline{\ell}_m$ $(m = N_1 - 1)$ of the main effects and interactions $\underline{\gamma}_1$, $\gamma_2, \ldots, \gamma_m$ (where γ_1 may denote α^A, γ_2 may denote α^{AB}, etc.) A second smaller 2^{k-q} fractional is added with $N_2 = 2^{k-q}$. From these N_2 runs a single least squares estimator $\underline{L}_2$ is obtained for the s confounded effects in the linear function λ $(s \leq m)$

$$\lambda = \delta_1\gamma_1 + \cdots + \delta_i\gamma_i + \cdots + \delta_s\gamma_s \tag{192}$$

where the δ's are plus or minus one depending on the aliases. The least squares estimator for λ based on the first experiment would be

$$\underline{L}_1 = \delta_1\underline{\ell}_1 + \cdots + \delta_i\underline{\ell}_i + \cdots + \delta_s\underline{\ell}_s \tag{193}$$

If $\underline{L}_1$ and $\underline{L}_2$ differ quite a bit, we shall correct our estimate of γ_i accordingly. Actually Box shows that the least squares estimator based on <u>both</u> experiments is

$$\hat{\underline{\gamma}}_i = \underline{\ell}_i + \frac{\delta_i N_2}{N_1 + N_2 s}(\underline{L}_2 - \underline{L}_1) \qquad (i = 1, 2, \ldots, s) \tag{194}$$

An example of the application of Box's formulas is given in Appendix IV.4. Box further derived

$$\operatorname{var}(\hat{\underline{\gamma}}_i) = \frac{4\sigma^2}{N_1}\left[\frac{N_1 + N_2(s - 1)}{N_1 + N_2 \cdot s}\right] \tag{195}$$

where the factor in square brackets indicates the variance reduction obtained from the second experiment; compare Eq. (98). If γ_i and γ_j both occur in λ, then

$$\mathrm{cov}(\hat{\underline{\gamma}}_i, \hat{\underline{\gamma}}_j) = \pm \frac{4N_2\sigma^2}{N_1(N_1 + N_2\ s)} \qquad (196)$$

where the minus sign is used in Eq. (196) if both γ_i and γ_j have the same sign in the λ relation (192); otherwise the plus sign is used. We can test hypotheses concerning the effects and determine confidence intervals using Eq. (195) and the familiar t-statistic

$$\underline{t}_{N_2} = \frac{\hat{\underline{\gamma}}_i - \underline{\gamma}_i}{[\mathrm{var}(\hat{\underline{\gamma}}_i)]^{1/2}} \qquad (197)$$

with the degrees of freedom for t equal to the degrees of freedom on which the estimate of σ^2 is based, i.e. $N_2 = 2^{k-q}$. For a joint test on more than one effect we suggest using the general formulas in, e.g. Johnston (1963, pp. 115-133) for testing least squares estimators; Dykstra (1959, p. 65) proposes another procedure which seems rather arbitrary.[18)]

An alternative, possibly biased, procedure for estimating σ^2 does not duplicate observations but assumes that certain effects are _nonexistent_. In Eq. (33) we saw that the expected value of the mean square of an effect is σ^2 if that effect is zero. If m effects are assumed to be nonexistent then we can pool their independent sums of squares, i.e.

$$\frac{\underline{SS}_1 + \cdots + \underline{SS}_j + \cdots + \underline{SS}_m}{df_1 + \cdots + df_j + \cdots + df_m} \quad (m \geq 1) \qquad (198)$$

is an estimator of the variance σ^2 with degrees of freedom equal to $(df_1 + \cdots + df_m)$. For a 2^{k-p} $(p \geq 0)$ design relation (198) reduces to relation (103) since $df_j = 1$. The problem remains as to how to _select_ the effects that go into relation (198). If we had _duplicated_ experimental points then we could test whether an effect is insignificant as we observed in our comment on Eq. (33). If we have no duplicates then we cannot test the significance of an effect in this way. Therefore Daniel (1956, pp. 93-95; 1959) developed a graphical procedure called "half-normal plots" that might be used to judge which effects are significant. Unfortunately as Daniel (1959, pp. 338, 339) points out "the use of half-normal plots...is still full of subjective biases" and only if "a small proportion of the totality of contrasts have effects, this plot can be used to make judgments about the reality of the largest effects found." Birnbaum (1959) gives more statistical theory on the half-normal plotting procedure. His results again show that this procedure is applicable only if we suppose that just a few effects are nonzero. Therefore, instead of half-normal plotting, it is usually assumed *a priori* that certain effects are nonexistent. (Remember that such an assumption forms the basis of all incomplete designs.) Most often the high-order interactions are taken to be zero and are pooled in Eq. (198). Once we have somehow obtained an estimate of σ^2 we can test the significance of effects using the F-ratio. If an effect is found to be insignificant at the selected significance level we may either continue to use the original estimate of σ^2, or pool the insignificant effect with the original estimator; compare Hunter (1959b, p. 9) and also Cohen (1968). We refer to Holms and Berrettoni (1969) for a possible procedure for determining when to stop pooling sums of squares.

From Eq. (33) it follows that relation (198) gives an *overestimated* σ^2 if a pooled effect is actually nonzero. Further there is a higher chance that σ^2 will be overestimated in a fractional factorial design. For in a fractional design each degree of freedom used to obtain an estimate of σ^2 implies that more than one effect is assumed zero. For instance, in the resolution IV design for seven factors, it is not enough to assume that the interaction between factors 1 and 4 is zero since Eq. (167) shows that we must also assume the interactions between 3 and 6 and between 5 and 7 to be zero. The consequences of an overestimated error variance are:

(i) The confidence intervals for the effects are too long.

(ii) When testing the hypothesis that an effect is not significant, we accept this hypothesis more often, i.e. more often than the nominal α-level of the test indicates. This might lead to the elimination of a factor from future experimentation if the factor is found, erroneously, to be insignificant. In our opinion this is less disastrous than it looks at first sight, for the amount of time and money available for experimentation is limited so that the elimination factor means that we can spend more time on the factors that are probably more important.[19)]

We can look at the pooling of effects in another way. When pooling effects we assume that these effects are zero. This assumption means that the regression model has fewer parameters. If the number of parameters, say J, is smaller than the number of observations N, then we can determine the "residual" mean square $\underline{S}_R$, i.e.

$$\underline{S}_R = \sum_{i=1}^{N} \frac{(\underline{y}_i - \hat{\underline{y}}_i)^2}{N - J} \tag{199}$$

where $\hat{\underline{y}}_i$ is the variable predicted by the regression model with least squares estimators of the J parameters and $\underline{y}_i$ is the true variable. In, e.g. Johnston (1963, pp. 106, 112) it is shown that

$$E(\underline{S}_R) = \sigma^2 \tag{200}$$

provided that the regression model used for the predictor $\hat{\underline{y}}_i$ is correct. For example, in a 2^3 design without replication we might assume that the linear model is adequate, i.e., $\beta_{12} = \beta_{13} = \beta_{23} = \beta_{123} = 0$. Hence, in Eq. (199) we have $N = 2^3 = 8$ and $J = 4$ (compare β_0, β_1, β_2, β_3); $\underline{S}_R$ has $(8-4) = 4$ degrees of freedom. Notice that because of the orthogonality of the design we could split the sum of squares in the numerator of Eq. (199) into four independent sums of squares, corresponding with β_{12}, β_{13}, β_{23}, and β_{123}. Pooling an effect emphasizes the assumption that some effect has a particular value (namely zero), whereas $\underline{S}_R$ emphasizes the assumption that the model has a certain order (e.g. first-order is assumed). This leads us to the problem of testing the adequacy of a particular model.

Suppose we start an experiment assuming that an adequate model contains only k main effects, i.e. in regression terminology we have a first-order model. If we take a saturated resolution III design the model will fit exactly and no check on the adequacy of the model is possible. However, if (k+1) is not a multiple of four we have an unsaturated resolution III design, or if (k+1) is a multiple of four we may take a resolution IV design. In both cases we can estimate some (confounded) two-factor interactions. Further, if one or more experimental points are duplicated we have an independent estimator of σ^2 and we can test the significance of these two-factor interactions. Suppose that some two-factor interactions turn out to be significant and some to be insignificant. Then it may be sensible to use a model with all interactions still included. For even while some interactions are not significant their minimum variance unbiased least squares estimators will be nonzero (albeit small). So if, e.g. all variables are quantitative we may decide to use a second-order polynomial (with all two-factor interactions plus pure quadratic effects) instead of a first-order model. Compare also the discussion by Box (1954, p. 57) and Hunter (1959b, p. 9) concerning the practice of testing individual parameters. Summarizing, instead of testing effects separately we may test them jointly by pooling their sums of squares and comparing the resulting mean square with an independent estimator of σ^2.[20)]

If we are not interested in the individual sums of squares we may also proceed as follows. Suppose that the N observations contain N_2 duplicates. Then we can determine the residual sum of squares, i.e. the numerator of Eq. (199), say

$$SS_{\underline{R}} = \sum_{i=1}^{N} (\underline{y}_i - \hat{\underline{y}}_i)^2 \qquad (201)$$

and an estimator of σ^2 based on duplication, say $\hat{\underline{\sigma}}_D^2$

$$\hat{\underline{\sigma}}_D^2 = SS_{\underline{D}}/N_2 \tag{202}$$

where $SS_{\underline{D}}$ denotes the numerator of Eq. (190). The estimator $\hat{\underline{\sigma}}_D^2$ does not depend on our model specification; the estimator $\underline{S}_R = SS_{\underline{R}}/(N-J)$, however, does depend on the correctness of the model used to determine $\hat{\underline{y}}_i$. If the model is correct then both estimators give (approximately) the same estimate for σ^2. If the model is incorrect then $\hat{\underline{y}}_i$ deviates from $\underline{y}_i$ not only because of experimental error but also because of specification error. So a value of $\underline{S}_R$ much higher than that of $\hat{\underline{\sigma}}_D^2$ indicates a model specification error. If we want to use an F-test to compare $\underline{S}_R$ with $\hat{\underline{\sigma}}_D^2$ we have to remember that the F-statistic has a numerator independent of the denominator. However, $\underline{S}_R$ also contains the duplicated observations that go into $\hat{\underline{\sigma}}_D^2$. In e.g. Mendenhall (1968, p. 201) it is shown how to proceed: Determine the "lack of fit" sum of squares $SS_{\underline{L}}$:

$$SS_{\underline{L}} = SS_{\underline{R}} - SS_{\underline{D}} \tag{203}$$

and its mean square

$$MS_{\underline{L}} = SS_{\underline{L}}/(N - J - N_2) \tag{204}$$

In this way $SS_{\underline{R}}$ is split into two independent sums of squares, viz. $SS_{\underline{L}}$ and $SS_{\underline{D}}$. Hence to test the lack of fit compare $MS_{\underline{L}}/\hat{\underline{\sigma}}_D^2$ with the tabulated F-ratio. If the critical F-level is surpassed then we reject the hypothesis that our model specification is correct. (Also compare Draper and Herzberg (1971, pp. 231-232) and Hunter (1958, pp. 20-22; 1959a, pp. 10, 12; 1959b, p. 9). We shall give an application in Chap. VI.

If we reject the hypothesis that our model is correct then we usually proceed to a higher-order model.[21] This leads us to the sequentialization of designs. We might start with a design involving only a few runs. We have seen before that resolution III designs are available for studying k factors in only $N = k+1$ runs if N is a multiple of four; otherwise we take the next multiple of four. If N is not a multiple of four, or otherwise if some additional experimental points are taken, then we are able to examine whether the fitted first-order model is adequate. For we can calculate some interaction sums of squares or the residual sum of squares. An F-test is possible if we have an independent estimator of σ^2 (from duplication of some design points or from previous experiments). If interactions are found to be significant, we might proceed to a resolution IV design.

Fortunately, we have seen that the construction of a _resolution IV design_, once we have a resolution III design, is not difficult at all. We just replicate the resolution III design with reversed signs, i.e. besides the N_1 runs of the resolution III design[22)] we have already obtained, we take N_1 more runs. By definition the resolution IV design provides estimators of main effects that are not biased by two-factor interactions. Hence, from the resolution IV design we may safely conclude whether a factor has no main effect (provided that there are no interactions among three or more factors; this condition may be tested by a lack of fit test for the resolution IV design). Assuming that factors with no main effects have no interactions either, it is possible that after the resolution IV design we eliminate certain factors. Having fewer factors means that the number of observation needed in the experiment reduces considerably; compare, e.g. Table 8. The remaining factors may be examined in a resolution V design.

When one or more factors are dropped after the resolution IV design, it may be that this design provides a design of _resolution V_ (or higher) for the _remaining_ factors. Consider the example in Table 25. From the construction of the design it follows immediately that if factor 4 is not significant, then the 2^{4-1}_{IV} design is a full factorial in the three factors, 1, 2, and 3. (We shall return to this "screening" for the important factors among the k factors in Sec. IV.7.) If, from the very beginning of the experiment involving k factors, we want to include the possibility of proceeding to a resolution V design in k factors, then more care is needed. For the resolution III and IV designs should be chosen such that they form a subset of the resolution V design. For 2^{k-p} designs this can be

TABLE 25

A 2_{IV}^{4-1} Fractional Design

$\vec{1}$	$\vec{2}$	$\vec{3}$	$\vec{4} = \vec{123}$
-	-	-	-
+	-	-	+
-	+	-	+
+	+	-	-
-	-	+	+
+	-	+	-
-	+	+	-
+	+	+	+

achieved by keeping all words in the defining relation of the 2_V^{k-p} design in the defining relation of the preceding resolution III and IV 2^{k-p} designs; an example of this procedure is given by Daniel (1956, pp. 96-97). However, in Table 20 we saw that 2_V^{k-p} designs are unattractive since they require many more observations than there are effects to be estimated. Therefore we may prefer Webb's saturated resolution V designs. In general his designs do not contain the preceding resolution III and IV designs as a subset. Yet his designs may be attractive since Table 23 shows that for high values of k the number of additional runs is smaller than for the 2_V^{k-p} designs. Note that when all factors (remaining after the resolution IV design) are quantitative, it is simple to augment the resolution IV design to a central composite design. Figure 3 demonstrated that such a design comprises a 2^{k-p} design $(p \geq 0)$ to which axial points and center points are added.

Instead of the above procedure, where we go from a resolution III design to a resolution IV and next to a resolution V design, we may proceed in smaller steps. In each step we then add a small fraction to the observations we have available. Addelman (1969) categorizes sequences of two-level fractional factorial designs for three to eleven factors and not more than 256 observations all together.

His additional fractions are restricted to belong to the same family. For example, in each step we may add $2^{7-4} = 8$ runs from one of the sixteen fractions belonging to the family characterized by Eq. (142). Addelman's catalogue shows how many main effects and two-factor interactions are estimable after each step assuming that there are no interactions among three or more factors. To demonstrate the usefulness of proceeding in small steps we briefly consider the 2_{IV}^{7-3} design that yielded Eq. (167). If in Eq. (167) only the first contrast for the two-factor interactions is large, whereas all other two-factor contrasts are small, then we may not want to proceed to a resolution V design. For in a resolution V design all two-factor interactions can be estimated while we may assume that only the interactions α^{24}, α^{35}, and α^{67} are important (neglecting the possibility of the other confounded interactions canceling out exactly). For such cases special procedures have been developed. As it is difficult to formulate general rules for these special cases, we do not discuss these procedures here but refer to Addelman (1969) and also to Daniel (1962), Jacoby and Harrison (1962, pp. 123-126), and John (1966). The analysis of sequentially augmented designs may be performed using the procedures developed by Hunter (1964) or Box (1966). The latter procedure was presented in Eqs. (192) through (197). Proceeding in small steps may also be guided by the Dykstra (1971) procedure which we discussed in relation with the augmentation of designs (with old and new points) for estimating experimental error and effects.

One final comment on sequentialization: We saw in Eq. (152) that we can duplicate a design with the signs of a single column switched. This procedure permits the examination of the main effect of a particular factor and its interactions with other factors.

Before turning to the next type of designs let us briefly consider the quality of the above designs. Hunter (1968, p. 4) and Naylor and Hunter (1969, p. 3) list the following requirements for experimental designs.

(1) Small number of observations.

(2) Minimum variance estimators.

(3) Providing a measure of the adequacy of the model.

(4) Desirable confounding patterns.

(5) Ease of computation.

To these requirements we would further add the possibility of sequentialization. Regarding condition (1) we remark that the number of experimental points, N_1, should be at least equal to the number of parameters, say J, to be estimated. Concerning (2), the covariance matrix of the least squares (minimum variance, unbiased) estimators is

$$\sigma^2(\vec{X}'\vec{X})^{-1} \tag{205}$$

Box (1952, p. 50) proved that the diagonal elements of relation (205) are minimized[23)] if

$$\vec{X} = N^{1/2}\,\vec{G} \tag{206}$$

where all elements in the first column of $\vec{G}$ are equal to one and moreover $\vec{G}$ is <u>orthogonal</u>, i.e.

$$\vec{G}'\vec{G} = \vec{I} \tag{207}$$

and N is the number of observations ($N = N_1$ if there are no duplicates). If $N > N_1 > J$ then a lack of fit test can be performed so that condition (3) is met. A desirable confounding pattern means that low-order effects are not confounded with other low-order effects but with high-order effects. Calculations are simplified if in the least squares estimators

$$\vec{\underline{b}} = (\vec{X}'\vec{X})^{-1}\,\vec{X}'\vec{\underline{y}} \tag{208}$$

and in relation (205), we have a diagonal matrix $(\vec{X}'\vec{X})$. In that case the inverse is obtained by simply taking the reciprocals of the diagonal elements of $(\vec{X}'\vec{X})$. Moreover computer rounding errors of the usual inversion programs are then avoided. Most of the above requirements are met simultaneously by the designs presented above. A different view is given by Box and Draper (1971). They use as a criterion maximization of the determinant of $\vec{X}'\vec{X}$. A very restrictive condition is their assumption that the model is true (so no check for possible higher order terms is possible); see Box and Draper (1971, p. 733). If the experimenter a priori assumes that a linear model is true then he may use the designs tabulated by Box and Draper (1971, pp. 736-737) for k factors ($k = 2, \ldots, 7$) and $N = k+1$ runs. (For $k = 2, 3, 7$ their designs are identical to 2^{k-p} designs!) We refer to Srivastava and Chopra (1971, pp. 258-259) for a brief discussion and references concerning various possible design criteria (e.g. determinant, trace, or largest root of $\vec{X}'\vec{X}$ in the covariance matrix of the estimated effects).

IV.7. SCREENING DESIGNS

At the beginning of an investigation there may be many, say k, conceivably important factors. We may suppose that not all k factors are important but only a few, say k', factors. We do not know the value of k' and neither do we know which k' factors among the k factors are important, and therefore we have to screen for these factors. This situation is also discussed by Jacoby and Harrison (1962, p. 128). They believe that such screening situations are quite frequently encountered in simulation experiments. We shall discuss several types of screening designs, namely (1) 2^{k-p} designs; (2) random designs; (3) supersaturated designs; (4) group-screening designs.

2^{k-p} Designs

The use of 2^{k-p} designs when screening for important factors is mentioned, e.g. by Hunter (1959a, p. 15) and Box and Hunter (1961a, pp. 318, 341-344). As an example we shall discuss the design in Table 25 repeated again in Table 26. From the construction of the design it follows that if factor 4 has no influence, then the 2_{IV}^{4-1} design is a full factorial in the three factors 1, 2, and 3. But also if one of the factors 1, 2, or 3 is not important, the above table shows that the design is a full factorial in the remaining three factors. (The design in these remaining factors is not listed

TABLE 26

A 2_{IV}^{4-1} Fractional Design

$\vec{1}$	$\vec{2}$	$\vec{3}$	$\vec{4} = \vec{12}$
-	-	-	-
+	-	-	+
-	+	-	+
+	+	-	-
-	-	+	+
+	-	+	-
-	+	+	-
+	+	+	+

in the standard order given in Table 6.) That table 26 results in a full design is also very simply shown by the generator for the above 2^{4-1} design, viz. $\vec{I} = \vec{1234}$. For we have remarked before that dropping a factor from the design, means that all words in the defining relation containing that factor vanish; compare with Eq. (143). Hence in this example the generator $\vec{1234}$ vanishes, no generators remain, and therefore no words in the defining relation remain and the effects are not confounded anymore, i.e. the design has become a full factorial. If not one but two factors are unimportant, then the 2^{4-1} design is a full factorial in the remaining two factors, replicated twice. If a single factor is important, then the 2^{4-1} design is a design for one factor replicated four times.

Generalizing, we know that by definition a resolution R design has a defining relation with words of R or more characters. So if only (R-1) factors are important, then all words in the defining relation contain at least one unimportant factor. Dropping unimportant factors means that all words with these factors vanish. Hence all words in the resolution R design vanish if only (R-1) factors are important. Consequently a resolution R design is a full factorial in (R-1) factors; a repeated full factorial in less than (R-1) factors; a repeated full factorial in (R-1) factors, if the number of observations is a multiple of $2^{(R-1)}$. As an example of the last possibility consider a 2^{8-4} design. It follows from, e.g. Eq. (170) that R = IV and therefore this design is a full factorial in three factors. The number of observations is $2^{8-4} = 2^4 = 2 \times 2^3$, i.e. the full factorial in three factors is replicated twice. If we want to screen R factors in a resolution R design, then not all words in the defining relation vanish. The remaining words show which combinations of factors are incomplete factorials. For instance, the word $\vec{1248}$ appears in the defining relation for the 2^{8-4} design. Hence for the factors 1, 2, 4, and 8 the design is an incomplete (but replicated) design. The word $\vec{1234}$ does not appear in the defining relation, so the design is a full factorial for the factors 1, 2, 3, and 4.

Box and Hunter (1961a, pp. 344-345) give an interesting example of factor screening. They have the 2_{IV}^{8-4} design with the responses given in Table 27. From this table the estimates of Table 28 follow. This table shows that a plausible interpretation of the results of this experiment is: There are only three important factors, viz. the factors 3, 5, and 8 since their main effects and their interactions can account for the large values in Table 28 (compare the circled values in the table). We saw above that the 2_{IV}^{8-4} design is a full factorial in three factors, replicated twice. This is shown in Table 29. If the interpretation of only three important factors is correct then the difference between two "replicates" is caused by experimental error and not by the difference between the levels of the nonsignificant factors 1, 2, 4, 6, and 7. As Box and Hunter (1961a, p. 345) remark, this can be tested if we have an independent estimator of experimental error available, obtained by duplication. The latter estimator can be compared with the estimator of σ based on "replication" following from Table 29.

Next we shall discuss a particular kind of screening situation. Suppose we are looking for, say, k' important factors among k conceivably important factors, where now k' is much smaller than k and k is very large, say 100 or more. Using a resolution

TABLE 27

Results for a Particular 2_{IV}^{8-4} Experiment

Run	Factor $\vec{1}$	$\vec{2}$	$\vec{3}$	$\vec{8}$	$\vec{4}$	$\vec{5}$	$\vec{6}$	$\vec{7}$	Response
1	-	-	-	-	-	-	-	-	60.4
2	+	-	-	-	+	+	-	+	66.0
3	-	+	-	-	+	-	+	+	62.1
4	+	+	-	-	-	+	+	-	63.3
5	-	-	+	-	-	+	+	+	82.9
6	+	-	+	-	+	-	+	-	75.4
7	-	+	+	-	+	+	-	-	82.4
8	+	+	+	-	-	-	-	+	73.0
9	-	-	-	+	+	+	+	-	68.1
10	+	-	-	+	-	-	+	+	61.2
11	-	+	-	+	-	+	-	+	71.3
12	+	+	-	+	+	-	-	-	59.6
13	-	-	+	+	+	-	-	+	67.3
14	+	-	+	+	-	+	-	-	75.3
15	-	+	+	+	-	-	+	-	66.7
16	+	+	+	+	+	+	+	+	77.1

TABLE 28

Screening Eight Factors in a 2_{IV}^{8-4} Design

Effect	Estimate
1	- 1.3
2	- 0.1
3	11.0
4	0.5
5	7.6
6	0.2
7	1.2
8	- 2.4
12 + 37 + 48 + 56	- 1.1
13 + 27 + 58 + 46	1.7
14 + 28 + 36 + 57	0.8
15 + 38 + 26 + 47	- 4.5
16 + 78 + 34 + 25	0.6
17 + 23 + 68 + 45	- 0.3
18 + 24 + 35 + 67	1.2

TABLE 29

The 2^{8-4} Design as a Twice Replicated Design in Three Factors

Run	Factor			Response	
	$\vec{3}$	$\vec{8}$	$\vec{5}$		
1,3	-	-	-	60.4	62.1
6,8	+	-	-	75.4	73.0
10,12	-	+	-	61.2	59.6
13,15	+	+	-	67.3	66.7
2,4	-	-	+	66.0	63.3
5,7	+	-	+	82.9	82.4
9,11	-	+	+	68.1	71.3
14,16	+	+	+	75.3	77.1

IV design as in Table 27 would mean a number of observations equal to 2k. So if the number of factors k is high then the number of observations may be prohibitively high. Even in resolution III designs the number of runs equals (k+1) and may be too large. The next three types of design may be used to solve the problem of too many observations because of too many factors.

Random Designs

An elaborate discussion of random designs was presented in Technometrics in 1959. Satterthwaite (1959, p. 112) defined random designs as designs where all or some of the elements of the design matrix are chosen by a random sampling process. Several random sampling processes are possible. Usually we shall sample the various levels of a particular factor with equal probabilities. However, if we have some prior knowledge as to which are the most promising levels of the factor, then we may sample from a nonuniform distribution. We may sample with or without replacement. The latter technique can achieve that all levels of a factor appear equally often in the experiment. It is customary to reject any selections that create correlation coefficients between explanatory variables that are outside prechosen limits. Suppose that the sampling process results in a correlation coefficient for x_1 and x_2 that is plus one. This means that if in a run $x_1 = +1$ then $\underline{x}_2 = +1$, and if $\underline{x}_1 = -1$ then $\underline{x}_2 = -1$. Then it is impossible to disentangle the effect of $\underline{x}_1$ and $\underline{x}_2$ on the response; the two effects are completely confounded. Observe that in fractional or full factorials the "correlation coefficient" is zero. For obviously the "correlation coefficient" between the nonstochastic variables x_1 and x_2 is defined as

$$\rho = \frac{\Sigma_{i=1}^{N} (x_{i1} - \mu_1)(x_{i2} - \mu_2)}{\sigma_1 \sigma_2} \tag{209}$$

where

$$\mu_j = N^{-1} \sum_{i=1}^{N} x_{ij} \qquad (j = 1, 2) \tag{210}$$

$$\sigma_j^2 = N^{-1} \sum_{i=1}^{N} (x_{ij} - \mu_j)^2 \tag{211}$$

In 2^{k-p} designs $(p \geq 0)$ we have $\mu_j = 0$, so Eq. (209) reduces to

$$\rho = \frac{\Sigma_i^N x_{i1} x_{i2}}{\sigma_1 \sigma_2} \tag{212}$$

Since the designs are orthogonal Eq. (212) shows that the "correlation coefficient" is zero. This property of factorials gives independent estimators of the main effects. However, in random designs the sample correlation coefficient $\underline{r}$ for $\underline{x}_1$ and $\underline{x}_2$, defined analogous to Eq. (209) as

$$\underline{r} = \frac{\Sigma_{i=1}^{N} (\underline{x}_{i1} - \bar{\underline{x}}_1)(\underline{x}_{i2} - \bar{\underline{x}}_2)}{\underline{s}_1 \underline{s}_2} \tag{213}$$

is a stochastic variable not necessarily zero. We shall return to this correlation when treating supersaturated designs. For a further discussion of the various sampling processes for random designs we refer to Anscombe (1959, pp. 195-196), Budne (1959a), Jacoby and Harrison (1962, p. 129), and Satterthwaite (1959, pp. 112-113). Several case studies employing random designs are presented by Budne (1959a); for a simulation experiment using random designs we refer to Cyert and March (1963, pp. 173-178).

The attractive feature of random designs is that the number of observations can be determined independently of the number of factors. In nonrandom designs there is a mathematical relation between N and k, e.g. $N = 2^{k-p}$. In random designs there is no such relation. (Of course we obtain better estimates if N/k increases.) So random designs can be used to examine a large number of factors in a moderate number of experiments; N may even be smaller than k.

Regarding the analysis of random designs Satterthwaite (1959, p. 126) remarks that there are no specific techniques for these designs; any technique familiar from the traditional designs may be used with random designs; or "any desired analysis can be made on any (sufficiently small) set of variables ignoring all the others. The ignored variables are valid residual error for the analysis being made." He does not elaborate on this "sufficiently small" subset. We can interpret his remark as follows. Satterthwaite (1959, p. 127) mentions the possibility of using ANOVA for the analysis of random designs. We know that the degrees of freedom in such an analysis cannot exceed the number of observations; hence the number of effects analyzed in the ANOVA must be smaller than the number of observations. So if the number of factors in the experiment is larger than the number of runs, then we have to restrict the ANOVA to a "sufficiently small" subset of the variables.

An analysis technique familiar from the traditional designs is regression analysis which comprises ANOVA as we saw in Sec. IV.2. Sattherthwaite shows that both multiple regression and simple regression can be applied. The latter technique means that in the model with several variables we estimate a particular effect ignoring all other effects and using the familiar simple regression formula, given in, e.g. Johnston (1963, p. 12). For instance, in

$$\underline{y}_i = \beta_0 + \sum_{j=1}^{k} \beta_j \underline{x}_{ij} + \underline{e}_i \qquad (i = 1, \ldots, N) \tag{214}$$

we estimate β_j $(j = 1, \ldots, k)$ by

$$\underline{b}_j = \frac{\Sigma_{i=1}^{N} (\underline{x}_{ij} - \bar{\underline{x}}_j)(\underline{y}_i - \bar{\underline{y}})}{\Sigma_j (\underline{x}_{ij} - \bar{\underline{x}}_j)^2} \tag{215}$$

If $k \geq N$ then multiple regression estimates do not exist, but in random designs simple regression estimates "do exist, are valid, are associated with $(N-2)$ degrees of freedom, and have finite confidence limits;" Satterthwaite (1959, p. 135). Note that in Eq. (215) the explanatory variable is stochastic. In most textbooks, e.g. Johnston (1963), this variable is nonstochastic. Fisz (1967, p. 96), however, uses a stochastic variable $\underline{x}$.

Another means of analysis is a graphical one. Though this technique is also applicable in traditional designs, it is the authors on random designs who have emphasized a simple graphical analysis. We can make a scatter diagram per factor, i.e. per factor the y-axis corresponds with the response and the x-axis with the various levels of that factor. If the factor has no effect then the average response per level will be the same for all levels.[24] An example of a scatter diagram is given in Fig. 4 (each level appears an equal number of times in this example). Next we eliminate the factors that are obviously effective. In the example of Fig. 4 factor A is effective. We calculate the average response per level of the effective factor, say $\bar{\underline{y}}_1^A$, $\bar{\underline{y}}_2^A$, and $\bar{\underline{y}}_3^A$, and in each run where factor A appears at level h $(h = 1, 2, 3)$ we subtract $\bar{\underline{y}}_h^A$ from the response. These corrected responses are plotted against the levels of the factors which in the first round were not detected as being significant. We stop this procedure as soon as no factor is found effective. An elaborate example of such a graphical analysis (where interactions are also examined) is given by Budne (1959b) and (1959d, pp. 143-154). Obviously scatter diagrams can be a wise first step in the analysis of any experiment,

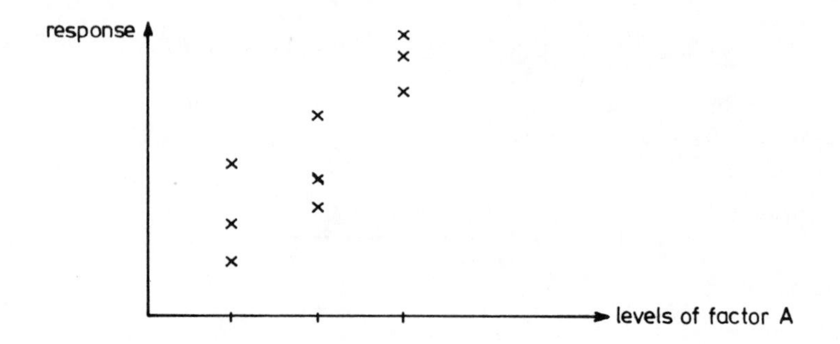

Fig. 4. Scatter diagram for factor A.

either with random or classic designs. Depending on the type of questions and the responses such a simple analysis may give the answer without any further sophisticated analysis. These diagrams are also attractive for presenting the results of the experiment to "the public".

To supplement the visual inspection of the scatter diagrams, Anscombe (1959) has given several significance tests again applied per factor.

1. The ANOVA F-test

Because each factor is considered separately we use the formula for the F-statistic in a one-way ANOVA presented before in Eq. (31), i.e.

$$\underline{F}_{J-1,J(I-1)} = \frac{(J-1)^{-1} \Sigma_j I_j(\underline{y}_{.j} - \underline{y}_{..})^2}{(N-J)^{-1} \Sigma_i \Sigma_j (\underline{y}_{ij} - \underline{y}_{.j})^2} \tag{216}$$

where the factor has J levels and per level I_j observations are made; $\underline{y}_{ij}$ is the original observation or the "corrected" observation after elimination of some effective factors in previous rounds as described above. This test checks whether $\underline{b}_j$ in Eq. (215) is zero; compare our discussion of the relations between ANOVA and regression analysis.

2. Welch's Randomization Test

Welch does not specify a parent distribution whereas the F-statistic is based on normality. He uses a randomization approach: We have N observations which are grouped corresponding with the J levels of the factor. If that factor has no effect the association of the observations with these levels is random, i.e. the particular grouping may be considered a random drawing from all possible groupings of N observations into groups of size $I_1, I_2, \ldots, I_J$. If we define some statistic we can calculate the distribution of that statistic, i.e. we can calculate which value that statistic has under each of the possible groups. [There are $N!(I_1!\, I_2! \cdots I_J!)^{-1}$ ways of grouping; because of these permutations randomization tests are also called permutation tests.] The number of different groupings is limited, so the distribution of the statistic is discrete. The statistic Welch (1938, p. 149) used is

$$\underline{E}^2 = \frac{\underline{SS}_{\text{between levels}}}{\underline{SS}_{\text{total}}} = \frac{\Sigma_{j=1}^{J} \Sigma_{i=1}^{I_j} (\underline{y}_{.j} - \underline{y}_{..})^2}{\Sigma_j \Sigma_i (\underline{y}_{ij} - \underline{y}_{..})^2} \tag{217}$$

Welch (1938, p. 152) derived the mean and variance of $\underline{E}^2$ under the hypothesis of no effect. $\underline{E}^2$ does not have a standard distribution with tabulated critical values. To find the critical, say, 5% level we have to calculate the various possible values of $\underline{E}^2$, tabulate these values and calculate the 5%-point. Instead Anscombe (1959, pp. 199-200) determines the expectation and the variance of $\underline{E}^2$ derived by Welch under the hypothesis of an ineffective factor; if the observed value of $\underline{E}^2$ is much larger than, say, two or three times its standard deviation, then Anscombe rejects the hypothesis. However, we would suggest that if we do not calculate a critical level for $\underline{E}^2$, we might as well stick to the ANOVA F-test (to which Welch's procedure approaches) or to the graphical analysis; we may also use the next test.

3. Tukey's Randomization Test for a Factor at Two Levels

The two levels of the factor divide the observations into two groups. Tukey (1959a, p. 32) requires

> "(i) to count the number of values in the one group exceeding all values in the other, (ii) to count the number of values in the other group falling below all those in the one, and (iii) to sum these two counts (we require that neither count be zero)."

As an example let us consider Fig. 4 above, deleting the observations at the third level. Then step (i) gives one, (ii) gives two, and (iii) gives three. If the number of observations for each level is the same (a condition which we can impose on the sampling procedure in random designs), then the critical values of the count in (iii) are given in Table 30 (based on Tukey's Table 2A). Tukey also gives critical values for unequal group sizes (in his Table 2A). These critical values differ only slightly from those in our Table 30. (For the 5% level they range from 7 to 11 as the difference between the group sizes ranges from 1 to 10 and N ranges from 6 to higher than 86.) For the sake of completeness we mention that Westlake (1971) further developed Tukey's test for special experimental situations.

TABLE 30

Two-Sided Critical Values for Tukey's Test

Total number of observations N	Critical value		
	5%	1%	0.1%
8-16	7	9	13
18-42	7	10	13
44-48	7	10	14
50-	8	10	14

We have seen that in random designs the number of observations, N, does not depend on the number of factors k. Neither does N depend on the number of levels of the factors. Nevertheless Anscombe (1959, pp. 200-201) suggests that it is not wise to use many levels. To strengthen the power of Welch's test (which is related to the F-test in ANOVA) he suggests using only two levels if one feels sure that the response varies linearly with the levels of the (quantitative) factor and using three or four levels if the response is "a low-degree polynomial regression curve with not more than one maximum" (if at least the factor does have any effect at all). He also derives that to reduce the possibility of confounding effects (caused by nonorthogonal design-columns) the number of observations N should satisfy

$$N \geq 8J \tag{218}$$

where J is the maximum number of levels of the k factors in the experiment, i.e.

$$J = \max_h(J_h) \qquad (h = 1, \ldots, k) \tag{219}$$

So Eq. (218) shows that a large number of levels would require a much larger number of observations; also compare Budne (1959b, p. 11).

To conclude our discussion of random designs we shall present the advantages of random designs as listed by Satterthwaite (1959, pp. 119-121) and we shall evaluate these alleged advantages at the same time.

(i) "Design simplicity." However, we think that this chapter shows that even the nonprofessional statistician can construct a classical design like a 2^{k-p} design.

(ii) "Analysis simplicity." Nevertheless Satterthwaite (1959, p. 126) himself states elsewhere in his paper, that most analysis techniques in random designs originate from classical designs (cf. ANOVA, regression, scatter diagrams). So in classical designs simple analysis techniques are also possible.

(iii) "Feedback and sequentialization." "It is not difficult to revise a random balance design during the course of the investigation as preliminary analyses of the early experiments indicate opportunities to concentrate later experiments more closely in the domain of major interest." Yet in Sec. IV.6 we saw that classical designs also permit sequentialization.

(iv) "Number of experiments." We consider this characteristic to be the major advantage of random designs. In classical designs the number of runs is determined by the number of factors and the number of levels. Incomplete factorials may often mitigate this disadvantage. Unfortunately, if the number of levels of the various factors varies, these designs are nonexistent or not readily available. Further even if we consider all factors at only two levels a classical design may be too large for high values of k. So the merit of the authors on random designs was that they emphasized the need for developing designs with a moderate number of runs for investigating situations with very many factors. We shall see that their work stimulated the development of nonrandom designs with few observations for a large number of factors.

(v) "Efficiency." Satterthwaite presents a table for the efficiency of random designs. However, this table seems to have no scientific justification, as was also pointed out by several discussants of Satterthwaite's paper; compare Kempthorne (1959, p. 161) and Box (1959, p. 175). Therefore we shall not present Satterthwaite's table but we shall discuss Box's results, which compare the (statistical) efficiency of random balance and orthogonal designs such as 2^{k-p} designs. Box (1959, pp. 175-177) restricts his comparisons to the first-order model

$$\underline{y}_i = \beta_0 + \sum_{j=1}^{k} \beta_j x_{ij} + \underline{e}_i \qquad (i = 1, \ldots, N) \tag{220}$$

and assumes that $N > k$. [Actually the condition $N > k$ means that Box restricts his comparisons to the situation which we do not see as most appropriate for random designs as we suggested under (iv) above.] In the random design the factor $\underline{x}_j$ is at L_j levels and each level is assumed to be repeated N/L_j times in the experiment. The efficiency is measured by

$$E_j = \frac{\text{variance of estimator of } \beta_j \text{ in orthogonal design}}{\text{var}(\underline{b}_j) \text{ in random balance design}} \tag{221}$$

In orthogonal designs the variance of the estimator of β_j can be shown to be

$$\text{var}(\underline{b}_j) = \frac{\sigma^2}{\Sigma_{i=1}^{N} (x_{ij} - \bar{x}_j)^2} \qquad (j = 1, 2, \ldots, k) \tag{222}$$

For random designs Box considers two alternative analyses.

(a) Least squares analysis. This yields

$$E_j = \frac{N-k}{N-1} \qquad (j = 1, 2, \ldots, k) \tag{223}$$

So especially for large k the inefficiency of random designs is large, and it are exactly situations with large k where random designs are recommended. (However, we repeat that Box assumes $N > k$, even for large k.)

(b) Scatter diagrams per factor. Then

$$E_j = \left\{ 1 + (N-1)^{-1} \sum_{h \neq j}^{k} [\sum_{i}^{N} (x_{ih} - \bar{x}_h)^2] \beta_h^2 \sigma^{-2} \right\}^{-1} \tag{224}$$

which is smaller than one, since all terms in Eq. (224) are positive. Equation (224) reduces to a simpler form if all factors are at only two levels. We then have

$$E_j \approx \left(1 + \sum_{h \neq j}^{k} \frac{\beta_h^2}{\sigma^2} \right)^{-1} \tag{225}$$

So Box has proven that for $N > k$ random designs are inefficient because the columns in the design matrix are not mutually orthogonal.

For the case $N \leq k$ Box suggests using systematic, i.e. non-random designs also. Not all columns in $\vec{X}$, the matrix of independent variables, can be mutually orthogonal if $N \leq k$. Consider

$$\vec{X}_{N\times(k+1)} = (\vec{I}_{N\times 1}, \vec{D}_{N\times k}) \tag{226}$$

where $\vec{I}_{N\times 1}$ is an N-dimensional vector of plus ones and $\vec{D}_{N\times k}$ is the design matrix so $\vec{X}$ is an $N \times (k+1)$ matrix. In an N-dimensional space not more than N orthogonal vectors can exist. Consequently if $N \leq k$ not all $(k+1)$ columns of $\vec{X}$ can be orthogonal. Or in more statistical terminology, independent estimators of k main effects and the grand mean based on N $(\leq k)$ observations would mean that the number of degrees of freedom becomes negative. If all $\vec{X}$-columns can not be mutually orthogonal, Box suggests achieving as much orthogonality as possible rather than having the degree of orthogonality be determined by a random sampling process. This hint was used by Booth and Cox (1962) who derived "supersaturated designs."

Supersaturated Designs

Booth and Cox considered systematic two-level factorials, where each of the k factors is at the "low" level N/2 times and the "high" level N/2 times (N being even). It is supposed that $N \leq k$. Orthogonality of the design columns $\vec{d}_h$ and $\vec{d}_j$ means

$$d_h'\vec{d}_j = 0 \qquad (h \neq j;\ j = 1, \ldots, k) \tag{227}$$

If $N \leq k$ then Eq. (227) cannot be satisfied for all h and j. Nevertheless we can try to satisfy Eq. (227) "as well as possible." This was quantified by Booth and Cox (1962, pp. 489-490) as minimizing

$$\max_{h\neq j} |d_h'd_j| \tag{228}$$

If more designs have the same value for Eq. (228), then that design is chosen in which the number of pairs of columns attaining Eq. (228) is smallest. In this way the authors derived designs for various N and k. The designs are tabulated by Booth and Cox (1962, pp. 490-492). Their results are reproduced in Table 31. For values of k not given in Table 31 designs are still available taking the design for the next higher value of k that is tabulated and dropping the final columns in that design.

To compare the efficiency of supersaturated and random designs the authors compare the variance of $\underline{z}$, where $\underline{z}$ is the inner product of two design columns, i.e.

$$\underline{z} = \vec{\underline{d}}_h' \cdot \vec{\underline{d}}_j \qquad (h \neq j) \tag{229}$$

Before proceeding to the results of the authors we observe that the expected value of $\underline{z}$ in random designs is equal to that in supersaturated designs, viz. zero. When in a random design each factor is sampled independently of the other factors and the levels plus and minus one have the same probability then we have

$$\begin{aligned} E(\underline{z}) &= E(\vec{\underline{d}}_h' \cdot \vec{\underline{d}}_j) = E(\sum_{i=1}^{N} \underline{d}_{ih}\underline{d}_{ij}) \\ &= \sum_i E(\underline{d}_{ih}\underline{d}_{ij}) = \sum_i E(\underline{d}_{ih})\, E(\underline{d}_{ij}) \\ &= \sum_i 0 \cdot 0 = 0 \end{aligned} \tag{230}$$

In a supersaturated design the elements plus and minus one are not sampled as in random designs. Nevertheless, whether we denote the "high" level of a particular factor by plus or minus one is decided randomly.[25] This means that, if for the tabulated design Eq. (229) yields the value z for two particular columns, then the value $-z$ is equally likely. Hence its expected value is zero. Next let us return to Booth and Box (1962, p. 494), who calculated the variance of $\underline{z}$. We have reproduced their results in Table 32.

TABLE 31

Supersaturated Designs for k Factors Using N Observations

Design I. k = 16, N = 12

+	+	+	+	+	+	+	+	+	+	+	-	-	-	-	-
+	-	+	+	+	-	-	-	+	-	-	-	-	-	-	-
-	+	+	+	-	-	-	+	-	-	+	+	+	-	+	+
+	+	+	-	-	-	+	-	-	+	-	-	+	+	+	+
+	+	-	-	-	+	-	-	+	-	+	+	+	-	+	-
+	-	-	-	+	-	-	+	-	+	+	+	+	+	-	+
-	-	-	+	-	-	+	-	+	+	+	+	-	+	+	+
-	-	+	-	-	+	-	+	+	+	-	+	-	+	-	+
-	+	-	-	+	-	+	+	+	-	-	+	+	+	+	-
+	-	-	+	-	+	+	+	-	-	-	-	+	+	-	-
-	-	+	-	+	+	+	-	-	-	+	-	-	-	+	+
-	+	-	+	+	+	-	-	-	+	-	-	-	-	-	-

Design II. k = 20, N = 12

+	-	-	+	+	+	+	-	+	+	-	-	-	+	-	-	-	-	+	-
-	-	-	-	+	+	-	+	-	-	+	-	+	-	-	+	-	-	-	-
-	-	+	+	+	-	+	+	-	+	-	-	-	-	-	-	+	+	+	+
-	+	-	-	-	+	+	-	+	-	-	+	-	-	-	+	-	+	+	+
+	+	+	-	+	+	-	+	+	-	-	+	+	+	-	+	+	-	-	+
+	+	+	+	+	-	+	-	-	-	+	-	+	-	+	-	-	+	-	-
+	-	-	-	-	+	-	-	-	+	+	+	-	-	+	-	+	-	+	-
-	+	-	+	+	-	-	-	-	+	+	+	+	+	+	-	-	-	-	+
+	+	-	-	-	-	-	+	-	+	-	-	-	+	+	+	+	+	-	-
+	-	+	-	-	-	+	+	+	-	+	-	-	-	-	-	-	-	-	+
-	-	+	+	-	+	+	+	+	+	+	+	+	+	+	+	+	+	+	-
-	+	+	+	-	-	-	-	+	-	-	+	+	+	+	+	+	+	+	+

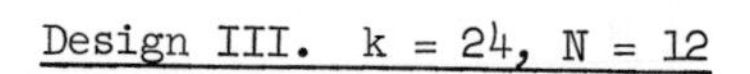

Design III. k = 24, N = 12

+	-	-	+	+	+	+	-	+	+	-	-	-	+	-	-	-	-	-	-	-	-	+	+
-	-	-	-	+	+	-	+	-	-	+	-	+	-	-	+	-	-	-	-	+	-	-	-
-	-	+	+	+	-	+	+	-	+	-	-	-	-	-	-	+	+	+	-	-	+	-	-
-	+	-	-	-	+	+	-	+	-	-	+	-	-	-	+	-	+	+	+	-	-	-	-
+	+	+	-	+	+	-	+	+	-	-	+	+	+	-	+	+	-	+	-	+	-	-	-
+	+	+	+	+	-	+	-	-	-	+	-	+	-	+	-	-	+	-	+	-	-	+	-
+	-	-	-	-	+	-	-	-	+	+	+	-	-	+	-	+	-	-	-	+	+	-	-
-	+	-	+	+	-	-	-	-	+	+	+	+	+	+	-	-	-	+	+	-	+	-	+
+	+	-	-	-	-	-	+	-	+	-	-	-	+	+	+	+	+	-	-	-	-	+	+
+	-	+	-	-	-	+	+	+	-	+	-	-	-	-	-	-	-	+	+	+	+	+	+
-	-	+	+	-	+	+	+	+	+	+	+	+	+	+	+	+	+	-	+	+	+	+	+
-	+	+	+	-	-	-	-	+	-	-	+	+	+	+	+	+	+	+	+	+	+	+	+

TABLE 31 (Continued)

Design IV. k = 24, N = 18

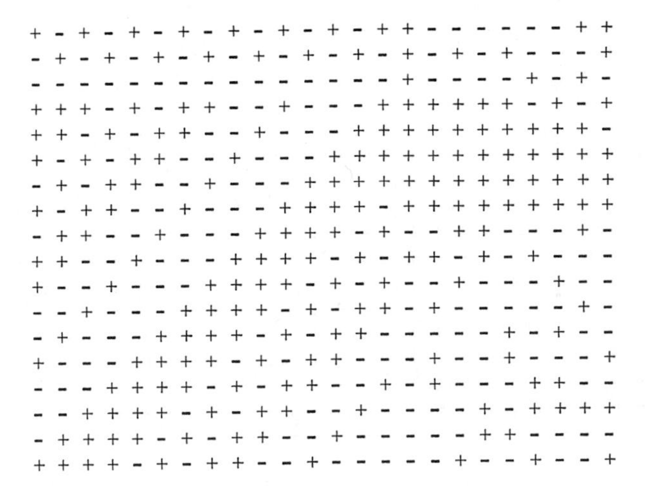

+	-	+	-	+	-	+	-	+	-	+	-	+	-	+	+	-	-	-	-	-	-	+	+
-	+	-	+	-	+	-	+	-	+	-	+	-	+	-	+	-	+	-	+	-	-	-	+
-	-	-	-	-	-	-	-	-	-	-	-	-	-	-	+	-	-	-	-	+	-	+	-
+	+	+	-	+	-	+	+	-	-	+	-	-	-	+	+	+	+	+	+	-	+	-	+
+	+	-	+	-	+	+	-	-	+	-	-	-	+	+	+	+	+	+	+	+	+	+	-
+	-	+	-	+	+	-	-	+	-	-	-	+	+	+	+	+	+	+	+	+	+	+	+
-	+	-	+	+	-	-	+	-	-	-	+	+	+	+	+	+	+	+	+	+	+	+	+
+	-	+	+	-	-	+	-	-	-	+	+	+	+	-	+	+	+	+	+	+	+	+	+
-	+	+	-	-	+	-	-	-	+	+	+	+	-	+	-	-	+	+	-	-	-	+	-
+	+	-	-	+	-	-	-	+	+	+	+	-	+	-	+	+	-	+	-	+	-	-	-
+	-	-	+	-	-	-	+	+	+	+	-	+	-	+	-	-	+	-	-	-	+	-	-
-	-	+	-	-	-	+	+	+	+	-	+	-	+	+	-	+	-	-	-	-	-	+	-
-	+	-	-	-	+	+	+	+	-	+	-	+	+	-	-	-	-	-	+	-	+	-	-
+	-	-	-	+	+	+	+	-	+	-	+	+	-	-	-	+	-	-	+	-	-	-	+
-	-	-	+	+	+	+	-	+	-	+	+	-	-	+	-	+	-	-	-	+	+	-	-
-	-	+	+	+	+	-	+	-	+	+	-	-	+	-	-	-	-	+	-	+	+	+	+
-	+	+	+	+	-	+	-	+	+	-	-	+	-	-	-	-	-	+	+	-	-	-	-
+	+	+	+	-	+	-	+	+	-	-	+	-	-	-	-	-	+	-	-	+	-	-	+

Design V. k = 30, N = 18

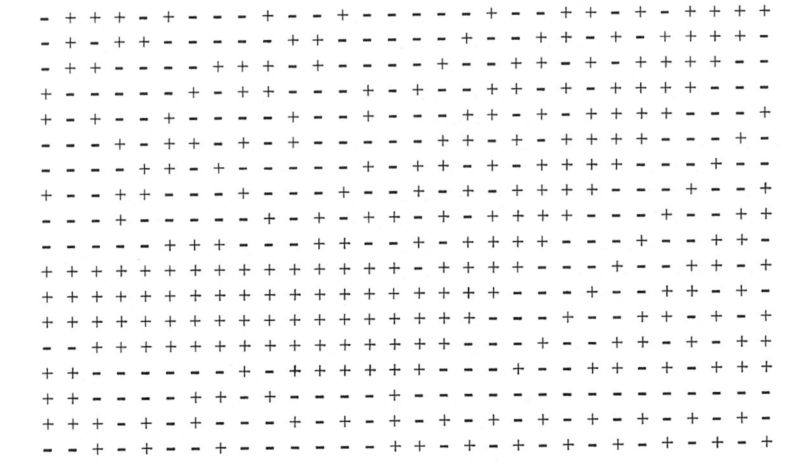

-	+	+	+	-	+	-	-	-	+	-	-	+	-	-	-	-	-	+	-	-	+	+	-	+	-	+	+	+	+
-	+	-	+	+	-	-	-	-	-	+	+	-	-	-	-	-	+	-	-	+	+	-	+	-	+	+	+	+	-
-	+	+	-	-	-	-	+	+	+	-	+	-	-	-	-	+	-	-	+	+	-	+	-	+	+	+	+	-	-
+	-	-	-	-	-	+	-	+	+	-	-	-	+	-	+	-	-	+	+	-	+	-	+	+	+	+	-	-	-
+	-	+	-	-	+	-	-	-	-	+	-	-	+	-	-	-	+	+	-	+	-	+	+	+	+	-	-	-	+
-	-	-	+	-	+	+	-	+	-	+	-	-	-	-	-	+	+	-	+	-	+	+	+	+	-	-	-	+	-
-	-	-	-	+	+	-	+	-	-	-	-	-	+	-	+	+	-	+	-	+	+	+	+	-	-	-	+	-	-
+	-	-	+	+	-	-	-	+	-	-	-	+	-	-	+	-	+	-	+	+	+	+	-	-	-	+	-	-	+
-	-	-	+	-	-	-	-	-	+	-	+	-	+	+	-	+	-	+	+	+	+	-	-	-	+	-	-	+	+
-	-	-	-	-	+	+	+	-	-	-	+	+	-	-	+	-	+	+	+	+	-	-	-	+	-	-	+	+	-
+	+	+	+	+	+	+	+	+	+	+	+	+	+	+	-	+	+	+	+	-	-	-	+	-	-	+	+	-	+
+	+	+	+	+	+	+	+	+	+	+	+	+	+	+	+	+	+	+	-	-	-	+	-	-	+	+	-	+	-
+	+	+	+	+	+	+	+	+	+	+	+	+	+	+	+	+	+	-	-	-	+	-	-	+	+	-	+	-	+
-	-	+	+	+	+	+	+	+	+	+	+	+	+	+	+	+	-	-	-	+	-	-	+	+	-	+	-	+	+
+	+	-	-	-	-	-	-	+	-	+	+	+	+	+	+	-	-	-	+	-	-	+	+	-	+	-	+	+	+
+	+	-	-	-	-	+	+	-	+	-	-	-	-	+	-	-	-	-	-	-	-	-	-	-	-	-	-	-	-
+	+	+	-	+	-	+	-	-	-	+	-	+	-	+	-	+	-	+	-	+	-	+	-	+	-	+	-	+	-
-	-	+	-	+	-	-	+	-	-	-	-	-	-	+	+	-	+	-	+	-	+	-	+	-	+	-	+	-	+

TABLE 31 (Continued)

Design VI. k = 36, N = 18

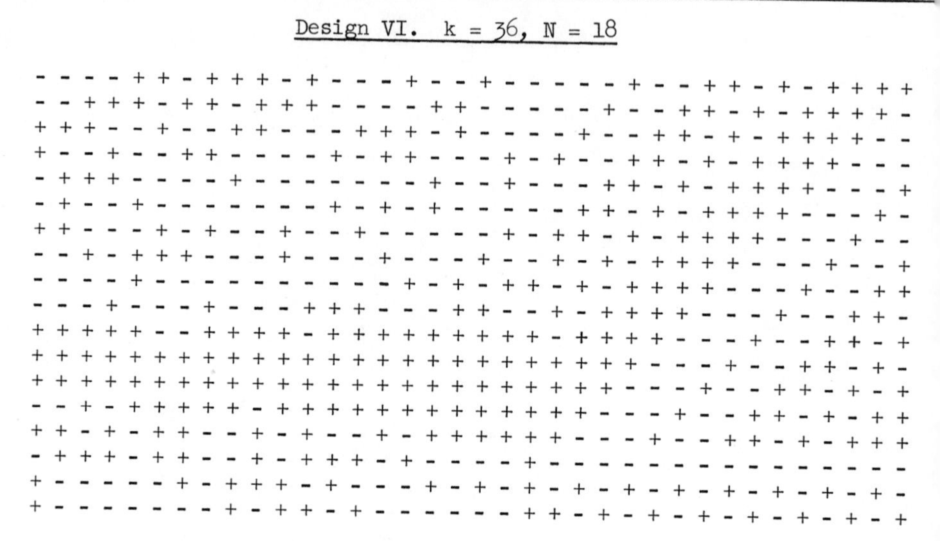

-	-	-	-	+	+	-	+	+	+	-	+	-	-	-	+	-	-	+	-	-	-	-	-	+	-	-	+	+	-	+	-	+	+	+	+
-	-	+	+	+	-	+	+	-	+	+	+	-	-	-	-	+	+	-	-	-	-	-	+	-	-	+	+	-	+	-	+	+	+	+	-
+	+	+	-	-	+	-	-	+	+	-	-	-	+	+	+	-	+	-	-	-	-	+	-	-	+	+	-	+	-	+	+	+	+	-	-
+	-	-	+	-	-	+	+	-	-	-	-	+	-	+	+	-	-	-	+	-	+	-	-	+	+	-	+	-	+	+	+	+	-	-	-
-	+	+	+	-	-	-	-	+	-	-	-	-	-	-	-	+	-	-	+	-	-	-	+	+	-	+	-	+	+	+	+	-	-	-	+
-	+	-	-	+	-	-	-	-	-	-	-	+	-	+	-	+	-	-	-	-	-	+	+	-	+	-	+	+	+	+	-	-	-	+	-
+	+	-	-	-	+	-	+	-	-	+	-	-	+	-	-	-	-	-	+	-	+	+	-	+	-	+	+	+	+	-	-	-	+	-	-
-	-	+	-	+	+	+	-	-	-	+	-	-	-	+	-	-	-	+	-	-	+	-	+	-	+	+	+	+	-	-	-	+	-	-	+
-	-	-	-	+	-	-	-	-	-	-	-	-	-	-	+	-	+	-	+	+	-	+	-	+	+	+	+	-	-	-	+	-	-	+	+
-	-	-	+	-	-	-	+	-	-	-	+	+	+	-	-	-	+	+	-	-	+	-	+	+	+	+	-	-	-	+	-	-	+	+	-
+	+	+	+	+	-	-	+	+	+	+	-	+	+	+	+	+	+	+	+	+	-	+	+	+	+	-	-	-	+	-	-	+	+	-	+
+	+	+	+	+	+	+	+	+	+	+	+	+	+	+	+	+	+	+	+	+	+	+	+	+	-	-	-	+	-	-	+	+	-	+	-
+	+	+	+	+	+	+	+	+	+	+	+	+	+	+	+	+	+	+	+	+	+	+	+	-	-	-	+	-	-	+	+	-	+	-	+
-	-	+	-	+	+	+	+	+	-	+	+	+	+	+	+	+	+	+	+	+	+	+	-	-	-	+	-	-	+	+	-	+	-	+	+
+	+	-	+	-	+	+	-	-	+	-	+	-	-	+	-	+	+	+	+	+	+	-	-	-	+	-	-	+	+	-	+	-	+	+	+
-	+	+	+	-	+	+	-	-	+	-	+	+	+	-	+	-	-	-	-	+	-	-	-	-	-	-	-	-	-	-	-	-	-	-	-
+	-	-	-	-	-	+	-	+	+	+	-	+	-	-	-	+	-	+	-	+	-	+	-	+	-	+	-	+	-	+	-	+	-	+	-
+	-	-	-	-	-	-	-	+	-	+	+	-	+	-	-	-	-	-	-	+	+	-	+	-	+	-	+	-	+	-	+	-	+	-	+

Design VII. k = 30, N = 24

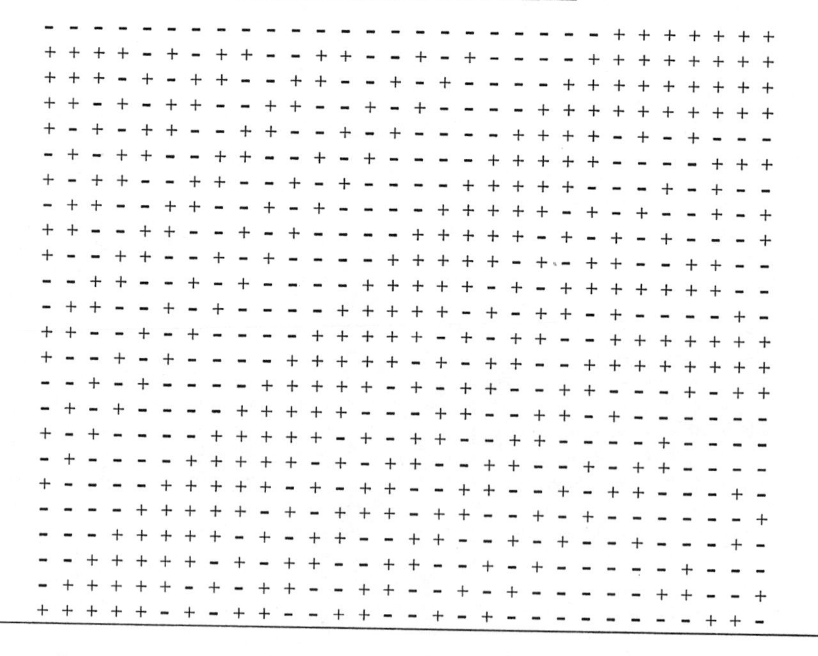

-	-	-	-	-	-	-	-	-	-	-	-	-	-	-	-	-	-	-	-	-	-	-	+	+	+	+	+	+	+
+	+	+	+	-	+	-	+	+	-	-	+	+	-	-	+	-	+	-	-	-	-	+	+	+	+	+	+	+	+
+	+	+	-	+	-	+	+	-	-	+	+	-	-	+	-	+	-	-	-	-	+	+	+	+	+	+	+	+	+
+	+	-	+	-	+	+	-	-	+	+	-	-	+	-	+	-	-	-	-	+	+	+	+	+	+	+	+	+	+
+	-	+	-	+	+	-	-	+	+	-	-	+	-	+	-	-	-	-	+	+	+	+	-	+	-	+	-	-	-
-	+	-	+	+	-	-	+	+	-	-	+	-	+	-	-	-	-	+	+	+	+	+	-	-	-	-	+	+	+
+	-	+	+	-	-	+	+	-	-	+	-	+	-	-	-	-	+	+	+	+	+	-	-	-	+	-	+	-	-
-	+	+	-	-	+	+	-	-	+	-	+	-	-	-	-	+	+	+	+	+	-	+	-	+	-	-	+	-	+
+	+	-	-	+	+	-	-	+	-	+	-	-	-	-	+	+	+	+	+	-	+	-	+	-	+	-	-	-	+
+	-	-	+	+	-	-	+	-	+	-	-	-	-	+	+	+	+	+	-	+	-	+	+	-	-	+	+	-	-
-	-	+	+	-	-	+	-	+	-	-	-	-	+	+	+	+	+	-	+	-	+	+	+	+	+	+	+	-	-
-	+	+	-	-	+	-	+	-	-	-	-	+	+	+	+	+	-	+	-	+	+	-	+	-	-	-	-	+	-
+	+	-	-	+	-	+	-	-	-	-	+	+	+	+	+	-	+	-	+	+	-	-	+	+	+	+	+	+	+
+	-	-	+	-	+	-	-	-	-	+	+	+	+	+	-	+	-	+	+	-	-	+	+	+	+	+	+	+	+
-	-	+	-	+	-	-	-	-	+	+	+	+	+	-	+	-	+	+	-	-	+	+	-	-	-	+	-	+	+
-	+	-	+	-	-	-	-	+	+	+	+	+	-	-	-	+	+	-	-	+	+	-	+	-	-	-	-	-	-
+	-	+	-	-	-	-	+	+	+	+	+	-	+	-	+	+	-	-	+	+	-	-	-	-	+	-	-	-	-
-	+	-	-	-	-	+	+	+	+	+	-	+	-	+	+	-	-	+	+	-	-	+	-	+	+	-	-	-	-
+	-	-	-	-	+	+	+	+	+	-	+	-	+	+	-	-	+	+	-	-	+	-	+	+	-	-	-	+	-
-	-	-	-	+	+	+	+	+	-	+	-	+	+	+	-	+	+	-	-	+	-	+	-	-	-	-	-	-	+
-	-	-	+	+	+	+	+	-	+	-	+	+	-	-	+	+	-	-	+	-	+	-	-	+	-	-	-	+	-
-	-	+	+	+	+	+	-	+	-	+	+	-	-	+	+	-	-	+	-	+	-	-	-	-	-	+	-	-	-
-	+	+	+	+	+	-	+	-	+	+	-	-	+	+	-	-	+	-	+	-	-	-	-	-	+	+	-	-	+
+	+	+	+	+	-	+	-	+	+	-	-	+	+	-	-	+	-	+	-	-	-	-	-	-	-	-	+	+	-

TABLE 32

The Variance of $\underline{z}$, the Measure of Nonorthogonality

Design	I	II	III	IV	V	VI	VII
Random	13.1	13.1	13.1	19.1	19.1	19.1	25.0
Systematic	7.07	9.68	10.1	13.6	15.3	17.4	11.4

Whereas the mean value of $\underline{z}$ is the same for both types of designs, supersaturated designs have a smaller spread, which means that less high values of (the absolute value of) $\underline{z}$ occur. High values of $|\underline{z}|$ are undesirable, as nonorthogonality means that the variance of the estimated factor effects is large; even if $N \leq k$ random designs are less efficient than the systematic designs derived by Booth and Cox. If, however, $k > 36$ or we want to make fewer runs than given in Table 31, then no tabulated designs are available. We could then derive a supersaturated design using the iterative procedure on a computer, as described by Booth and Cox (1962, pp. 492-494). The time needed to write and to run the computer program for the generation of such a design might very well be prohibitive. In that case we might still use a random design.

There is one more case where random designs seem attractive. This is demonstrated by the following example based on Satterthwaite (1959, pp. 114-115). First suppose that the response is approximated by

$$y = \beta_0 + \beta_1 x_1 + \beta_2 x_2 \tag{231}$$

Using a 2^2 factorial design, the columns in the matrix of independent variables are mutually orthogonal, as shown by Table 33.

Next suppose that it is discovered that not Eq. (231) but Eq. (232) is an adequate model.

$$y = \gamma_0 + \gamma_1 (x_1 x_2) + \gamma_2 (x_1 / x_2) \tag{232}$$

TABLE 33

Matrix of Independent Variables in 2^2 Factorial Design

$\vec{x}_0$	$\vec{x}_1$	$\vec{x}_2$
+1	-1	-1
+1	+1	-1
+1	-1	+1
+1	+1	+1

Then Table 34 gives the matrix of fundamental independent variables corresponding to the 2^2 design in the original variables. This table shows that the columns of the fundamental variables are no longer orthogonal; it is even impossible to estimate the parameters γ_1 and γ_2 separately. Especially in the initial phase of an investigation an adequate model may not be known at all. In that case the advantages of orthogonality between the original variables may vanish. (Of course the random design will also show nonorthogonality for the new fundamental variables.)

Summarizing, because of the flexibility of random designs it may be attractive to use a random design with a large number of factors and a moderate number of runs in order to get some impression

TABLE 34

Matrix of Fundamental Independent Variables

$\vec{x}_0$	$\overrightarrow{(x_1, x_2)}$	$\overrightarrow{(x_1/x_2)}$
+1	+1	+1
+1	-1	-1
+1	-1	-1
+1	+1	+1

about the form of the response function, about the experimental region of interest, and about the important factors. Without such a pilot-phase certain factors are arbitrarily assumed to have no effect or, alternatively, certain factors are fixed at a particular level. (The latter practice would restrict the conclusions of the experiment to that particular level.) In the next phase we can at least approximate the true response function in the limited area of future experimentation and concentrate on the important factors. Then we can use either a systematic orthogonal design for the variables in the approximate response function, a systematic supersaturated design, or a systematic "group-screening" design. The latter type of designs will be discussed next.

Group-Screening Designs

In these designs the k factors are grouped into g groups, each group being considered as a single factor. These g group-factors are tested in a systematic design. The assumptions of group-screening imply that if a group-factor is found to be insignificant, then all original factors within that group-factor are insignificant and can be dropped from further investigation. If a group-factor is found to be significant, then one or more original factors in that group are significant. So in the next stage these original factors should be further investigated. In a two-stage procedure the g group-factors are tested in the first-stage and the original factors of the significant groups only are tested individually in the second stage. Such a procedure was introduced by Watson in 1961. In multistage group-screening the groups found to be significant in the first stage (or more generally in stage t) are repartitioned into smaller groups that are tested in the next stage. These multistage designs were introduced independently of each other by Patel and Li in 1962. Let us consider two-stage and multistage group-screening in some more detail.

a. Two-Stage Group-Screening

Watson (1961, p. 372) used the following assumptions (the possibility to relax certain assumptions will be discussed later):

(i) all factors have, independently, the same prior possibility of being effective, p ($q = 1-p$),

(ii) effective factors have the same effect, $\Delta > 0$,

(iii) there are no interactions present,

(iv) the required designs exist,

(v) the directions of possible effects are known,

(vi) the errors of all observations are independently normal with a constant known variance, σ^2,

(vii) $k = gf$ where g = number of groups, and f = number of factors per group."[26]

Because of assumption (v) we can define the upper level of a factor as the level producing the highest response. The upper level of a group-factor is then defined as the level where all factors in that group are at their high level. Together with assumption (iii) this ensures that no cancellation of effects can occur. So a group-factor containing one or more effective factors gives a nonzero effect. We shall discuss an example of group-screening before investigating these assumptions.

Suppose there are nine original factors, A, B, ..., I, partitioned into three groups labeled X, Y, Z. Hence, each group consists of three original factors. Testing three (group-)-factors for their main effects can be done in four runs: $N_1 = 2^{3-1}$. A possible fraction is given in Table 35. In run 1, X and Y are at their low levels and Z is at its high level. If X consists of A, B, and C; Y consists of D, E, and F, and Z of H, J, and I, then in run 1 the original factors A, B, ..., I are at the levels -1, -1, -1, -1, -1, -1, 1, 1, 1. Suppose the first stage shows that only group-factor X is effective. [Because of assumption (vi) ANOVA can be used, provided we have an independent estimator of σ^2.] In the second stage we then have to test only the factors within X, i.e. A, B, and C. These three factors can be tested in four runs:

TABLE 35

A 2_{III}^{3-1} Design in the Three Group-Factors X, Y, and Z

Run	$\vec{X}$	$\vec{Y}$	$\vec{Z} = \vec{X}\vec{Y}$
1	-1	-1	+1
2	+1	-1	-1
3	-1	+1	-1
4	+1	+1	+1

$N_2 = 2^{3-1}$. If, e.g. we use the analog of Table 35 we run the combinations c, a, b, and abc, respectively.

Watson (1961, p. 374) points out that we can save runs by a good choice of the fractions run in each stage and by a good choice of the levels of the original factors found to be insignificant in the first stage. For choosing the fraction listed in Table 35 implies that in run 4 all original factors are at their high level. In stage 2 we run combination abc, i.e. the combination where the factors A, B, and C are at the high level. If we set all insignificant factors D, E, ..., I at the high level in the second stage then combination abc is the same as combination xyz in run 4 of stage 1. So this run has already been performed. Hence, to test A, B, and C for their main effects we need only three additional runs. We would comment on Watson's procedure that if the insignificant factors are truly unimportant then their levels do not matter. Because of experimental error, however, we may declare a factor to be insignificant, whereas that factor is actually important. As we demonstrate in Appendix IV.5 it is best not to vary an insignificant factor, i.e. it is best to set an insignificant factor either at its low level in all runs of stage 2 or at its high level in all those runs. In this way a possible main effect of this factor is confounded with the grand mean. Setting the factor at its low level in some runs and at its high level in some other runs of stage 2 confounds its main effect with the main effect of the significant factors.

We shall next show that the assumptions (i) through (vii) are not very restrictive. Assumption (iv) was introduced in order to derive the group-size that minimizes the number of runs in stages 1 and 2. This assumption means that in the first stage a design exists for testing g group-factors in $(g+1)$ runs. Actually we have seen that a resolution III design in $N_1 = (g+1)$ runs is available only if N_1 is a multiple of four. Moreover this assumption means that in the second stage a design exists with $N_2 = fs$ runs, where s is the number of group-factors found to be significant in the first stage. This implies that one run of the previous stage is used and that $(fs+1)$ is a multiple of four. In practice N_1 and (N_2+1) may not be a multiple of four. Therefore the derived optimal values of the group-size will not be exactly optimal. However, this does not invalidate the procedure. Further if $(g+1)$ and $(fs+1)$ are not a multiple of four then the number of observations is larger than the number of parameters, and degrees of freedom remain to estimate the experimental error variance. Assumption (ii) is also needed to derive the optimal group-size and therefore is not crucial. Assumption (v) can be weakened, as Watson (1961, p. 385) derived: "As optimum sized groups, for $p \leq .15$, have only a chance of 0.06 or less of containing two or more effective factors, we will in practice not need to know the directions of all the possible effects;" also compare our comment on Eq. (234) below. Assumption (vi) is again needed for the derivation of the optimal design and also makes ANOVA possible. We have seen before that ANOVA is robust with regard to nonnormality and heteroscedasticity. Assumption (i) should be interpreted as follows. We need some prior rough estimate as to how many factors are thought to be effective among the total of k factors. Then p is equal to the ratio of the likely number of effective factors and the total number of factors. This p determines the optimal group-size f, derived by Watson (1961, p. 381) to be

$$f_0 \approx \frac{1}{[(1 - \alpha_1)p]^{1/2}} \tag{233}$$

where α_1 is the significance level in the first stage. Equation (233) shows that the group-size decreases as p increases; for high p the optimal group-size becomes one, i.e. the factors are tested individually. This is a reasonable result since a high p means that many groups contain effective factors, so that in the second stage all or nearly all original factors must be tested and nothing has been gained by the grouping procedure. Of course a high value of p is contradicting the definition of a screening situation. If we have no firm estimate of p this would mean that our grouping will not be optimal. However, Eq. (233) for the optimal group-size is only approximate and because the group-size f must be an integer f_0 will not fluctuate much with varying p (and α_1). For instance, if there is no experimental error, i.e. $\alpha_1 = 0$, then for $0.03 \leq p \leq 0.30$, Watson's Table 2 shows $6 \geq f_0 \geq 3$. If we do not suppose that all factors have the same prior probability p, we can form classes of factors, each class having its own prior probability. Equation (233) shows that factors with a high probability, p, should be placed in small groups; if $p > .30$ we take a group-size of one, i.e. these factors are tested individually. So different estimates of p (which are more realistic) can be incorporated into the group-screening method. Moreover they make the grouping more flexible. The group-size is no longer a constant, but can be varied, and assumption (vii) is replaced by the less restrictice assumption

$$k = g_1 \cdot f_1 + g_2 \cdot f_2 + \cdots + g_J \cdot f_J \tag{234}$$

where g_j is the number of groups of size f_j $(j = 1, \ldots, J)$. Further, the number of group-factors tested in the first stage, i.e. Σg_j, can be so chosen that $N_1 = \Sigma g_j + 1$ is a multiple of four in which case a saturated design is possible; compare Watson (1961, pp. 383-385). We remark that unequal group-sizes make is possible to test a factor individually when we do not know the direction of its effect (and are afraid of cancellations of various effects notwithstanding the remark we made above concerning the low probability of such cancellations). Finally in Appendix IV.6 we weakened assumption (iii) about <u>interactions</u>. For we derived that a two-factor interaction β_{zw} biases the estimator of the main effect of a factor p, only if the factors z, w, and p belong to three different group-factors (z, w, p = 1, ..., k). (Pure quadratic effects β_{jj} never bias the estimator of the main effect.) Therefore, if we assume that two-factor interactions exist only between particular (related) factors then we should place those <u>related</u> factors in the <u>same</u> group. Then their two-factor interactions will not bias any estimated main effect. We further derived that if we examine the g <u>group</u>-factors in a <u>resolution IV</u> design, then main effects are not biased by any two-factor interaction (but of course are still confounded with each other within the same group-factor).

Finally, we briefly consider the influence of the <u>significance levels</u> in the first and second stage, i.e. α_1 and α_2, where α_i is the probability of falsely rejecting the hypothesis of no effect in stage i (i = 1, 2). If α_1 increases we shall declare many group-factors to be significant and this increases the number of runs in the second stage. If α_2 increases, N $(= N_1 + N_2)$ is not influenced; of course it means that more unimportant factors may be declared significant. A truly optimal choice of α_1 would have to be based on cost considerations and would require a "decision theory" approach, but such a solution is not available. [For the sake of completeness we mention that Curnow (1965) detected a nonserious error in Watson's derivations.]

b. <u>Multistage Group-Screening</u>

Patel generalized Watson's procedure to more than two stages in order to achieve a further reduction of the number of runs. He used, explicitly or implicitly, the following <u>assumptions</u>: Watson's assumptions (i), (ii), (iii), (iv), and (v). [Assumption (iv) is to be interpreted as the number of runs in each stage is equal to the number of group-factors in that stage except for stage 1, where the runs are one unit larger than the number of group-factors.] Watson's assumption (vi) is replaced by the assumption of no experimental error. His assumption (vii) is replaced by its "sequentialized" version as we shall show below.

Let us consider the following example.[27)] Suppose there are two hundred factors ($k = 200$). In the first stage we can form ten groups ($g_1 = 10$), so each group consists of twenty factors ($f_1 = 20$). In the second stage we can split each group of the first stage into, say, five groups ($g_2 = 5$; note that g_2 is not the total number of groups in the second stage, that number is g_1g_2). Then each group in the second stage consists of $f_2 = 20/5 = 4$ factors. If we have a three-stage procedure the number of groups into which each group of the second stage is split is four; so $g_3 = f_2 = 4$. Hence, $k = g_1g_2g_3 = 10 \times 5 \times 4 = 200$. In each stage only the group-factors found significant in the previous stage are investigated further. So if s_i factors are found to be significant in stage i, then in stage (i+1) we shall investigate $s_i \cdot g_{i+1}$ group-factors.

Patel (1962, p. 214) derived that the expected number of runs over all $(n+1)$ stages is minimized by choosing the <u>number of groups</u> as in Eqs. (235) and (236).

$$g_1 \approx kp^{n/(n+1)} \tag{235}$$

and

$$g_2 = g_3 = \cdots = g_n = g_{n+1} \approx p^{-1/(n+1)} \tag{236}$$

Remember that g_i does not denote the total number of groups in stage i, but the number of groups into which g_{i-1} is split. With Eqs. (235) and (236) the following <u>group-sizes</u> correspond.

$$f_i \approx p^{-[n-(i-1)]/(n+1)} \tag{237}$$

Equation (237) shows that the group-sizes are geometrically decreasing with the ratio $p^{1/(n+1)}$.

The next question is: <u>How many stages</u> n should the procedure have? Patel (1962, p. 215) showed that the total number of runs is minimized by choosing an n-stage procedure instead of an (n-1)-stage procedure if

$$p < \left(1 - \frac{1}{n}\right)^{n(n-1)} \tag{238}$$

Equation (238) yields Table 36, which is duplicated from Li (1962, p. 463).

Li derived a similar multistage procedure. Though he does not refer to Watson or Patel, we can easily compare his assumptions and results with those of the latter authors. Li (1962, p. 456) uses the equivalent of Watson's assumption (i); further he uses Watson's assumptions (iii) and (iv) in the sense that in each stage the number of runs is equal to the number of group-factors. Assumption (vi) is replaced by the assumption of "small" experimental error; actually Li neglects experimental error exactly as Patel did. Instead of assumption (v), which Watson used because of possible cancellations within a group, Li (1962, p. 458) assumes that the P

TABLE 36

Optimal Number of Stages as a Function of Prior Probability p

p higher than	Optimal number of stages
	1
2.5000×10^{-1}	
	2
8.7792×10^{-2}	
	3
3.1675×10^{-2}	
	4
1.1529×10^{-2}	
	5
4.2131×10^{-3}	
	6
1.5423×10^{-3}	
	7
5.6516×10^{-4}	
	8
2.0758×10^{-4}	
	9
7.6248×10^{-5}	
	10

important variables show up in exactly P subgroups at every stage.[28)] This means that per group there is not more than one important factor, so that cancellations cannot occur. Moreover Li (1962, p. 456) assumes that the P "important variables have much greater effects than all the unimportant variables combined." This assumption can be compared with Watson's assumption (ii) stating that important factors have an effect $\Delta > 0$ and unimportant factors have an effect zero, for this latter assumption implies that Li's assumption is satisfied. Finally Li used the multistage equivalent of Watson's assumption (vii): $k_i = g_i f_i$, where the subscript i denotes the stage, k_i is the total number of orginal factors considered at stage i, and f_i is the group-size. Notice that here g_i is the total number of groups in stage i, and this definition differs from Patel's definition.

For the two-stage procedure Li derives that $f_1 = p^{-1/2}$ which agrees with Watson's result for the case of no experimental error, i.e. $\alpha_1 = 0$ in Eq. (233). For the procedure with $(n+1)$ stages it can be easily shown that Li derived the same optimal group-sizes f_i as Patel derived.[29)] The optimal number of groups in stage i follows immediately from $k_i = f_i g_i$. Because Li assumes that each important factor shows up in a separate group, his total number of groups per stage remains constant:

$$g_i = k \cdot p^{n/(n+1)} \tag{239}$$

It can be easily shown that with Li's assumption Patel also arrives at a constant total number of groups per stage. For the optimal number of stages Li derived the same critical values of p as Patel did.

Li (1962, p. 461) points out that it may be attractive to choose a procedure with less stages than the optimum if the increase in the total number of runs is only small. He derives that the percent change in the number of runs when choosing an (n+1)-stage procedure instead of an n-stage procedure is

$$\frac{N_{n+1} - N_n}{N_n} = \left(\frac{n+1}{n}\right) p^{1/n(n+1)} - 1 \qquad (240)$$

We would suggest using Eq. (240) as follows. Given p we determine the optimal number of stages, n_0. Put $n_0 = n+1$. If Eq. (240) gives only a small negative percentage then it is attractive to use (n_0-1) stages.

Next we shall consider the problem of <u>estimating the prior probability</u> p.

(i) Li assumes that all important factors show up in separate groups and neglects experimental error. Then after the first stage we know exactly P, the number of important factors. For

$$P = s_1 \qquad (241)$$

where s_1 is the number of significant groups in stage 1. So after stage 1 in an n-stage procedure we have to go through $(n-1)$ more stages, where the number of factors is now $k_1 = s_1 f_1$ and p is exactly known:

$$p = \frac{P}{s_1 f_1} = \frac{s_1}{s_1 f_1} = \frac{1}{f_1} \qquad (242)$$

So the optimal number of groups in the remaining (n-1) stages follows from our Eq. (239) (which holds for an $n+1$ stage procedure):

$$\begin{aligned} g_i &= k_1 \cdot (f_1^{-1})^{(n-2)/(n-1)} \\ &= s_1 \cdot f_1^{1/(n-1)} \end{aligned} \qquad (243)$$

Li (1962, pp. 462-464) shows that the number of observations over all n stages, increases only slightly when our estimate $\hat{p}$ is wrong. For instance, if our initial estimate of p is wrong by a factor a, i.e.

$$\hat{p} = a \cdot p \qquad (244)$$

and $a = 2$ or 0.5 then the total number of observations increases less than 6%.

(ii) Suppose, contrary to Li, that we take into account the, albeit small, probability that more than one important factor occurs within a group. Then we might use the analog of Eq. (241) in every stage:

$$\hat{\underline{P}}_i = \underline{s}_i \tag{245}$$

where $\underline{s}_i$ is the number of significant groups in stage i. As i increases, the group-size decreases, and Eq. (245) will be a better approximation, for the smaller the group-size, the smaller the probability that two or more important factors occur together in a single group. A summary of the way various quantities influence each other is given in Fig. 5. Figure 6 shows how after each stage we can recalculate the optimal number of stages n_0, the group-size f_0 and the resulting number of groups g_0. When calculating these quantities we use the most recent information on P, and act as if we were to start with the first stage of an n-stage procedure characterized by the current P and the current k, which is the number of factors after elimination of insignificant groups. A closer look at the blocks 8, 9, and 2 in Fig. 6 reveals that in stage i, p is estimated by Eq. (246) where $\underline{P}_{i-1}$ is the number of significant factors in the previous stage.

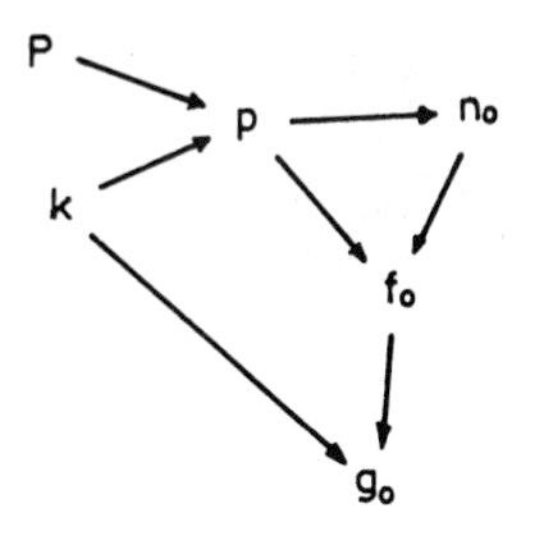

P = number of important factors

k = number of factors

p = probability of being an important factor

n_0 = optimal number of stages

f_0 = optimal group-size

g_0 = optimal number of groups

Fig. 5. Interdependence of the various parameters of a multistage group-screening design.

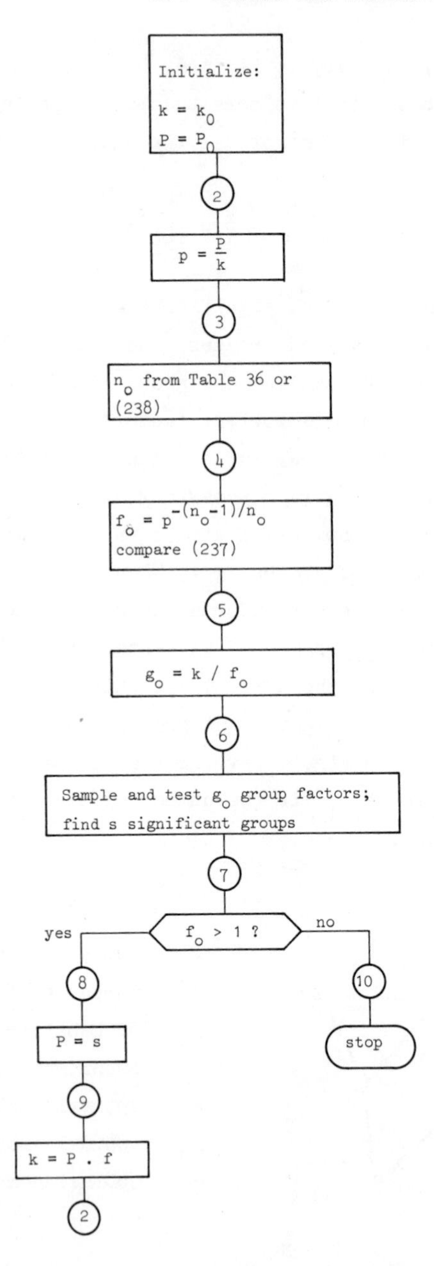

Fig. 6. Flow diagram for the sequentialized estimation of the parameters of a multistage group-screening design.

$$\hat{\underline{p}}_i = \frac{\underline{P}_{i-1}}{\underline{k}_{i-1}} = \frac{\underline{P}_{i-1}}{\underline{P}_{i-1} \cdot \underline{f}_{i-1}} = \frac{1}{\underline{f}_{i-1}} \tag{246}$$

The true p in stage i is equal to Eq. (247) (assuming that none of the P important factors is erroneously eliminated).

$$\underline{p}_i = \frac{P}{\underline{k}_i} = \frac{P}{\underline{P}_{i-1} \cdot \underline{f}_{i-1}} \tag{247}$$

So $\hat{p}_i$ will be a correct estimate if $\underline{P}_{i-1} = P$. The probability that the latter equality holds, increases as the group-size f_{i-1} decreases. The smaller the group-size, the smaller the probability that two or more important factors occur together within one group. We can prove that the group-size does decrease as the procedure goes on, i.e $f_i < f_{i-1}$. For block 4 in Fig. 6 gives:

$$f_i = \hat{p}_i^{-(n_i-1)/n_i} \tag{248}$$

Using Eq. (246) we have

$$\begin{aligned} f_i &= (f_{i-1}^{-1})^{-(n_i-1)/n_i} \\ &= f_{i-1}^{(n_i-1)/n_i} \end{aligned} \tag{249}$$

Because $(n_i-1) < n_i$, Eq. (249) shows that indeed $f_i < f_{i-1}$. So Fig. 6 implies that the prior probability in stage i is set equal to the reciprocal of the group-size in the previous stage. This procedure works better as we proceed from stage to stage.

Li (1962, pp. 465-466) shows that the total number of observations is not affected when we divide the total group of k factors into b <u>subgroups</u>, estimate the prior probability in each subgroup by the <u>same</u> quantity P/k, and use his procedure for each subgroup separately. We would go one step further and estimate the prior probability for each subgroup by its <u>own</u> estimator if we used prior knowledge to put related factors together in one subgroup.

Then in the first stage we could save observations as our estimate $\hat{p}_j$ $(j = 1, \ldots, b)$ is more correct.

To conclude this discussion of group-screening, the most crucial assumptions are that there are no interactions and that the directions of the main effects are known. Otherwise a group-factor may show no effect, whereas actually effects do exist. Elimination of this group-factor means that the effects of the original factors in that group are not detected. Regarding the assumption of known directions of the main effects, we repeat that factors with unknown directions may be put into separate groups of size one, or we may rely on the low probability of more than two important factors occurring within one group, and moreover having opposite effects that cancel out exactly (or having opposite effects of nearly the same magnitude such that the experimental error and the combined effects of minor factors mask the net result of the important factors with opposite signs). Regarding the assumption about the interactions we repeat that main effects are not masked by two-factor interactions if these interactions occur only between factors of the same group, or if we examine the group-factors in a resolution IV design. We observe that, independently of Li, Patel and Watson, group-screening (or sequential bifurcation) for simulation experiments was proposed by Jacoby and Harrison (1962, pp. 131-133). At each stage they propose to split each significant group into exactly two subgroups (i.e. in Patel's notation $g_i = 2$). For group-screening applied to simulation we also refer to Mihram (1972, pp. 399-400) and Nolan and Sovereign (1972, p. 684). A related type of design for testing (2^N-1) factors in N runs, is presented by Ott and Wehrfritz (1972).

Comparing these group-screening designs with the supersaturated and random designs we would recommend the first designs if there are very many factors. Li (1962, p. 455) himself speaks of "cases involving up to 100,000 or more variables." For so many factors no tabulated supersaturated designs are available; random designs with a moderate number of runs and very many factors would result in the confounding of many factors with each other.

Finally, we mention that Jacoby and Harrison (1962) briefly discuss several varieties of the above designs for screening situations. Anscombe (1963) presents a different approach to screening situations. His approach implies the use of prior distributions for certain parameters (like β_0, β_1, and σ) and loss functions. We refer to Anscombe (1963, pp. 725, 726) for the specification of the particular prior distributions and loss functions he used, so that the reader can check whether in a particular application he is willing to use the same distributions and functions. There remains the difficulty that this analysis requires complicated numerical integration.

IV.8. MORE ASPECTS OF EXPERIMENTAL DESIGN

Above we mentioned the different type of approach followed by Anscombe (1963), namely _Bayesian_ analysis and _decision theory_. Loss functions were also used by Last (1962) to derive "truly" optimal designs (instead of designs with minimum variance or another purely statistical criterion). The problem of decision theory and experimental design is further discussed by Herzberg and Cox (1969, pp. 36-37) who give more references. Unfortunately this theoretically very attractive approach does not lead to generally applicable simple designs.

Another aspect of experimental design is the possibility of _multiple responses_. As we observed in Sec. II.10 for the selection of a system we need to bring together various aspects ("responses") of the system in a _criterion_ function. Such a criterion function might be the profit of the simulated firm, the cost of a production process, or a subjective weighing of the various responses (the weights being given, for instance, by the manager). The last possibility was used, e.g. by Knight and Ludeman (1968, pp. 9-14) who constructed a "performance index" in which the various responses of a simulated computer job-shop were incorporated; Fromm (1969) discussed "utility functions" for the evaluation of economic policies; also see Kotler (1970) for a general discussion of deriving such weights. The system variant with the highest (or lowest) value of the criterion is selected. In this case the multiple response problem vanishes. Nevertheless situations do exist with multiple responses. Especially in the initial phase of an investigation, experimentation might aim at giving a survey of various possible system variants, each variant being associated with several (probably conflicting) responses. In that case we are not yet selecting a system but we are trying only to explain how the various factors influence the system in order to get more insight into the behavior of the system. Naylor and Hunter (1969, p.22) state: "Unfortunately, experimental designs for multiple response experiments are virtually nonexistent." As exceptions we mention, e.g. the designs derived by Daniel (1960), Draper and Hunber (1966), and Roy et al. (1971, pp. 113-145), but their designs are applicable only in very special situations. Therefore most authors design the experiment as if there were only a single response; compare, e.g. Kaczka and Kirk (1967) and Sasser (1969, p. 13). Even then we are still confronted with the analysis of multiple responses. We usually want to realize a particular confidence level α that simultaneously holds for all our statements about the several responses. This may require a multivariate analysis. A survey of multivariate analysis is given by, e.g. Anderson (1958), Roy (1957) and Roy et al. (1971); nonparametric multivariate analysis is developed by Puri and Sen (1971). Roy's results are applied to simulation experiments by Dear (1961). Mihram (1972, pp. 393-397) discusses the analysis of the correlation coefficients among the multivariate responses. In Chap. V we shall further discuss simultaneous inference making. In Chap. VI we shall apply such an analysis to the Monte Carlo study of a multiple ranking procedure. (For other aspects of experimental design we also refer to the next section.)

IV.9. LITERATURE

Even this rather voluminous chapter gives only a part of the picture. More incomplete designs exist, for instance, for situations with all factors at more than two levels, or with not all factors at the same number of levels (i.e. mixed designs), or with nonorthogonal estimators of effects, etc. A survey of such designs can be found in Addelman (1963) who gives a list of 48 references. Webb (1969, pp. 429-430 and 1971) constructed a catalogue with designs for factors at two and three levels with no more than twenty runs. He applied ad hoc methods to construct his designs and used a computer to evaluate candidate designs, the criterion being the "averaged" variance of the estimated response. Webb (1969, pp. 430-432) applied his type of design in a simulation experiment with a rocket engine. Other types of design and analysis exist: random effects (i.e. the factor levels are sampled and not fixed to their extreme values as in, e.g. 2^{k-p} and random designs), nested designs (i.e. not all factor levels are crossed but some kind of hierarchy exists), etc. We also refer to the very extensive bibliography by Herzberg and Cox (1969). Good textbooks on experimental design are, e.g. Cochran and Cox (1957), Davies (1963), Hicks (1966), and John (1971).

If the experimenter is confronted with a simulation study where he prefers, e.g. a confounding pattern or factor level combinations different from the designs discussed in detail in this chapter, then he is advised to consult an expert statistician. We hope that this chapter has made the reader aware of what is at stake in the choice of an experimental design.

APPENDIX IV.1. ANALYSIS OF VARIANCE FOR A ONE-FACTOR EXPERIMENT

Consider an experiment with one factor at J levels and I observations per level, as in Table 2 of Sec. IV.2. Let $\underline{y}_{ij}$ denote the ith response for the jth level (i = 1, ..., I and j = 1, ..., J). Obviously the following identity holds.

$$y_{ij} - \underline{y}_{..} \equiv (\underline{y}_{.j} - \underline{y}_{..}) + (\underline{y}_{ij} - \underline{y}_{.j}) \tag{1.1}$$

where

$$\underline{y}_{.j} = I^{-1} \sum_{i} \underline{y}_{ij} \tag{1.2}$$

$$\underline{y}_{..} = J^{-1} \sum_j \underline{y}_{.j} = (IJ)^{-1} \sum_i \sum_j \underline{y}_{ij} \tag{1.3}$$

Squaring both sides of Eq. (1.1) and summing over i and j gives

$$\sum_i \sum_j (\underline{y}_{ij} - \underline{y}_{..})^2$$

$$= \sum_i \sum_j (\underline{y}_{.j} - y_{..})^2 + \sum_i \sum_j (\underline{y}_{ij} - \underline{y}_{.j})^2 + \sum_i \sum_j 2(\underline{y}_{.j} - \underline{y}_{..})(\underline{y}_{ij} - \underline{y}_{.j}) \tag{1.4}$$

The last term in the right-hand side of Eq. (1.4) reduces to

$$2 \sum_j (\underline{y}_{.j} - \underline{y}_{..}) \sum_i (\underline{y}_{ij} - \underline{y}_{.j})$$

$$= 2 \sum_j (\underline{y}_{.j} - \underline{y}_{..})(\sum_i \underline{y}_{ij} - I\underline{y}_{.j})$$

$$= 2 \sum_j (\underline{y}_{.j} - \underline{y}_{..})(I\underline{y}_{.j} - I\underline{y}_{.j})$$

$$= 0 \tag{1.5}$$

Combining Eqs. (1.4) and (1.5) yields

$$\sum_i \sum_j (\underline{y}_{ij} - \underline{y}_{..})^2 = \sum_i \sum_j (\underline{y}_{.j} - \underline{y}_{..})^2 + \sum_i \sum_j (\underline{y}_{ij} - \underline{y}_{.j})^2 \tag{1.6}$$

If the hypothesis of no factor effect holds, then all observations come from a single population. Then the well-known formula for estimating the variance holds, i.e.

$$(IJ - 1)^{-1} \sum_i \sum_j (\underline{y}_{ij} - \underline{y}_{..})^2 = \underline{\hat{\sigma}}^2 \tag{1.7}$$

Under this hypothesis an estimator of the variance of the <u>means</u>, $\underline{y}_{.j}$, is given by

$$\frac{\Sigma_j (\underline{y}_{.j} - \underline{y}_{..})^2}{J - 1} = \underline{\hat{\text{var}}}(\underline{y}_{.j}) \tag{1.8}$$

Combining Eq. (1.8) with a well-known result for the variance of an average, viz.

$$\text{var}(\underline{y}_{.j}) = \frac{\text{var}(\underline{y}_{ij})}{I} = \frac{\sigma^2}{I} \tag{1.9}$$

gives

$$\frac{\Sigma_i \, \Sigma_j \, (\underline{y}_{.j} - \underline{y}_{..})^2}{(J - 1)} = \underline{\hat{\sigma}^2} \tag{1.10}$$

Whether the above hypothesis does or does not hold, Eq. (1.11) is always valid

$$\sum_i (\underline{y}_{ij} - \underline{y}_{.j})^2 = (I - 1) \, \underline{\hat{\text{var}}}(y_{ij}) = (I - 1) \, \underline{\hat{\sigma}^2} \tag{1.11}$$

This is equivalent to

$$\frac{\Sigma_i \, \Sigma_j \, (\underline{y}_{ij} - \underline{y}_{.j})^2}{J(I - 1)} = \underline{\hat{\sigma}^2} \tag{1.12}$$

The eqs. (1.7), (1.10), and (1.12) are summarized in Table 3 of Sec. IV.2. We refer, e.g. to Scheffé (1964) for a proof of the independence of the mean squares in Eqs. (1.10) and (1.12). Because of this independence we can use an F-test to see if both mean squares are equal.

APPENDIX IV.2. ESTIMATION OF INTERACTIONS IN A 2^k EXPERIMENT

In this appendix we shall show how the two-factor interaction α^{AB} in a 2^3 experiment can be estimated, using $x_{gs} = -1$ if factor s is at its low level in factor combination g $(g = 1, \ldots, N)$, and $x_{gs} = +1$ if factor s is at its high level. From Eq. (21) in Sec. IV.2 it follows that

$$\alpha^{AB}_{22} = \eta_{22.} - \eta_{2..} - \eta_{.2.} + \eta_{...} \tag{2.1}$$

Because of the relations in Eq. (26) we know that

$$\alpha_{22}^{AB} = - \alpha_{21}^{AB} = - \alpha_{21}^{AB} = \alpha_{11}^{AB} \tag{2.2}$$

In the 2^3 experiment we define "the" interaction between A and B as plus or minus two times the AB-interactions in the general definition, i.e.

$$\alpha^{AB} = 2\alpha_{22}^{AB} = 2\alpha_{11}^{AB} = - 2\alpha_{21}^{AB} = - 2\alpha_{12}^{AB} \tag{2.3}$$

By definition we have

$$\eta_{22.} = \frac{\eta_{221} + \eta_{222}}{2}$$

$$\eta_{2..} = \frac{\eta_{211} + \eta_{212} + \eta_{221} + \eta_{222}}{4}$$

$$\eta_{.2.} = \frac{\eta_{121} + \eta_{122} + \eta_{221} + \eta_{222}}{4}$$

$$\eta_{...} = \frac{\eta_{111} + \eta_{112} + \eta_{121} + \eta_{122} + \eta_{211} + \eta_{212} + \eta_{221} + \eta_{222}}{8} \tag{2.4}$$

From Eqs. (2.1), (2.3), and (2.4) it follows that

$$\alpha^{AB} = \frac{\eta_{111} - \eta_{211} - \eta_{121} + \eta_{221} + \eta_{112} - \eta_{212} - \eta_{122} + \eta_{222}}{4} \tag{2.5}$$

So the estimated interaction can be calculated from

$$\underline{\alpha}^{AB} = \frac{2}{N} \sum_{g=1}^{N} (x_{g1}x_{g2})\, \underline{y}_g \tag{2.6}$$

where x_{g1} and x_{g2} are the gth element in the column of factor A and B, respectively in Table 6.

APPENDIX IV.3. THE ANALYSIS OF WEBB'S DESIGNS

The matrix of the independent variables $\vec{X}$ in Webb's fold over design is

$$\vec{X} = \begin{bmatrix} \vec{I}_k, & \vec{U}, & \vec{V} \\ \vec{I}_k, & -\vec{U}, & \vec{V} \end{bmatrix} \tag{3.1}$$

where $\vec{I}_k$ corresponds with the grand mean and is a vector of size $(N/2) = k$ consisting of one's; U corresponds with the main effects and is defined by

$$\vec{U} = (x_{us}), \qquad u = 1, \ldots, N/2, \; s = 1, \ldots, k \tag{3.2}$$

and $\vec{V}$ corresponds to the two-factor interactions and is defined by

$$\vec{V} = (x_{us}x_{uz}),$$

$$u = 1, \ldots, N/2, \quad s = 1, \ldots, k-1, \quad z = (s+1), \ldots, k \tag{3.3}$$

The model is

$$\vec{\underline{y}} = \vec{X}\vec{\beta} + \vec{\underline{e}} \tag{3.4}$$

where $\vec{\underline{y}}$ is the vector of N observations; $\vec{\beta}$ is the vector of the grand mean β_0, the k main effects β_s and the $k(k-1)/2$ two-factor interactions β_{sz}; $\vec{\underline{e}}$ is the vector of experimental errors.[30] An equivalent expression for Eq. (3.4) is

$$\underline{y}_i = \beta_0 + \sum_{s=1}^{k} \beta_s x_{is} + \sum_{s=1}^{k-1} \sum_{z=s+1}^{k} \beta_{sz} x_{is} x_{iz} + \underline{e}_i$$

$$(i = 1, \ldots, N) \tag{3.5}$$

Because of Eq. (3.4) the least squares estimators $\vec{\underline{b}}$ should satisfy

$$\vec{X}'\vec{\underline{y}} = \vec{X}'\vec{X}\,\vec{\underline{b}} \tag{3.6}$$

The matrix $(\vec{X}'\vec{X})$ in Eq. (3.6) cannot be inverted for $k \geq 3$, as it is singular. $(\vec{X}'\vec{X})$ is a square matrix with $1 + k + k(k-1)/2$ rows and columns. However, its rank is not larger than $N = 2k$ since it follows from, e.g. Scheffé (1964, p. 394) that

$$\text{rank}(\vec{X}'\vec{X}) = \text{rank}(\vec{X}) \leq \min[N,\ 1 + k + k(k-1)/2] = 2k \tag{3.7}$$

From the definition of $\vec{X}$ in Eq. (3.1), we see that

$$(\vec{X}'\vec{X}) = \begin{bmatrix} 2\vec{I}_k'\vec{I}_k & \vec{0} & 2\vec{I}_k'\vec{V} \\ \vec{0} & 2\vec{U}'\vec{U} & \vec{0} \\ 2\vec{V}'\vec{I}_k & \vec{0} & 2\vec{V}'\vec{V} \end{bmatrix} \tag{3.8}$$

From relations (3.6) and (3.8) we can derive that the main effects, represented by $\vec{\beta}_M' = (\beta_1, \ldots, \beta_k)$, are estimated by

$$\vec{\underline{b}}_M = \frac{1}{2}\,(\vec{U}'\vec{U})^{-1}\,(\vec{U}', -\vec{U}')\,\vec{\underline{y}} \tag{3.9}$$

as $(U'U)$ is nonsingular. It can be shown that these estimators are indeed unbiased. Further, we can derive that the matrix of variances and covariances of $\vec{\underline{b}}_M$ is

$$\Omega_M = \frac{1}{2}\,\sigma^2(\vec{U}'\vec{U})^{-1} \tag{3.10}$$

The two-factor interactions cannot be uniquely determined since there are not enough observations. Some interactions can be assigned arbitrary values which then determine the values of the remaining interaction estimates. This is a well-known result for a consistent system of linear equations with more variables than equations.[31] The general solution of such a system can be found using a generalized inverse, as shown in John (1971, pp. 23-26), Rao (1965, p. 26), and Healy (1968a, b, and c). For a full discussion of the generalized inverse we refer to Rao and Mitra (1971). A particular solution can be found by setting some two-factor interactions to zero. Applying the formulas for this particular solution derived by Webb (1968a, p. 294) we give the following procedure.[32] Rearrange the matrix $\vec{X}$ as in (3.11).

$$\vec{X} = (\vec{X}_1, \vec{X}_2, \vec{X}_3) \tag{3.11}$$

where $\vec{X}_1$ corresponds with the k main effects, $\vec{X}_2$ corresponds with (N-k) two-factor interactions and the grand mean (N is the rank of the matrix $\vec{X}$); $\vec{X}_3$ corresponds with the remaining interactions. (There are $k(k-1)/2 - [(N-k)-1]$ remaining interactions.) Comparing Eq. (3.11) with Eq. (3.1) yields

$$X_1 = \begin{bmatrix} \vec{U} \\ -\vec{U} \end{bmatrix} \tag{3.12}$$

and

$$X_2 = \begin{bmatrix} \vec{I}_k, & \vec{V}_1 \\ \vec{I}_k, & \vec{V}_1 \end{bmatrix} \tag{3.13}$$

where $\vec{V}_1$ is that part of $\vec{V}$ which makes $[\vec{X}_1, \vec{X}_2]$ nonsingular. Applying Webb's results gives the same solution for the main effects as we have already derived in Eq. (3.9). The grand mean and those interactions that correspond with $\vec{X}_2$ are estimated by

$$\vec{\underline{b}}_2 = (\vec{X}_2'\vec{X}_2)^{-1} \vec{X}_2' \vec{\underline{y}} \tag{3.14}$$

The interactions corresponding with $\vec{X}_3$ are put equal to zero, i.e.

$$\vec{\underline{b}}_3 = \vec{0} \tag{3.15}$$

It can be easily shown that $\vec{\underline{b}}_2$ is unbiased if the interactions in $\vec{\beta}_3$ are zero. Further, the matrix of variances and covariances for $\vec{\underline{b}}_2$ is given by

$$\vec{\Omega}_2 = \sigma^2(\vec{X}_2'\vec{X}_2)^{-1} \tag{3.16}$$

APPENDIX IV.4. AN EXAMPLE OF THE APPLICATION OF BOX'S FORMULA FOR AUGMENTED DESIGNS

Box (1966, p. 185) demonstrates the use of his formula given in Eq. (194) above with the following example. Suppose that we started with a 2_V^{8-2} design with generators -ACDFH and -BDEGH and that we added a 2^{8-4} design with the additional generators -ABC and -CDE; compare Table 24. Consequently the four generators for the added 2^{8-4} design are

$$-ACDFH, -BDEGH, -ABC, -CDE \tag{4.1}$$

From the generators in (4.1) the defining relation follows:

$$I = -ACDFH = - BDEGH = -ABC = -CDE = - AGH = \cdots \tag{4.2}$$

where the last word is obtained by multiplying the last three generators -BDEGH, -ABC, and -CDE. From the defining relation it follows that, for instance

$$A = -BC = -GH \tag{4.3}$$

where we have omitted the aliased interactions among three or more factors. Hence, a linear function λ, defined in Eq. (192), is, e.g.

$$\lambda = A - BC - GH \tag{4.4}$$

So in this case we have

$$\gamma_1 = A,\ \delta_1 = +1,\ \gamma_2 = BC,\ \delta_2 = -1,\ \gamma_3 = GH,\ \delta_3 = -1 \tag{4.5}$$

Applying Eq. (194) gives an estimator for the main effect of factor A

$$\hat{\underline{\gamma}}_1 = \underline{\ell}_1 + \frac{N_2}{N_1 + N_2 \cdot 3} (\underline{L}_2 - \underline{L}_1)$$

$$= \underline{\ell}_1 + \frac{16}{64 + 48} (\underline{L}_2 - \underline{L}_1)$$

$$= \underline{\ell}_1 + \frac{1}{7} (\underline{L}_2 - \underline{L}_1) \tag{4.6}$$

where $\underline{\ell}_1$ is the estimator of A based on the first 2^{8-2} runs; $\underline{L}_1$ is the estimator $(\hat{\underline{A}} - \hat{\underline{BC}} - \hat{\underline{GH}})$, where $\hat{\underline{A}}$, $\hat{\underline{BC}}$ and $\hat{\underline{GH}}$ are independent estimators from the first 2^{8-2} design, and $\underline{L}_2$ is the single estimator of the confounded effects (A - BC - GH) in the second 2^{8-4} design. In the same way we have

$$\underline{\gamma}_2 = \underline{\ell}_2 - \frac{1}{7} (\underline{L}_2 - \underline{L}_1) \tag{4.7}$$

and

$$\underline{\gamma}_3 = \underline{\ell}_3 - \frac{1}{7} (\underline{L}_2 - \underline{L}_1) \tag{4.8}$$

APPENDIX IV.5. SELECTION OF THE LEVELS OF THE INSIGNIFICANT FACTORS IN GROUP-SCREENING

Consider the example of Table 35 with three group-factors X, Y, and Z each consisting of three individual factors labeled A, B, C, D, etc. In the first stage only X is significant. Hence, in the second stage the three factors A, B, and C are investigated in four runs. The insignificant factor D may be set at its (i) low level in all four runs; (ii) high level in all four runs; (iii) low level in two runs, high level in the other two runs; (iv) high level in one run, low level in the other three runs. In the cases (i) and (ii) a possible main effect of D is confounded with the grand mean; the A, B, and C-effects can be estimated without bias. In case (iii)

D is confounded with one of the main effects A, B, or C as it follows from Table 5.1. In this table the first column of asterisks denotes that D is either at level +1 in the runs 1 and 2 and at level -1 in the runs 3 and 4, or at level -1 in the runs 1 and 2 and at level +1 in the runs 3 and 4. Assuming that A, B, and C are run as in Table 35, D is confounded with B if D is taken as in the first column of asterisks. In case (iv) D will also disturb the estimates of the main effects. For instance, suppose D is at the levels +, -, -, - in runs 1, 2, 3, and 4, respectively. B is estimated by

$$\underline{b} = \frac{-(\underline{y}_1 + \underline{y}_2) + (\underline{y}_3 + \underline{y}_4)}{4} \tag{5.1}$$

where

$$\begin{aligned} \underline{y}_1 &= \mu - \alpha - \beta + \gamma + \delta + \underline{e}_1 \\ \underline{y}_2 &= \mu + \alpha - \beta - \gamma - \delta + \underline{e}_2 \\ \underline{y}_3 &= \mu - \alpha + \beta - \gamma - \delta + \underline{e}_3 \\ \underline{y}_4 &= \mu + \alpha + \beta + \gamma - \delta + \underline{e}_4 \end{aligned} \tag{5.2}$$

TABLE 5.1

Possible Level-Settings of D and Resulting Confounding

Run	Possible settings of the levels of D (* denotes either + or -, a blank notes the opposite sign)					
1	*	*	*			
2	*			*	*	
3		*		*		*
4			*		*	*
D confounded with:	B	A	C	C	A	B

Substitution of Eq. (5.2) into Eq. (5.1) gives

$$E(\underline{b}) = \beta - \delta/2 \tag{5.3}$$

The conclusion is that the best policy is to set all factors, found to be insignificant in a previous stage, at one level. This implies that if these factors are actually effective they bias only the grand mean, not the main effects of the other factors.

APPENDIX IV.6. THE EFFECT OF INTERACTIONS IN GROUP-SCREENING

Suppose the true model is

$$E(\underline{y}_i) = \beta_0 + \sum_{j=1}^{k} \beta_j x_{ij} + \sum_j \sum_h \beta_{jh} x_{ij} x_{ih} \tag{6.1}$$

where if all the factors are quantitative we may permit $j = h$, i.e. pure quadratic effects β_{jj}. Let the k factors x_j be grouped into g groups $X_1, \ldots, X_g$. Test these g groups in a resolution III design. Then we know that

$$\sum_{i=1}^{N} x_{ij} = 0 \qquad j = 1, \ldots, k \tag{6.2}$$

and if the factors j and j' belong to the <u>same</u> group

$$\sum_i (x_{ij} x_{ij'}) = \sum_i (+1) = N \tag{6.3}$$

If the factors j and j' belong to two different group-factors then

$$\sum_i (x_{ij} x_{ij'}) = 0 \tag{6.4}$$

since the two group-factors are orthogonal in a resolution III design. Next let us consider the sum of products of three factors, i.e. $\Sigma_i \, x_{ij} x_{ij'} x_{ij''}$. There are three possibilities:

(i) The factors j, j', and j'' belong to the <u>same</u> group-factor. Then

$$(x_{ij}x_{ij'}) = +1 \qquad (i = 1, \ldots, N) \tag{6.5}$$

and

$$\sum_i (x_{ij}x_{ij'})\, x_{ij''} = \sum_i x_{ij''} = 0 \tag{6.6}$$

where the last equality follows from Eq. (6.2).

(ii) The factors j and j' belong to one group-factor and j'' belongs to another group. Then as in (6.5) and (6.6) we find

$$\sum_i x_{ij}x_{ij'}x_{ij''} = 0 \tag{6.7}$$

(iii) The three factors j, j' and j'' all belong to different group-factors, say X_1, X_2, and X_3. In a 2^{g-p}_{III} design one group-factor must be confounded with the interaction between two other group-factors, say $\vec{X}_3 = \vec{X_1X_2}$. Consequently

$$x_{ij''} = x_{ij}x_{ij'} \qquad (i = 1, \ldots, N) \tag{6.8}$$

So

$$\sum_i x_{ij}x_{ij'}x_{ij''} = \sum_i (x_{ij''})^2 = \sum_i (+1) = N \tag{6.9}$$

In the discussion of Table 14 we saw that in a Plackett-Burman design, not a 2^{g-p}_{III} design, the two-factor interaction can be expressed as a linear combination of the main effects and the grand mean. Hence, the column of interaction between j and j', say $\vec{x}_{jj'}$, satisfies

$$(\vec{x}_{jj'}) = \sum_{\ell=0}^{g} a_\ell \vec{x}_\ell \tag{6.10}$$

or

$$x_{ij}x_{ij'} = \sum_{\ell=0}^{g} a_\ell x_{i\ell} \tag{6.11}$$

Consequently

$$\sum_i x_{ij}x_{ij'}x_{ij''} = \sum_i \sum_{\ell=0}^{g} a_\ell x_{i\ell}x_{ij''}$$

$$= \sum_\ell a_\ell \sum_i x_{i\ell}x_{ij''}$$

$$= N\, a_{j''} \qquad (6.12)$$

where the last equality follows from Eqs. (6.3) and (6.4).

The main effect of factor p $(p = 1, \ldots, k)$ is estimated by

$$\underline{\hat{\alpha}}_p = \frac{2}{N} \sum_{i=1}^{N} x_{ip}\underline{y}_i \qquad (6.13)$$

so

$$E(\underline{\hat{\alpha}}_p) = \frac{2}{N} \sum_i x_{ip}\, E(\underline{y}_i)$$

$$= \frac{2}{N} \sum_i x_{ip}(\beta_0 + \sum_{j=1}^{k} \beta_j x_{ij} + \sum_j \sum_h \beta_{jh} x_{ij} x_{ih})$$

$$= \frac{2}{N} (\beta_0 \sum_i x_{ip} + \sum_j \beta_j \sum_i x_{ip}x_{ij} + \sum_j \sum_h \beta_{jh} \sum_i x_{ip}x_{ij}x_{ih}) \qquad (6.14)$$

Consider the three sum terms within the brackets.

(i) Because of Eq. (6.2) the first term reduces to zero.

(ii) Because of Eqs. (6.3) and (6.4) the second term reduces to $N\,\Sigma_s\,\beta_s$ where factor s belongs to the same group as p (or is factor p itself if $p = s$).

(iii) Because of Eqs. (6.6), (6.7), (6.9), and (6.12) the last term reduces to $N\,a_p\,\Sigma_z\,\Sigma_w\,\beta_{zw}$, where the factors z, w, and p belong to three different group-factors (and $a_p = 1$ in a 2^{g-p}_{III} design). So

$$E(\underline{\hat{\alpha}}_p) = 2 \sum_s \beta_s + 2a_p \sum_z \sum_w \beta_{zw} \qquad (6.15)$$

Hence, if all factors have two-factor interactions then the (confounded) main effect of a factor is biased. However, if we assume that two-factor interactions exist only between particular (related) factors then we can place these factors in the same group. Their two-factor interactions do not bias the main effect of a factor in that group nor do these interactions bias main effects of factors outside that group since z and w in Eq. (6.15) must correspond with two different groups. The last argument also implies that pure quadratic effects β_{jj} do not bias main effects.

If we test the g group-factors in a <u>resolution IV</u> design then main effects of the group-factors are orthogonal to interactions between two group-factors. Hence, Eqs. (6.9) and (6.12) are replaced by

$$\sum_i x_{ij} x_{ij'} x_{ij''} = 0 \tag{6.16}$$

and Eq. (6.15) reduces to

$$E(\underline{\hat{\alpha}}_p) = 2 \sum_s \beta_s \tag{6.17}$$

EXERCISES

1. An experimenter wants to investigate the effects of varying symmetry of the distribution of a stochastic variable. He likes to make the factor "symmetry" a quantitative variable. Propose a family of distributions that yields varying symmetry as its parameter changes.

2. In a one-way layout level j is replicated r_j times ($j = 1, \ldots, J$). What is the mean square for error? Show that for a constant number of replications this mean square reduces to the averaged estimated variance $\underline{\hat{\sigma}}^2 = \Sigma_1^J \underline{\hat{\sigma}}_j^2 / J$, where

$$\underline{\hat{\sigma}}_j^2 = \sum_{i=1}^{r_j} (\underline{y}_{ij} - \underline{y}_{.j})^2 / (r_j - 1).$$

3. (a) A 2^{7-4} experiment with defining relation $\vec{I} = \overrightarrow{1235} = \overrightarrow{1246} = \overrightarrow{2347}$ yielded the following responses (each response is actually the average of 400 replications): .9975 .9662 .9937 1.000 .9987 .9712 .9925 .9987 .9512 .9850 .9912 .9837 .9925 .9987 .9975 .9912

where $\underline{y}_1 = .9975$ is the (average) response of run 1, the runs being listed in standard order (i.e. column 1 of the design matrix is + - + - + - etc., column 2 is + + - - + + etc.); $\underline{y}_2 = .9662$ is the response of run 2, etc. Determine which main effects and two-factor interactions are not significant when tested at the 5% level.

(b) The standard deviations of the sixteen responses could also be measured and are as follows: .0018 .0063 .0028 .0000 .0013 .0058 .0030 .0013 .0074 .0043 .0033 .0044 .0030 .0013 0.0052 .0033.

Each standard deviation is based on 400 replications. Assume that all responses are normally distributed with a common variance (i.e. the sixteen standard deviations are assumed to have a common mean σ). Apply a goodness of fit test to check if terms not higher than two-factor interactions give good fit.

Note: The numbers in this example are taken from the Monte Carlo experiment with a multiple ranking procedure for $P^* = 0.99$ in Chap. VI.

4. Derive that for Webb's nonorthogonal resolution IV designs:

   (a) The main effects $\vec{\beta}_M$ are estimated by Eq. (174), given $\vec{U}$ is nonsingular.

   (b) The covariance matrix of $\underline{\vec{b}}_M$ is given by Eq. (177).

   (c) The variance of $\underline{\hat{b}}_1$ for $k = 3$ is $\sigma^2/4$. (What would the variance have been if the design matrix were orthogonal?)

   (d) The estimated main effects $\underline{\vec{b}}_M$ are unbiased by grand mean and two-factor interactions.

5. Patel's group-size in stage i $(i = 2, 3, \ldots)$ is $p^{-1/(n+1)}$; see Eq. (236). Assume that each important factor shows up in a separate group. Prove that the total number of groups in stage i remains constant, viz. $kp^{n/(n+1)}$; see Eq. (239).

6. Simulate the single-server queueing system. Investigate the effects of varying mean interarrival time, varying mean service time, and FIFO vs. LIFO queueing discipline.

7. Repeat exercise 6 with common random numbers, antithetic variates and both techniques combined.

8. Simulate a system (e.g. a multi-item inventory or a multichannel queueing system) that has a great many input variables and parameters (but that has a known analytical solution so that you can check your results). Screen for the most important factors applying various screening designs.

   Note: Many ANOVA exercises can be found in the standard textbooks on ANOVA, e.g. Cochran and Cox (1957), Davies (1963), Hicks (1966), and Scheffé (1964). Many (viz. 43) exercises requiring the techniques of this chapter and/or the next chapter are given in Schmidt and Taylor (1970, pp. 558-574); also see the many exercises in Mihram (1972, for instance pp. 383-393).

NOTES

1. We hope that our selection is based on a complete survey of the relevant literature. We examined Sec. 9 on the design of experiments in Statistical Theory and Method Abstracts (1959 through 1972), all issues of Technometrics (1959 through 1972, no. 3) and bibliographies given in other articles.

2. Jacoby and Harrison (1962, pp. 126-127) also discuss the advantages that the one-factor-at-a-time approach may have.

3. It is customary in experimental design terminology to denote the the number of factors by k.

4. If all factors were quantitative then a more general model could be used. We shall return to this problem.

5. For those readers familiar with the distinction between fixed and random factors we point out that in the simulation by Sasser et al. "plans" is a fixed factor while "starting conditions" and "random numbers" are random factors. It is not clear from Sasser et al. (1970, p. 290) whether they indeed analyzed their experiment as a mixed design.

6. Matrices, including vectors, are denoted by using the symbol $\rightarrow$. The transpose of a matrix X is indicated by X'.
7. Scheffé (1964, p. 83) rejects Bartlett's test since it is extremely sensitive to nonnormality, and he derived an alternative test procedure.
8. A detailed discussion of the appropriateness of approximating a response curve by main effects only is given by Yntema and Torgerson (1961, p. 22).
9. E.g., Table 10 shows a 2^{3-1} design.
10. This yields eight rows for the design matrix but only three columns.
11. Except for N = 92. Webb (1969, p. 427) states that Golomb and Baumert (1963) succeeded in deriving a matrix for N = 92; see also John (1971, p. 184). However, we found that their matrix gives design columns that are not orthogonal to the identity column $\vec{I}_{92}$ of the grand mean.
12. The above may clarify the statement by Tukey (1959b, p. 170). "If only a few main effects are large, the chance of concealment of a modestly large side effect (= interaction) by an accumulation of weakly confounded parts of main effects (is diminished)."
13. Remember that Yntema and Torgerson (1961) have argued that ignoring all interactions can yield valid conclusions.
14. The design for nine factors has been derived from the 2^{11-4} design and Box and Hunter (1961b, pp. 454-456) explain why dropping factors 3 and 11 leads to a favorable alias structure.
15. Compare Dykstra (1959, p. 63): "the experimenter's ability to reset and reattain the desired experimental conditions leads to real experimental error, as opposed to the errors of repeated observations taken under a particular setting of the factors under study."

16. Box (1966, pp. 183-186) gives an example with $G = 16$ and $J_g = 2$. So obviously he errs when he claims "an estimate of pure replication error having fifteen degrees of freedom would be available from the differences between replicated runs." Actually there are sixteen degrees of freedom.

17. $\mathrm{var}(\hat{\underline{y}}|x_0) = \sigma^2 \vec{x}_0' (\vec{X}'\vec{X})^{-1} \vec{x}_0$ where $\vec{X}$ comprises only the old experimental points.

18. Remember that Eq. (196) shows that estimated effects are correlated in general. In the usual unaugmented designs the estimated effects are independent and the well-known ANOVA procedure of pooling their sums of squares can be applied; compare our comment on Eq. (33).

19. In RSM another consequence of erroneously accepting the hypothesis of no effect, can be that we use a first degree polynomial as regression equation while actually a higher-order polynomial gives a better representation. We expect that this ill fitting first degree polynomial will imply that it takes more steps, i.e. more observations, to reach the optimum.

20. Observe that in a 2^{k-p} design each effect is either completely orthogonal to another effect or completely confounded with another effect. Hence, their sums of squares are either independent or identical. However, we saw, e.g. in Tables 14 and 18 that other designs yield effects that are not completely orthogonal to each other. Yet sums of squares per effect can be determined for any design [compare Johnston (1963, p. 124)] but these sums of squares are not independent; compare also Hunter (1959a, p. 10).

21. For quantitative factors an alternative is to fit a low-order model to <u>transformed</u> factors; compare Box and Tidwell (1962).

22. Here N_1 does not comprise possible duplicates and extra observations for the residual sum of squares.

23. Actually Box assumed a first order model, but his proof is valid for any model that is linear in its parameters, i.e. $E(\vec{\underline{y}}) = \vec{X}\vec{\beta}$.

24. Budne (1959d, p. 143) uses the median instead of the mean to measure the "average" response per level since the median is less sensitive to extreme responses. Such an extreme value might be an erroneous observation.

25. This randomization rule is not mentioned explicitly by Booth and Cox (1962) but it is a standard procedure in experimental design (and was affirmed by Cox in private communication).

26. We have used our symbol k to denote the number of original factors rather than Watson's symbol. We use the symbol f for the group size.

27. We stick to our familiar symbols. This means that Patel's k and f become our f and k, respectively.

28. We stick to our familiar symbols. So Li's symbols v, p, c, and s_i are replaced by our symbols k, P, n, and f_i, respectively.

29. Substitute our symbols p and $(n+1)$ for Li's symbols p/v, and c, respectively in Li's formula (26). The result is the same as in our Eq. (237).

30. β_s and β_{sz} are defined as only one half of the main effects and two-factor interactions in the standard definition for designs with all factors at only two levels; compare our comment on Eq. (95).

31. The system is consistent since $\vec{X}'\vec{X}$ spans the same space S as $\vec{X}'$ does. So $(\vec{X}'\vec{X})b$ is a vector in that space S and so is $\vec{X}'\vec{y}$.

32. When applying Webb's formulas we can use Eq. (3.8) which shows that Webb's matrix $\vec{B}$ is a zero-matrix in our case.

BIBLIOGRAPHY ON RESPONSE SURFACE METHODOLOGY

This is a preliminary bibliography on RSM and related maximum seeking methods. Obviously many publications could be assigned to more than one category. We have assigned such publications to the one category we thought most relevant. We have placed an asterisk in front of those publications that seem most suited for those readers that are not familiar with the fundamentals of RSM.

A. General

Andersen, S.L. (1959). "Statistics in the strategy of chemical experimentation," Chem. Eng. Prog. 55, 61-67.

*Box, G.E.P. (1954). "The exploration and exploitation of response surfaces: some general considerations and examples." Biometrics, 10, 16-60.

Bradley, R.A. (1958). "Determination of optimum operating conditions by experimental methods, Part I, mathematics and statistics fundamental to the fitting of response surfaces," Ind. Qual. Control, 15, 16-18.

*Burdick, D.S. and T.H. Naylor (1969). "The use of response surface methods to design computer simulation experiments with models of business and economic systems," in The Design of Computer Simulation Experiments (T.H. Naylor, ed.) Duke University Press, Durham, N.C.

*Davies, O.L. (editor)(1960). The Design and Analysis of Industrial Experiments, Hafner, New York, pp. 495-578.

Draper, N.R. and A.M. Herzberg (1971). "On lack of fit," Technometrics, 13, 231-241.

Dutton, J.M. and W.H. Starbuck (1971). Computer Simulation of Human Behavior, Wiley, New York, pp. 593, 706.

*Herzberg, A.M. and D.R. Cox (1969). "Recent work on the design of experiments: a bibliography and review," Roy. Stat. Soc., Ser. A, 132, 29-67.

*Hill, W.J. and W.G. Hunter (1966). "A review of response surface methodology: A literature survey," Technometrics, 8, 571-590.

*Hunter, J.S. (1958). "Determination of optimum operating conditions by experimental methods; Part II-1," Ind. Qual. Control, 15, 16-24.

*Hunter, J.S. (1959). "Determination of optimum operating conditions by experimental methods; Part II-2," Ind. Qual. Control, 15, 7-15.

*Hunter, J.S. (1959). "Determination of optimum operating conditions, Part II-3," Ind. Qual. Control, 15, 6-14.

John, P.W.M. (1971). Statistical Design and Analysis of Experiments, MacMillan, New York, pp. 193-218.

Mendenhall, W. (1968). Introduction to Linear Models and the Design and Analysis of Experiments, Wadsworth, Belmont, Cal., (pp. 267-305).

Meyer, D.L. (1963). "Response surface methodology in education and psychology," J. Exp. Ed., 31, 329-336.

Meyers, R.H. (1971). Response Surface Methodology, Allyn and Bacon, Boston.

Mihram, G.A. (1972). Simulation: Statistical Foundations and Methodology, Academic, New York, pp. 402-442.

Naylor, T.H. (1971). Computer Simulation Experiments with Models of Economic Systems, Wiley, New York, pp. 26-28 and 165-184.

Read, D.R. (1954). "The design of chemical experiments," Biometrics, 10, 1-15.

Roth, P.M. and R.A. Stewart (1969). "Experimental studies with multiple responses," Appl. Stat., 18, 221-228.

B. Designs

Bose, R.C. and R.V. Carter (1959). "Complex representation in the construction of rotatable designs," Ann. Math. Stat., 30, 771-780.

Bose, R.C. and N.R. Draper (1959). "Second order rotatable designs in three dimensions," Ann. Math. Stat., 30, 1097-1112.

*Box, G.E.P. (1952). "Multi-factor designs of first order," Biometrika, 39, 49-57.

Box, G.E.P. and D.W. Behnken (1958). A Class of Three Level Second Order Designs for Surface Fitting, Technical report no. 26, Statistical Techniques Research Group, Section of Mathematical Statistics, Department of Mathematics, Princeton University, New Jersey.

Box, G.E.P. and D.W. Behnken (1960). "Simplex-sum designs: a class of second order rotatable designs derivable from those of first order," Ann. Math. Stat., 31, 838-864.

Box, G.E.P. and D.W. Behnken (1960). "Some new three level designs for the study of quantitative variables, Technometrics, 2, 455-475.

*Box, G.E.P. and N.R. Draper (1959). "A basis for the selection of a response surface design," J. Amer. Stat. Assoc., 54, 622-654.

*Box, G.E.P. and N.R. Draper (1963). "The choice of a second order rotatable design," Biometrika, 50, 335-352.

*Box, M.J. and N.R. Draper (1971). "Factoral designs, the X'X criterion, and some related matters," Technometrics, 13, 731-742.

*Box, G.E.P. and J.S. Hunter (1957). "Multi-factor experimental designs for exploring response surfaces," Ann. Math. Stat., 28, 195-241.

Das, M.N. (1963). "On construction of second order rotatable designs through balanced incompletely block designs with blocks of unequal size," Calcutta Stat. Assoc. Bull., 12, 31-46.

Das, M.N. and A. Dey (1967). "Group-divisible rotatable designs," Ann. Inst. Stat. Math., 19, 331-347.

Das, M.N. and V.L. Narasimham (1962). "Construction of rotatable designs through balanced incomplete block designs," Ann. Math. Stat., 33, 1421-1439.

De Baun, R.M. (1959). "Response surface designs for three factors at three levels," Technometrics, 1, 1-9.

Dey, A. and A.K. Nigam (1968). "Group divisible rotatable designs --some further considerations," Ann. Inst. Stat. Math., 20, 477-481.

Doehlert, D.H. (1970). "Uniform shell designs," Appl. Stat., 19, 231-239.

Draper, N.R. (1960). "Second order rotatable designs in four or more dimensions," Ann. Math. Stat., 31, 23-33.

Draper, N.R. and W.E. Lawrence (1965). "Designs which minimize model inadequacies; cuboidal regions of interest," Biometrika, 52, 111-118.

Draper, N.R. and W.E. Lawrence (1966). "The use of second-order spherical and cuboidal designs in the wrong regions," Biometrika, 53, 596-599.

Draper, N.R. and W.E. Lawrence (1967). "Sequential designs for spherical weight functions," Technometrics, 9, 517-529.

Draper, N.R. and D.M. Stoneman (1968). "Response surface designs for factors at two and three levels and at two and four levels," Technometrics, 10, 177-179.

Dykstra, O. (1960). "Partial duplication of response surface designs," Technometrics, 2, 185-195.

Dykstra, O. (1971). "The augmentation of experimental data to maximize $|X'X|$," Technometrics, 13, 682-688.

Dykstra, O. (1971). Addendum to "The augmentation of experimental data to maximize $|X'X|$," Technometrics, 13, 927.

George, K.C. and M.N. Das (1966). "A type of central composite response surface designs," J. Indian Soc. Agricultural Stat., 18, 21-29.

Hartley, H.O. (1959). "Smallest composite designs for quadratic response surfaces," Biometrics, 15, 611-624.

Hartley, H.O. and P.G. Ruud (1969). "Computer optimization of second order response surface designs," in Statistical Computation (R.C. Milton and J.A. Nelder, eds.), Academic, New York.

Hebble, T.L. and T.J. Mitchell (1972). "Repairing response surface designs," Technometrics, 14, 767-779.

Herzberg, A.M. (1966). "Cylindrically rotatable designs," Ann. Math. Stat., 37, 242-247.

Herzberg, A.M. (1967). "The behaviour of the variance function of the difference between two estimated responses," J. Roy. Stat. Soc., Ser. B., 29, 174-179.

*Hunter, J.S. and T.H. Naylor (1969). "Experimental Design," in The Design of Computer Simulation Experiments (T.H. Naylor, ed.), Duke University Press, Durham, N.C.

Karson, M.J. (1970). "Design criterion for minimum bias estimation of response surfaces," J. Amer. Stat. Assoc., 65, 1565-1572.

Karson, M.J., A.R. Manson, and R.J. Hader (1969). "Minimum bias estimation and experimental design for response surfaces," Technometrics, 11, 6-17.

Mehta, J.S. and M.N. Das (1968). "Asymmetric rotatable designs and orthogonal transformations," Technometrics, 10, 313-322.

Nalimov, V.V., T.I. Golikova, and N.G. Mikeshina (1970). "On practical use of the concept of d-optimality," Technometrics, 12, 799-872.

Nigam, A.K. and M.N. Das (1966). "On a method of construction of rotatable designs with smaller number of points controlling the number of levels," Calcutta Stat. Assoc. Bull., 15, 174-157.

Rechtschaffner, R.L. (1967). "Saturated fractions of 2^n and 3^n factorial designs," Technometrics, 9, 569-575.

Shirafuji, M. (1959). "A two-stage sequential design in response surface analysis," Bull. Math. Stat., 8, 115-126.

Thaker, P.J. (1962). "Some infinite series of second order rotatable designs," J. Soc. Agricultural Stat., 14, 110-120.

Westlake, W.J. (1965). "Composite designs based on irregular fractions of factorials," Biometrics, 21, 324-336.

C. Maximum Seeking Techniques

*Brooks, S.H. (1959). "A comparison of maximum-seeking methods," Operations Res., 7, 430-457.

Brooks, S.H. and M.R. Mickey (1961). "Optimum estimation of gradient direction in steepest ascent experiments," Biometrics, 17, 48-56.

Carpenter, B.H. and H.C. Sweeny (1965). "Process improvement with simplex selfdirecting evolutionary operation," Chem. Eng., 72, 117-126.

Chow, W.H. (1962). "A note on the calculation of certain constrained maxima," Technometrics, 4, 135-137.

Doerfler, T.E. and O. Kempthorne (1963). The Compounding of Gradient Error in the Method of Parallel Tangents, ARL 63-144, Aerospace Research Laboratores, Wright-Patterson Air Force Base, Ohio.

Draper, N.R. (1963). "Ridge analysis of response surfaces," Technometrics, 5, 469-479.

Eldor, H. and L.B. Koppel (1971). "A generalized approach to the method of steepest ascent," Operations Res., 19, 1613-1618.

*Emshoff, J.R. and R.L. Sisson (1971). Design and Use of Computer Simulation Models, MacMillan, New York, second printing, 214-224.

Farlie, D.J. and J. Keen (1967). "Quick ways to the top: a game illustrating steepest ascent techniques," Appl. Stat., 16, 75-80.

Glass, H. and L. Cooper (1965). "Sequential search: a method for solving constrained optimization problems." J. ACM, 12, 71-82.

Hill, J.C. and J.E. Gibson (1966). "Hillclimbing on hills with many minima," in Theory of Self-Adaptive Control Systems (P. H. Hammond, ed.) Plenum, New York.

Johnson, C.H. and J.L. Folks (1964). "A property of the method of steepest ascent," Ann. Math. Stat., 35, 435-437.

Karr, H.W., E.L. Luther, H.M. Markowitz, and E.C. Russell (1965). Simoptimization Research Phase I. Report no. CACI 65-P2.0-1, Consolidated Analysis Center, Inc., Santa Monica, Calif.

Kiefer, J. (1953). "Sequential minimax search for a maximum," Proc. Amer. Math. Soc., 4, 502-506.

Krefting, J. and R.C. White (1971). "Adaptive Random Search," Report 71-E-24, Department of Electrical Engineering, Technische Hogeschool, Eindhoven (The Netherlands).

Luenberger, D.G. (1972). "The gradient projection method along geodesics," Management Sci., 18, 620-631.

Luther, E.L. and H.M. Markowitz (1965). Simoptimization Research Phase II. Report no. CACI, 65-P2-0-1, Consolidated Analysis Center, Inc., Santa Monica, Calif.

Luther, E.L. and N.H. Wright (1965). Simoptimization Research Phase III, Report no. CACI, 65-P2-0-1, Consolidated Analysis Center, Inc., Santa Monica, Calif.

McArthur, D.S. (1961). "Strategy in research--alternative methods for design of experiments," IRE Trans. Eng. Management, 1, 34-40.

McMurtry, G.J. (1971). "Adaptive optimization in learning control," in Pattern Recognition and Machine Learning (K.S. Fu, ed.), Plenum, New York.

Mayne, D.Q. (1966). "A gradient method for determining optimal control of nonlinear stochastic systems," in Proceedings of the Second IFAC Symposium; Theory of Self-Adaptive Control Systems (P. H. Hammond ed.), Plenum, New York.

Meier, R.C. (1967). "The application of optimum-seeking techniques to simulation studies: a preliminary evaluation," J. Financial Quant. Anal., 2, 31-51.

Meier, R.C., W.T. Newell, and H.L. Pazer, (1969). Simulation in Business and Economics, Prentice-Hall, Englewood Cliffs, N.J., second printing, pp. 313-327.

Molnar, G. (1968). "Self-optimizing simulation," in Simulation Programming Languages, Proceedings IFIP working conference on simulation programming languages, (J. N. Buxton, ed.), North Holland Publishing Co., Amsterdam.

Pierre, D.A. (1969). Optimization Theory with Applications, Wiley, New York, pp. 264-366.

Schmidt, J.W. and R.E. Taylor (1970). Simulation and Analysis of Industrial Systems, Richard D. Irwin, Inc., Homewood, pp. 529-553.

Spang, H.A. (1962). "A review of minimization techniques for non-linear functions," SIAM Rev., 4, 343-365.

Spendley, W., G.R. Hext, and F.S. Himsworth (1962). "Sequential application of simplex designs in optimisation evolutionary operation," Technometrics, 4, 441-461.

Törn, A. (1972). "Global Optimization as a Combination of Global and Local Search," in Working papers, volume 1, Symposium Computer Simulation Versus Analytical Solutions for Business and Economic Models, Graduate School of Business Administration, Gothenburg (Sweden).

Umland, A.W. and W.N. Smith (1959). "The use of Lagrangian multipliers with response surfaces," Technometrics, 1, 289-292.

Vajo, V.S. (1969). A Random Process for Optimization, Ecom-3179, U.S. Army Electronics Command, Fort Monmouth, New Jersey.

Westlake, W.J. (1962). "A numerical analysis problem in constrained quadratic regression analysis," Technometrics, 4, 426-430.

Wetherill, G.B. (1966). Sequential Methods in Statistics, Methuen, London, 144-161.

White, R.C. (1971). "A survey of random methods for parameter optimization," Simulation, 17, 197-205.

Wilde, D.J. (1964). Optimum Seeking Methods, Prentice Hall, Englewood Cliffs, New Jersey.

Wolfe, P. (1969). "Convergence conditions for ascent methods," SIAM Rev., 11, 226-235.

Wolfe, P. (1971). "Convergence conditions for ascent methods II: some corrections," SIAM Rev., 185-188.

Zedginidze, I.G. (1966). Optimizatsiya Plotnosti Sukhoy Zernovoi Smesi Metodom Svyazannogo Planirovaniya Eksperimenta, (Optimization of composition of dry grain mixture by the method of coherent planning of experiments), Trudy Gruzinskii Politekhnicheskii Institut (USSR), no. 4, pp. 197-201.

D. Applications

Boyd, D.F. (1964). The Emerging Role of Enterprise Simulation Models, Advanced Systems Development Division, IBM, Yorktown Heights, New York, pp. 74-78.

Davis, R.E., R.W. Faulkender, and W.W. Hines (1969). "A simulated port facility in a theatre of operations," Naval Res. Logistics Quart., 16, 259-269.

Dickey, J.W. and D.C. Montgomery (1970). "A simulation-search technique: an example application for left-turn phasing," Transportation Res., 4, 339-347.

Fine, G.H. and P.V. McIsaac (1966). "Simulation of a time-sharing system," Management Sci., Appl. Ser. 12, 180-194.

Hoggatt, A.C. (1971). "On stabilizing a large microeconomic simulation model," Logistics Rev. Military Logistics J., 1, 21-23. (Reprinted in Dutton, J.M. and W.H. Starbuck (1971). Computer Simulation of Human Behavior, Wiley, New York.)

Hoggatt, A.C. and B.J. Holtbrugge (1966). "Statistical techniques for the computer analysis of simulation models," in Studies in a Simulated Market (L.B. Preston and N.R. Collins eds.), Institute of Business and Economic Research, University of California, Berkeley.

Houston, B.F. and R.A. Huffman (1971). "A technique which combines modified pattern search methods with composite designs and polynomial constraints to solve constrained optimization problems," Naval Res. Logistical Quart., 18, 91-98.

Hufschmidt, M.M. (1966). "Analysis of simulation: Examination of response surface," in Design of Water-Resource Systems (Arthur Maass et al. eds.) Harvard University Press, Cambridge.

Klingel, A.R. (1966). "Bias in PERT project completion time calculations for a real network," Management Sci., 13, 194-207.

Luckie, P.T. and D.E. Smith (1968). Research applicable to problems of intelligence; final report. Report Number 4015, 11-F, HRB-Singer, Inc., State College, Penn.

Michaels, S.E. and P.J. Pengilly (1963). "Maximum yield for specified cost," Appl. Stat., 12, 189-193.

Mihram, G.A. (1971). An Efficient Procedure for Locating the Optimal Simular Response, University of Pennsylvania, Philadelphia.

Smith, D.E. (1968). Sensitivity Analysis and Optimization in Computer Simulation of Intelligence Situations; An Application of Response Surface Methodology, Report 4015.11-R-4, HRB-Singer, Inc., State College, Penn.

Spek, P. (1968). Toepassing van een Oplossingsalgoritme op het Scheepsontwerp (Application of an optimization algorithm to ship design.) Graduation thesis, Afdeling Algemene Wetenschappen, Technische Hogeschool, Delft, The Netherlands.

Taraman, K.S. and B.K. Lamert (1972). "Application of response surface methodology to the selection of machining variables," AIIE Trans., 4, 111-115.

Welch, L.F., W.E. Adams, and J.L. Carmon (1963). "Yield response surfaces, isoquants, and economic fertilizer optima for coastal Bermudagrass," Agron. J. 55, 63-67.

E. Miscellaneous

Atkinson, A.C. (1969). "Constrained maximization and the design of experiments," Technometrics, 11, 616-618.

Behnken, D.W. and N.R. Draper (1972). "Residuals and their variance patterns," Technometrics, 14, 101-111.

Bobis, A.H., and L.B. Andersen (1970). "An approach for economic discrimination between alternative chemical syntheses," Technometrics 12, 439-455.

Box, G.E.P. (1954). "Discussion on the symposium on interval estimation," J. Roy. Stat. Soc., Ser. B, 16, 211-212.

Box, G.E.P. and G.A. Coutie (1956). "Application of digital computers in the exploration of functional relationships," Proc. IEE, 103, Part B, Supplement No. 1, 100-107.

Box, G.E.P. and N.R. Draper (1969). Evolutionary Operation, Wiley New York.

Box, G.E.P. and J.S. Hunter (1954). "A confidence region for the solution of a set of simultaneous equations with an application to experimental design," Biometrika, 41, 190-199.

Box, G.E.P., and W.G. Hunter (1962). "A useful method for model-building," Technometrics, 4, 301-318.

Box, G.E.P. and W.G. Hunter (1965). "The experimental study of physical mechanisms," Technometrics, 7, 23-42.

Box, G.E.P. and P.W. Tidwell (1962). "Transformation of the independent variables," Technometrics, 4, 531-550.

Box, G.E.P. and P.V. Youle (1955). "The exploration and exploitation of response surfaces, an example of the link between the fitted surface and the basic mechanism of the system," Biometrics, 11, 287-323.

Clough, D.J. (1969). "An asymptotic extreme-value sampling theory for estimation of a global maximum," Can. Oper. Res. Soc. J., 7, 102-115.

Hill, W.J., W.G. Hunter, and D.W. Wichern (1960). "A joint design criterion for the dual problem of model discrimination and parameter estimation," Technometrics, 10, 145-161.

Hoerl, A.E. (1959). "Optimum solution of many variables equations," Chem. Eng. Prog., 55, 69-73.

Hunter, W.G. and J.R. Kittrell (1966). "Evolutionary operation: A review," Technometrics, 8, 289-297.

Fishman, G.S. and P.J. Kiviat (1967). Digital Computer Simulation: Statistical Considerations, Report no. RM-5387-PR, The Rand Corporation, Santa Monica, Calif., pp. 28-31. (Published as: The statistics of discrete-event simulation, Sci. Simul., 10, 185-195 (1968).)

Heuts, R.M.J. and P.J. Rens (1972). "A Numerical Comparison Among Some Algorithms for Unconstrained Non-Linear Function Minimization, Report EIT 34, Tilburg Institute of Economics, Department of Econometrics, Katholieke Hogeschool, Tilburg, The Netherlands.

Kitagawa, I. (1959). "Successive process of statistical inferences applied to linear regression analysis and its specializations to response surface analysis," Bull. Math. Stat., 8, 80-114.

Koster, H.J. (1970). "Analyse van functies van meer dan een variabele," (Analysis of functions of more than one variable), Informatie, 12, 15-18.

Kruskal, J.B. (1965). "Analysis of factorial experiments by estimating monotone transformations of the data," J. Roy. Stat. Soc., Ser. B, 27, 251-263.

Maarek, G. and A. Segond (1969). "L'estimation séquentiele dans les modèles linèaires," (Sequential estimation in linear models), Metra, 8, 553-578.

Marquardt, D.W. (1959). "Solutions of nonlinear chemical engineering models," Chem. Eng. Prog., 55, 65-70.

Scheffé, H. (1970). "Multiple testing versus multiple estimation; improper confidence sets; estimation of directions and ratios," Ann. Math. Stat., 41, 8.

Van der Vaart, H.R. (1960). "On certain types of bias in current methods of response surface estimation," Bull. L'Institut Inter. Stat., 37, 191-203.

Van der Vaart, H.R. (1961). "On certain characteristics of the distribution of the latent roots of a symmetric random matrix under general conditions," Ann. Math. Stat., 32, 864-873.

Van Horn, R.L. (1972). "An Optimizing Tree-Search Simulator," in Working papers, volume 4, Symposium Computer Simulation Versus Analytical Solutions for Business and Economic Models, Graduate School of Business Administration, Gothenburg (Sweden).

REFERENCES

Addelman, S. (1961). "Irregular fractions of the 2^n factorial experiments," Technometrics, 3, 479-496.

Addelman, S. (1963). "Techniques for constructing fractional replicate plans," J. Amer. Stat. Assoc., 58, 45-71.

Addelman, S. (1969). "Sequences of two-level fractional factorial plans," Technometrics, 11, 477-509.

Anderson, T.W. (1958). An Introduction to Multivariate Statistical Analysis, Wiley, New York.

Andrews, D.F. (1971). "A note on the selection of data transformations," Biometrika, 58, 249-254.

Anscombe, F.J. (1959). "Quick analysis methods for random balance screening experiments," Technometrics, 1, 195-209.

Anscombe, F.J. (1963). "Bayesian inference concerning many parameters with reference to supersaturated designs," Bull. Inter. Stat. Inst., 40, 721-733.

Anscombe, F.J. and J.W. Tukey (1963). "The examination and analysis of residuals," Technometrics, 5, 141-160.

Balderston, F.E. and A.C. Hoggatt (1962). Simulation of Market Processes, Institute of Business and Economic Research, University of California, Berkeley.

Beaton, A.E. (1969). "Algorithms for data maintenance and computation of analysis of variance," in, Statistical Computation (R.C. Milton and J.A. Nelder, eds.), Academic, New York.

Birnbaum, A. (1959). "On the analysis of factorial experiments without replication," Technometrics, 1, 343-359.

Bock, R.D. (1963). "Programming univariate and multivariate analysis of variance," Technometrics, 5, 95-117.

Bonini, C.P. (1967). Simulation of Information and Decision Systems in the Firm, Markam Publishing Co., Chicago.

Bonini, C.P. (1971). "Experimental design for a simulation model of the firm," in Computer Simulation of Human Behavior (J.M. Dutton and W.H. Starbuck, eds.) Wiley, New York.

Booth, K.H.V. and D.R. Cox (1962). "Some systematic supersaturated designs," Technometrics, 4, 489-495.

Box, G.E.P. (1952). "Multi-factor designs of first order," Biometrika, 39, 49-57.

Box, G.E.P. (1954). "The exploration and exploitation of response surfaces; some general considerations and examples," Biometrics, 10, 16-60.

Box, G.E.P. (1959). "Discussion of the papers of Messrs. Satterthwaite and Budne," Technometrics, 1, 174-180.

Box, G.E.P. (1966). "A note on augmented designs," Technometrics, 8, 184-188.

Box, G.E.P. and J.S. Hunter (1961a). "The 2^{k-p} fractional factorial designs, Part I," Technometrics, 3, 311-351.

Box, G.E.P. and J.S. Hunter (1961b). "The 2^{k-p} fractional factorial designs, Part II," Technometrics, 3, 449-458.

Box, G.E.P. and P.W. Tidwell (1962). "Transformation of the independent variables," Technometrics, 4, 531-550.

Box, G.E.P. and K.B. Wilson (1951). "On the experimental attainment of optimum conditions," J. Roy. Stat. Soc., Ser. B, 13, 1-38.

Box, M.J. and N.R. Draper (1971). "Factorial designs, the $|X'X|$ criterion and some related matters," Technometrics, 13, 731-742.

Box, M.J. and N.R. Draper (1972). "Corrigendum: factorial designs, the $|X'X|$ criterion and some related matters," Technometrics, 14, 511.

Boyd, D.F. (1964). The Emerging Role of Enterprise Simulation Models, Advanced Systems Development Division, IBM, Yorktown Heights, New York.

Budne, T.A. (1959a). "Random balance; part I: the missing statistical link in fact finding techniques," Ind. Qual. Control, 15, 5-10.

Budne, T.A. (1959b). "Random balance; part II: techniques of analysis," Ind. Qual. Control, 15, 11-16.

Budne, T.A. (1959c). "Random balance; part III: case histories," Ind. Qual. Control, 15, 16-19.

Budne, T.A. (1959d). "The application of random balance designs," Technometrics, 1, 139-155, 192-193.

Clough, D.J., J.B. Levine, G. Mowbray, and J.R. Walter (1965). "A simulation model for subsidy policy determination in the Canadian uranium mining industry," Can. Oper. Res. Soc. J., 3, 115-128.

Cochran, W.G. and G.M. Cox (1957). Experimental Designs, Wiley, New York, second edition.

Cohen, A. (1968). "A note on the admissibility of pooling in the analysis of variance," Ann. Math. Stat., 39, 1744-1746.

Curnow, R.N. (1965). "A note on G.S. Watson's paper "A study of the group screening method," Technometrics, 7, 444-446.

Cyert, R.M. and J.G. March (1963). A Behavioral Theory of the Firm, Prentice-Hall, Englewood Cliffs, New Jersey.

Daniel, C. (1956). "Fractional replication in industrial research," in Proceedings Third Berkeley Symposium on Mathematical Statistics and Probability, Volume 5 (J. Neyman, ed.), University of California Press, Berkeley.

Daniel, C. (1959). "Use of half-normal plots in interpreting factorial two-level experiments," Technometrics, 1, 311-341.

Daniel, C. (1960). "Parallel fractional replicates," Technometrics, 2, 263-268.

Daniel, C. (1962). "Sequences of fractional replicates in the 2^{p-q} series," J. Amer. Stat. Soc., 57, 403-429.

Davies, O.L. (1963), ed. The Design and Analysis of Industrial Experiments, Oliver and Boyd, London, 2nd edition.

Dear, R.E. (1961). Multivariate Analyses of Variance and Covariance for Simulation: Studies Involving Normal Time Series, Field note 5644, Systems Development Corporation, Santa Monica, Calif.

Dolby, J.L. (1963). "A quick method for choosing a transformation," Technometrics, 5, 317-327.

Donaldson, T.S. (1966). Power of the F-test for Nonnormal Distributions and Unequal Error Variances, Report no. RM-5072-PR, The Rand Corporation, Santa Monica, Calif.

Doornbos, P. (1965). "Optimale instelling van een pakmachine met behulp van simulatie op een computer," (Optimal setting of a packing-machine with the aid of simulation on a computer.) Statistica Neerl., 19, 206-212.

Draper, N.R. and A.M. Herzberg (1971). "On lack of fit," Technometrics, 13, 231-241.

Draper, N.R. and W.G. Hunter (1966). "Design of experiments for parameter estimation in multiresponse situations," Biometrika, 53, 525-533.

Draper, N.R. and W.G. Hunter (1969). "Transformations: some examples revisited," Technometrics, 11, 23-40.

Draper, N.R. and T.J. Mitchell (1967). "The construction of saturated 2_R^{k-p} designs," Ann. Math. Stat., 38, 1110-1126.

Draper, N.R. and T.J. Mitchell (1968). "Construction of the set of 256-run designs of resolution $\geq$ 5 and the set of even 512-run designs of revolution $\geq$ 6 with special reference to the unique saturated designs," Ann. Math. Stat., 39, 246-255.

Dykstra, O. (1959). "Partial duplication of factorial experiments," Technometrics, 1, 63-75.

Dykstra, O. (1960). "Partial duplication of response surface designs," Technometrics, 2, 185-195.

Dykstra, O. (1971a). "The augmentation of experimental data to maximize $|X'X|$," Technometrics, 13, 682-688.

Dykstra, O. (1971b). "Addendum," Technometrics, 13, 927.

Emshoff, J.R. and R.L. Sisson (1971). Design and Use of Computer Simulation Models, MacMillan, New York, 2nd printing.

Fisher, R.A. (1966). The Design of Experiments, Oliver and Boyd, Edinburgh, 8th edition.

Fisz, M. (1967). Probability Theory and Mathematical Statistics, Wiley, New York, 3rd printing.

Fowlkes, E.B. (1969). "Some operators for ANOVA calculations," Technometrics, 11, 511-526.

Fractional Factorial Designs for Factors at Two Levels, Statistical Engineering Laboratory. National Bureau of Standards. Distributed by Clearinghouse, Springfield, Virginia (1957).

Fromm, G. (1969). "The evaluation of economic policies," in The Design of Computer Simulation Experiments. (T.H. Naylor, ed.), Duke University Press, Durham, N.C.

Golomb, S.W. and L.D. Baumert (1963). "The search for Hadamard matrices," Amer. Math. Monthly, 70, 12-17.

Healy, M.J.R. (1968a). "Multiple regression with a singular matrix," Appl. Stat., 17, 110-117.

Healy, M.J.R. (1968b). "Triangular decomposition of a symmetric matrix," Appl. Stat., 17, 195-197.

Healy, M.J.R. (1968c). "Inversion of a positive semi-definite symmetric matrix," Appl. Stat., 17, 198-199.

Hebble, T.L. and T.J. Mitchell (1972). "Repairing response surface designs," Technometrics, 14, 767-779.

Herzberg, A.M. and D.R. Cox (1969). "Recent work on the design of experiments: a bibliography and a review," J. Roy. Stat. Soc., Ser. A, 132, 29-67.

Hicks, C.R. (1966). Fundamental Concepts in the Design of Experiments, Holt, Rinehart, and Winston, New York.

Hill, W.J. and W.G. Hunter (1966). "A review of response surface methodology: a literature survey," Technometrics, 8, 571-590.

Holms, A.G. and J.N. Berrettoni (1969). "Chain-pooling ANOVA for two-level factorial replication-free experiments," Technometrics, 11, 725-746.

Hunter, J.S. (1958). "Determination of optimum operating conditions by experimental methods, Part II-1," Ind. Qual. Control, 15, 16-24.

Hunter, J.S. (1959a), "Determination of optimum operating conditions by experimental methods; Part II-2," Ind. Qual. Control, 15, 7-15.

Hunter, J.S. (1959b). "Determination of optimum operating conditions by experimental methods; Part II-3," Ind. Qual. Control, 15, 6-14.

Hunter, J.S. (1964). "Sequential factorial estimation," Technometrics, 6, 41-55.

Hunter, J.S. (1968). "Experimental Designs in Simulation Analysis," presented at the Symposium on the Design of Computer Simulation Experiments, Duke University, Durham, N.C. (Also published in The Design of Computer Simulation Experiments (T.H. Naylor, ed.), Duke University Press, Durham, 1969.)

Ignall, E.J. (1972). "On experimental designs for computer simulation experiments," Management Sci., 18, 384-388.

Jacoby, J.E. and S. Harrison (1962). "Multi-variable experimentation and simulation models," Naval Res. Logistics Quart. 9, 121-136.

Jensen, R.E. (1966). "An experimental design for study of effects of accounting variations in decision making," J. Accounting Res., 4, 224-238.

John, P.W.M. (1966). "Augmenting 2^{n-1} designs," Technometrics, 8, 469-480.

John, P.W.M. (1971). Statistical Design and Analysis of Experiments, MacMillan, New York.

Johnston, J. (1963). Econometrics Methods, McGraw-Hill, New York.

Kaczka, E.E. and R.V. Kirk (1967). "Managerial climate, work groups, and organization performance," Administrative Sci. Quart., 12, 253-272.

Kempthorne, O. (1959). "Discussion of the papers of Messrs. Satterthwaite and Budne," Technometrics, 1, 159-166.

Knight, F.D. and M.M. Ludeman,(1968). Computer Job-Shop Simulation Model: A Decision Tool, Report no. DP-MS-67-100, Savannah River Laboratory, E.I. du Pont de Nemours and Co., Aiken, South Carolina.

Kotler, P. (1970). "A guide to gathering expert estimates," Business Horizons, 13, 79-87.

Last, K.W. (1962). Statistical Design of Complex Experimental Programs, Part I, Optimum Experimental Designs Obtained by Minimizing a Loss Function, Report no. ARL-62-373, Aeronautical Research Laboratories, Wright-Patterson Air Force Base, Ohio.

Li, C.H. (1962). "A sequential method for screening experimental variables," J. Amer. Stat. Assoc. 57, 455-477.

McQuie, R. (1969). "Experimental design and simulation in unloading ships by helicopter," Operations Res., 17, 785-799.

Marjolin, B. (1969). "Resolution IV fractional factorial designs," J. Roy. Stat. Soc., Ser. B, 31, 514-523.

Mendenhall, W. (1968). Introduction to Linear Models and the Design and Analysis of Experiments," Wadsworth, Belmont, Cal.

Mihram, G.A. (1970). "A cost-effectiveness study for strategic airlift," Transportation Sci., 4, 79-96.

Mihram, G.A. (1972). Simulation: Statistical Foundations and Methodology, Academic, New York.

Miller, J.R. (1968). Notes on a Simulation Investigation into the Relationship of Lateness to Queue Discipline and Labor Assignment Priority Rules in a Network of Waiting Lines, Western Management Science Institute Workshop, Graduate School of Business Administration, University of California, Los Angeles.

Nauta, F. (1967). Practical Problems in Digital Simulation, Technical notes, CEIR, The Hague.

Naylor, T.H., J.L. Balintfy, D.S. Burdick, and K. Chu (1967a). Computer Simulation Techniques, Wiley, New York, 2nd printing.

Naylor, T.H. and J.S. Hunter (1969). Experimental Designs for Computer Simulation Experiments, Econometric System Simulation Program Working Paper, No. 33, Duke University, Durham, N.C.

Naylor, T.H., K. Wertz and T.H. Wonnacott (1967b). "Methods for analyzing data from computer simulation experiments, Communications ACM, 10, 703-710.

Naylor, T.H., K. Wertz and T.H. Wonnacott (1968). "Some methods for evaluating the effects of economic policies using simulation experiments," Rev. Inter. Stat. Inst., 36, 184-200.

Nolan, R.L. and M.G. Sovereign (1972). "A recursive optimization and simulation approach to analysis with an application to transportation systems," Management Sci. Appl. Ser., 18, 676-690.

Ott, E.R. and F.W. Wehrfritz (1972). "A special screening program for many treatments," Statistica Neerl., 26, 165-170.

Overholt, J.L. (1968). "The problem of factor selection," presented at the Symposium on the Design of Computer Simulation Experiments, Duke University, Durham, N.C.

Overholt, J.L. (1970). Sensitivity Tests on SLAT Computer Simulations using Experimental Design, CNA Research Contribution No. 142, Naval Warfare Analysis Group, Center for Naval Analyses, Arlington, Va.

Patel, M.S. (1962). "Group-screening with more than two stages," Technometrics, 4, 209-217.

Patel, M.S. (1963). "Partially duplicated fractional factorial designs," Technometrics, 5, 71-83.

Peng, K.C. (1967). The Design and Analysis of Scientific Experiments (An Introduction with Some Emphasis on Computation), Addison-Wesley, Reading, Pa.

Plackett, R.L., and J.P. Burman (1946). "The design of optimum multifactorial experiments," Biometrika, 33, 305-325.

Puri, M.L. and P.K. Sen (1971). Nonparametric Methods in Multivariate Analysis, Wiley, New York.

Rao, C.R. (1965). Linear Statistical Inference and Its Applications, Wiley, New York.

Rao, C.R. and S.K. Mitra (1971). Generalized Inverse of Matrices and Its Applications, Wiley, New York.

Rechtschaffner, R.L. (1967). "Saturated fractions of 2^n and 3^n factorial designs," Technometrics, 9, 569-575.

Roy, S.N. (1957). Some Aspects of Multivariate Analysis, Wiley, New York.

Roy, S.N., R. Gnanadesikan, and J.N. Srivastava (1970). Analysis and Design of Certain Quantitative Multiresponse Experiments. Pergamon, Oxford.

Sasser, W.E. (1969). A Causal Relationship Between a Model's Characteristics and the Performances of the Estimators of the Model's Parameters: A Pilot Study, Graduate School of Business Administration, Harvard University, Boston.

Sasser, W.E., D.S. Burdick, D.A. Graham, and T.H. Naylor (1970). "The application of sequential sampling to simulation: an example inventory model," Communications ACM, 13, 287-296.

Satterthwaite, F.E., (1959). "Random balance experimentation," Technometrics, 1, 111-137, 184-192.

Scheffé, H. (1964). The Analysis of Variance, Wiley, New York, 4th printing.

Schink, W.A. and J.S.Y. Chiu (1966). "A simulation study of effects of multicollinearity and autocorrelation on estimates of parameters," J. Financial Quant. Anal., 1, 36-67.

Schmidt, J.W. and R.E. Taylor (1970). Simulation and Analysis of Industrial Systems, Richard D. Irwin, Inc., Homewood, Ill.

Seeger, P. (1966). Variance Analysis of Complete Designs, Almqvist and Wiksell, Uppsala.

Smith, D.E. (1968). Sensitivity Analysis and Optimization in Computer Simulation of Intelligence Situations, An Application of Response Surface Methodology, Report 4015, 11-R-4, HRB-Singer, Inc., State College, Penn.

Smith, H. (1969). "Regression analysis of variance," in The Design of Computer Simulation Experiments (T.H. Naylor, ed.), Duke University Press, Durham, N.C.

Srivastava, J.N. and D.A. Anderson (1969). Fractional Factorial Designs for Estimating Main Effects Orthogonal to Two-factor Interactions: 3^n and $2^m \times 3^n$ Series, ARL Technical Report No. 69-0123 Aerospace Research Laboratories, Wright-Patterson Air Force Base, Ohio.

Srivastava, J.N. and D.A. Anderson (1970). "Optimal fractional plans for main effects orthogonal to two-factor interactions: 2^m series," J. Amer. Stat. Assoc., 65, 828-843.

Srivastava, J.N. and D.V. Chopra (1971). "Balanced optimal 2^m fractional factorial designs of resolution V, $m \leq 6$," Technometrics, 13, 257-269.

Statistical Theory and Method Abstracts (published until 1965 as Statistical Theory and Methods), Oliver and Boyd, Edinburgh, 1959-72.

Tukey, J.W. (1959a). "A quick compact two-sample test to Duckworth's specifications," Technometrics, 1, 31-48.

Tukey, J.W. (1959b). "Discussion of the papers of Messrs. Satterthwaite and Budne," Technometrics, 1, 166-174.

Watson, C.S. (1961). "A study of the group screening method," Technometrics, 3, 371-388.

Webb, S. (1968a). "Non-orthogonal designs of even resolution," Technometrics, 10, 29-299.

Webb, S.R. (1968b). "Saturated sequential factorial designs," Technometrics, 10, 535-550.

Webb, S. (1969). "Interactions between the experiment designer and the computer," Naval Res. Logistics Quart., 16, 423-433.

Webb, S.R. (1971). "Small incomplete factorial experiment designs for two- and three-level factors," Technometrics, 13, 243-256.

Welch, B.L. (1938). "On tests for homogeneity," Biometrika, 30, 149-158.

Westlake, W.J. (1971). "A one-sided version of the Tukey-Duckworth test," Technometrics, 13, 901-903.

Whithwell, J.C. and G.K. Morbey (1961). "Reduced designs of resolution five," Technometrics, 3, 459-477.

Yntema, D.B. and W.S. Torgerson (1961). "Man-computer cooperation in decisions requiring common sense," IRE Trans. Human Factors Electron, HFE-2, 20-26.

Chapter V

SAMPLE SIZE AND RELIABILITY

INTRODUCTION AND SUMMARY

In this chapter we shall investigate the relationships between sample size and reliability. The *sample size* is the number of observations from one particular population (or system variant). The *reliability* is the statistical *accuracy* of an estimate from a sample. This "accuracy" is expressed by, e.g. the length of the confidence interval and the confidence coefficient $(1 - \alpha)$. First we shall concentrate on the case of a single population; next on the *general* case of k (≥ 2) populations. In both cases we can further distinguish between *fixed* sample sizes and *fluctuating* sample sizes. For predetermined sample sizes we have to find the resulting reliability of the estimates. Conversely, we may fix the desired reliability and determine the required sample size.

We have decided to divide this chapter into three parts. Part A: Reliability for a single population. Part B: Fixed sample sizes and k populations: *Multiple comparison* procedures. Part C: Sample size determination for k populations: *Multiple ranking* procedures. In *Part A* we shall discuss how we can estimate the *variance* of the average response of a simulation run. This estimated variance will be used in formulas for the *confidence interval* for the mean response and in formulas for the *sample size* required to estimate the mean with predetermined reliability. In *Part B* we shall present *multiple comparison procedures* (MCP) that yield confidence intervals, e.g. for comparisons among the means of k (≥ 2) populations, the intervals being *simultaneously* valid with predetermined reliability. In that part we shall also discuss procedures that select a *subset* from the k populations in such a way that this subset contains, e.g. the best mean with predetermined reliability. In Part C we shall present multiple ranking procedures (MRP) that determine how many observations should be taken from each of the k (≥ 2) populations in order to select the best population. At the beginning of each part a more detailed summary is given. Each part contains its own references and exercises.

V.A. RELIABILITY FOR A SINGLE POPULATION

V.A.1. SUMMARY

In Sec. V.A.2 we shall consider the estimation of the variance of the average response in a simulation run. A central limit theorem is given for dependent observations from a stationary process. Replicated and continued runs are discussed. In a prolonged run one may use nearly independent subruns or estimate the individual serial correlations, or create independent cycles of observations. Several more references on the estimation of the variance are given.

In Sec. V.A.3 we shall present some well-known results for confidence intervals and tests for the mean of one normal population or the difference between the means of two normal populations. We discuss, e.g. the t-statistic for one or two populations with unknown and possibly different variances. The assumptions of the t-statistic and simulation are studied. Also the binomial distribution and quantile estimation are discussed.

In Sec. V.A.4 sample size determination will be investigated. For a confidence interval of predetermined length double-sampling and (asymptotically consistent and efficient) sequential sampling are discussed. Multistage applications in simulation and Monte Carlo experiments show that the stopping rules do work. We also present sample size determination when testing hypotheses with predetermined α and β errors using a double-sampling procedure. Alternatively a selection (indifference zone) approach may be followed which selects the correct population with predetermined reliability; the resulting heuristic sequential procedure was applied in a simulation experiment. Testing hypotheses with predetermined α and β errors and fully sequential sampling can be based on Wald's sequential probability ratio test (SPRT) (provided no nuisance parameters occur; hence for binomial populations an exact SPRT exists). Part A is concluded with appendixes, exercises, and references.

V.A.2. ESTIMATION OF THE VARIANCE IN SIMULATION

As we saw in Sec. II.8, for terminating systems we can increase the sample size by replication of the simulation runs, each run giving one independent estimator of the response of interest (e.g. mean waiting time or probability of "long" waiting times). For nonterminating systems we may also decide to replicate a run, nowwithstanding the computer time possibly wasted in the transient state. Then the analysis can be performed by simply applying traditional statistical techniques based on independent observations. Since these techniques often assume normality let us first discuss the "stationary r-dependent central limit theorem." A process is called "strictly stationary" if the joint probability function of the observations $\underline{x}_1, \underline{x}_2, \ldots, \underline{x}_t, \ldots, \underline{x}_N$ in a time series,is not a function of the point of time t. So this probability does not vary over time, but is fixed. (This agrees with our definition of the steady-state given in the Secs. I.2 and II.4.) Such a joint probability function implies that the marginal probability function is the same for each $\underline{x}_t$. This in turn implies that all moments for $\underline{x}_t$ do not vary with t. Or, in particular, the mean and variance are constant as Eqs. (1) and (2) specify.

$$E(\underline{x}_t) = \mu \qquad (t = 1, 2, \ldots, N) \tag{1}$$

and

$$\text{var}(\underline{x}_t) = \sigma^2 \tag{2}$$

Moreover, the covariance between $\underline{x}_t$ and $\underline{x}_{t+s}$ does not vary with t but only with s, i.e. the distance or the lag between the observations. Or

$$E[(\underline{x}_t - \mu)(\underline{x}_{t+s} - \mu)] = c_s \tag{3}$$

By definition the autocovariance with lag s divided by the variance σ^2 gives the autocorrelation or serial correlation ρ_s, as Eq. (4) shows.

$$\rho_s = \frac{c_s}{\sigma^2} \tag{4}$$

If not the probability function itself but only the first and second moments in Eqs. (1) through (3) are assumed constant then the process is called "wide-sense" or "<u>covariance</u>" <u>stationary</u>. If the probability function is (multi-) normal then both types of stationarity coincide. A discussion of stationary processes including references can be found in, e.g. Fishman (1968, pp. 13, 17) and (1971, p. 22) and Dear (1961, pp. 9-13). The relations between time series and simulation are also examined by Mihram (1972, pp. 146-180, 443-483).

The "<u>r-dependence</u>" means that $\underline{x}_t$ and $\underline{x}_{t+s}$ are autocorrelated only if $s \leq r$. The "<u>stationary r-dependent central limit theorem</u>" can be formulated as follows. Given an r-dependent strictly stationary sample $\underline{x}_1, \underline{x}_2, \ldots, \underline{x}_t, \ldots, \underline{x}_N$, with $E(\underline{x}_t) = \mu$, and $E(|\underline{x}_t|^3)$ existing, then the sample mean

$$\bar{\underline{x}} = \sum_1^N \underline{x}_t / N$$

is <u>asymptotically normally</u> distributed with mean μ and variance as given in Eq. (5).

$$\begin{aligned} \operatorname{var}(\bar{\underline{x}}) &= \frac{1}{N}\left[\sigma^2 + \frac{2}{N}\sum_{s=1}^{N}(N-s)c_s\right] \\ &= \frac{\sigma^2}{N}\left[1 + 2\sum_{s=1}^{N}\left(1 - \frac{s}{N}\right)\rho_s\right] \end{aligned} \tag{5}$$

Observe that the variance given in Eq. (5) is no asymptotic property but holds for any N. References on this limit theorem can be found in, e.g. Andréasson (1971, pp. 215-223), Fraser (1957, p. 219), and Mihram (1972, pp. 278-281).

(The latter author also discusses asymptotic Weibull distributions that apply to responses that are the minimum of a number of variates as in reliability and PERT studies.) For replicated runs we do not use the variance specified in Eq. (5) but only the normality property. Each replication gives another sequence

$$\underline{x}_1, \underline{x}_2, \ldots, \underline{x}_t, \ldots, \underline{x}_N$$

and its average $\bar{\underline{x}}$. So if there are n replications we have the averages

$$\bar{\underline{x}}_1, \bar{\underline{x}}_2, \ldots, \bar{\underline{x}}_i, \ldots, \bar{\underline{x}}_n \ .$$

Each $\bar{\underline{x}}_i$ is independent and is assumed to satisfy the above limit theorem so that each $\bar{\underline{x}}_i$ is asymptotically normal. Therefore we can apply traditional statistical analysis to the averages $\bar{\underline{x}}_i$. For instance, $(1 - \alpha)$ confidence limits for μ may be given by relation (6).

$$\bar{\bar{\underline{x}}} \pm t_{n-1}^{\alpha/2} \, \underline{s}_{\bar{x}} / \sqrt{n} \tag{6}$$

where

$$\bar{\bar{\underline{x}}} = \sum_{i=1}^{n} \bar{\underline{x}}_i / n \tag{7}$$

$$\underline{s}_{\bar{x}}^2 = \sum_{i=1}^{n} (\bar{\underline{x}} - \bar{\bar{\underline{x}}})^2 / (n-1) \tag{8}$$

and $t_{n-1}^{\alpha/2}$ is the upper $\alpha/2$ percentile of Student's t-statistic with $n-1$ degrees of freedom. Notice that the longer a run is, the better the asymptotic normality holds. This forms an argument in favor of a continued run as opposed to replicated runs.

Next we shall consider the analysis of continued runs. (A brief discussion has already been given in Sec. II.9.) Conway (1963, p. 55) points out that as starting conditions for run 2 we can use the end conditions of run 1, i.e. we just continue the simulation run.

The advantage of this procedure is that the transient period of run 2, or better subrun 2, is nonexistent since we assume that run 1 is terminated in its steady-state. The first technique for analyzing a prolonged run, suggested by Conway (1963, pp. 55-56), throws away the transient phase and proceeds as follows. Divide the remaining run into, say, m subruns. Denote the averages of these subruns by $\bar{x}_1, \bar{x}_2, \ldots, \bar{x}_{m-1}, \bar{x}_m$. Suppose we are able to determine the length of each even-numbered subrun such that the odd-numbered subruns $\bar{x}_1, \bar{x}_3, \ldots, \bar{x}_{m-3}, \bar{x}_{m-1}$ (m is even) are independent. Notice that in order to make these odd-numbered subruns exactly independent, each even-numbered subrun should comprise more than r observations, assuming r-dependent stationary stochastic variables. Obviously it is a problem in itself to determine whether the even-numbered subruns are long enough so that the odd-numbered subruns are independent, at least approximately. We shall return to this problem later on in this section. Now we can again apply traditional analysis techniques to the odd-numbered subruns only. The price we pay is that we throw away the even-numbered subruns $\bar{x}_2, \bar{x}_4, \ldots, \bar{x}_{m-2}, \bar{x}_m$. (It would be nonsense to generate the mth subrun since it is thrown away in this approach; however, we shall need this subrun in a moment.) Observe further that this procedure would be efficient compared with replicated runs, only if the transient period is longer than the length of each even-numbered subrun needed to make the odd-numbered runs independent. However, we will not dwell on this technique any longer and we proceed to the second technique, suggested by Conway (1963) and based on the first technique.

The second technique bases the estimator of μ on all m subruns. Conway (1963, p. 56) proved, assuming that the odd-numbered and even-numbered subruns have the same length, that the variance of this new estimator is smaller than the variance of the above estimator based on the odd-numbered subruns only. This conclusion is intuitively appealing since even while successive observations are positively correlated they do provide information, and therefore it seems more efficient to use the whole prolonged run instead of throwing away all even numbered subruns. The next problem is that Conway did show that the whole run gives a more efficient estimator, but that he did not show how the variance of this estimator can be estimated. Of course we can give an upper-bound for this variance. This upper-bound is the variance of the estimator based on the odd-numbered runs only. This variance can be simply estimated from the independent odd-numbered runs (and from the even-numbered subruns which have the same length as the odd-numbered ones and therefore have the same variance). The disadvantages of this technique are that it gives only an upper-bound so that the reliability is under-estimated, and that we must find the length of the subruns that creates the independence between nonadjacent subruns. So in the rest of this section we shall concentrate on the efficient estimator which uses the whole run (except for a possible initial transient phase), and we shall present several methods for estimating the variance of this estimator.

Approach 1: "Independent" Subruns

As Conway (1963) suggested we may divide the complete run into m subruns. Let the number of individual observations per subrun be denoted by the symbol a. In Appendix V.A.1 we derive that the correlation between two consecutive subruns is given by Eq. (9), where ρ_s $(s = 1, 2, \ldots, a)$ is the correlation between the individual observations lag s apart and these observations are assumed to be stationary variables so that Eq. (3) above holds.

$$\rho(\bar{x}_i, \bar{x}_{i+1}) = \frac{\Sigma_{s=1}^{a} \frac{s}{a} \rho_s + \Sigma_{s=1}^{a-1} \frac{s}{a} \rho_{(2a-s)}}{2(\Sigma_{s=1}^{a} \frac{a-s}{a} \rho_s) + 1} \qquad (i = 1, 2, \ldots, m-1) \qquad (9)$$

We assume that the serial correlation coefficient ρ_s is positive and decreases as the lag s increases. The first assumption seems reasonable in most simulations; the second assumption holds if there are no periodicities.[1)] The denominator in Eq. (9) increases as a increases since the weighing-coefficient $(a-s)/a$ increases with a

and moreover the denominator then comprises more positive terms. In the numerator the weighing-coefficient s/a decreases with a and this decreases the value of the numerator. This effect is mitigated by the increase of the number of terms in the numerator, but for large lags the correlation coefficient is assumed to be close to zero. So we assume that as a, the length of the subrun, increases the correlation between consecutive subruns can be neglected for practical purposes. Or as Hauser et al. (1966, p. 81) conclude more intuitively "if the intervals are sufficiently large (the averages of the subruns or intervals) will effectively be uncorrelated because the effects of the correlations of early values in the interval will be averaged with the large number of values which occur later in the interval and which have no correlation with the values in the previous interval." So if we have m subruns then we can establish confidence intervals for the population mean using Eqs. (6) through (8) where we replace n by m, the number of (approximately) independent subruns. This procedure is valid if a is so large that the subruns are indeed approximately independent. A procedure for selecting an appropriate value for a has been developed by Mechanic and McKay and will be considered in a moment. We observe that a subrun-procedure with intuitively chosen subrun length was applied in simulation by several experimenters. Andréasson (1971, p. 6) created subruns when estimating the variance of a certain probability in his telephone exchange simulation, and so did Adhikari (1967, p. 54) in his simulation of parallel service counters. Hauser et al. (1966, p. 83) also used subruns in their simulated feedback system. Huisman (1970) created 50 subruns comprising 50, 100, 200 and 400 individual observations, respectively, for the simple queuing problem (with utilization 80% and 90%) and gives pictures showing the estimated serial correlation of the subrun averages; also compare Huisman (1969).

A problem not adequately solved by these authors is the selection of the subrun length such that the subruns are nearly independent. Mechanic and MaKay (1966) derived an iterative procedure where the correlation between the averages of the subruns (or "batches" in their terminology) is estimated and larger subruns are taken until this correlation is "small" enough. The smallness of the correlation is judged against an experimentally derived criterion. Mechanic and McKay (1966, pp. 24-39) applied their procedure to several queuing problems and derived some theoretical relations, both with satisfying results for their method. We have reproduced their algorithm in Appendix V.A.2. Derriks (1971) gives flow charts for the implementation of Mechanic and McKay's procedure and applied the technique to several systems (simple time series and a queuing system with known solutions, simple job shop with no known solution). He found that the procedure is relatively fast and works even with serial correlations that are negative before decreasing to zero. Note that Bruzelius (1972) also tried to find an appropriate length for the subrun size. He takes successively larger subruns, and assumes that the variance of the subrun-averages first increases and then stabilizes. We think, however, that it is not proved that this variance increases steadily with increasing subrun length. (The empirical results in Bruzelius (1972, p. 25) seem to confirm our objection.) Moreover the author uses (subjective) judgment to decide whether the variance has "stabilized."

Approach 2: Estimation of Autocorrelation

This alternative procedure does not consider subruns but estimates the correlations or covariances among the individual observations. If the whole run consists of N individual observations, after discarding the transient phase, then the variance of their average is given by Eq. (5) above. For the estimation of this variance, Hauser et al. (1966, p. 81) used Eq. (10) as an approximation to Eq. (5).

$$\operatorname{var}(\bar{\underline{x}}) \cong \frac{\sigma^2}{N}\left(1 + 2 \sum_{s=1}^{k} p_s\right) \tag{10}$$

where k denotes the maximum lag accounted for. We may compare

Eq. (10) with an expression given in, e.g. Blomqvist (1967, p. 165) and Fishman (1967, p. 3) for "large" N:[2)]

$$\mathrm{var}(\underline{\bar{x}}) = \frac{\sigma^2}{N} (1 + 2 \sum_{s=1}^{\infty} \rho_s) \qquad (N \longrightarrow \infty) \tag{11}$$

Comparing Eq. (11) with (10) shows that the formula used by Hauser et al. is correct, if N is large and if the autocorrelation vanishes after k lags. The latter condition is not very restrictive unless the system shows periodicities. Instead of Eq. (10) we may use another formula applied by Fishman (1967, pp. 3, 16-18). He estimates the variance of the average from Eqs. (12) through (14).

$$\mathrm{var}(\underline{\bar{x}}) = \frac{\underline{m}}{N} \tag{12}$$

where N is large and m is estimated by Eq. (13).

$$\underline{\hat{m}} = [\underline{\hat{c}}_0 + 2 \sum_{s=1}^{k} (1 - s/k)\, \underline{\hat{c}}_s]/(1 - k/N) \tag{13}$$

with

$$\underline{\hat{c}}_s = N^{-1} \sum_{t=1}^{N-s} [(\underline{x}_t - \underline{\bar{x}})(\underline{x}_{t+s} - \underline{\bar{x}})]$$

$$s = 0, 1, \ldots, k < N \tag{14}$$

Comparing Eq. (13) with (5) we see that σ^2 and c_s in Eq. (5) are estimated by $\underline{\hat{c}}_0$ and $\underline{\hat{c}}_s$, respectively, that a factor $(1 - k/N)$ is introduced to compensate the bias in these estimators and that not N but only k covariances are used. We refer to Fishman (1967, pp. 17-18) for a discussion of the appropriate choice of k; Mihram (1972, pp. 460-467) also discusses the estimation of c_s and the effect that trend-elimination has on the estimates. The second approach was applied by Geisler (1964a and b) and Hauser et al. (1966); also see Clark et al. (1972). The extensive experimentation by Geisler (1964b) shows that this approach may also yield a valid estimator for $\mathrm{var}(\underline{\bar{x}})$. (We shall give a more detailed report on his results in Sec. V.A.4.)

Recently Fishman (1971) developed a variant of the second approach. He expresses the observation $\underline{x}_t$ of the time series as a moving average, i.e.

$$\underline{x}_t = \mu + \sum_{s=0}^{\infty} a_s \underline{y}_{t-s} \tag{15}$$

with

$$\underline{y}_t : \text{NID}(0, \sigma^2) \tag{16}$$

[For additional references on this representation Naylor (1971, p. 252) may be consulted.] At present it is not clear whether this approach is superior to the other approaches we have discussed above; also see Fishman (1972a).[3)]

Approach 3: Independent Cycles

(i) Kabak (1968) pointed out that in a single-server queuing system the arrival of a new customer into an empty system (i.e. the server is idle) starts a new history that is independent of the past history. This is the so-called renewal or regeneration property of the system. So in a simulation we partition the whole simulated history into epochs (also called cycles, tours, blocks), each epoch starting when a customer arrives into an empty system. Kabak studied the following estimator of the percentage of customers not served, based on M cycles:

$$\hat{\underline{p}} = \frac{\Sigma_1^M \underline{n}_i'}{\Sigma_1^M \underline{n}_i} = \sum_1^M \underline{g}_i \underline{p}_i \tag{17}$$

with

$\underline{n}_i$ = total number of customers in cycle i

$\underline{n}_i'$ = total number of customers in cycle i, not served

$\underline{p}_i = \underline{n}_i'/\underline{n}_i$

$\underline{g}_i = \underline{n}_i/\Sigma\, \underline{n}_i$

So $\hat{\underline{p}}$ is a weighted average of the fraction of nonserved customers in cycle i. He showed that the estimator $\hat{\underline{p}}$ is asymptotically unbiased (the unweighted average of the $\underline{p}_i$ is biased). Note that the $\underline{n}_i$ (and the $\underline{n}_i'$) are independent and identically distributed. Kabak estimates the variance of the ratio of the two terms in Eq. (17), using the following result derived by Kendall and Stuart (1963, p. 232) (based on Taylor series expansion):

$$\operatorname{var}\left(\frac{\underline{x}_1}{\underline{x}_2}\right) = \left(\frac{\mu_1}{\mu_2}\right)^2 \left[\frac{\sigma_1^2}{\mu_1^2} + \frac{\sigma_2^2}{\mu_2^2} - \frac{2\operatorname{cov}(\underline{x}_1, \underline{x}_2)}{\mu_1\mu_2}\right] \quad (\underline{x}_2 \geq 0) \tag{18}$$

For instance, σ_2^2 becomes $\operatorname{var}(\Sigma\, \underline{n}_i) = M \operatorname{var}(\underline{n}_i)$, where $\operatorname{var}(\underline{n}_i)$ can be easily estimated since the $\underline{n}_i$ are independent.

(ii) The idea of independent cycles was used again by Crane and Iglehart (1972a) and Fishman (1972a), independent of Kabak's work. They emphasized that this approach also eliminates the problem of selecting adequate starting conditions and determining transient-state length; compare our discussion in Sec. II.4. The approach further permits a thorough statistical analysis based on independent observations. Crane and Iglehart (1972a) discuss <u>multi</u>-server systems. A cycle starts when a customer arrives into an empty system, i.e. <u>all</u> servers are idle; also see Fishman (1973, p. 14) for an alternative definition of cycles. They estimate various responses like $E(\underline{w})$, $\sigma^2(\underline{w})$, $P(\underline{w} \geq a)$, etc. ($\underline{w}$ denoting steady-state waiting time), mean number of customers in the system, etc. For illustration purposes let us consider the following estimator of $\mu_w = E(\underline{w})$:

$$\hat{\underline{\mu}}_w = \frac{\Sigma_1^M \underline{y}_i}{\Sigma_1^M \underline{n}_i} \tag{19}$$

with

$$\underline{y}_i = \sum_{k=1}^{\underline{n}_i} \underline{w}_{ik}$$

$\underline{w}_{ik}$ = waiting time of customer k within cycle i.

To compare this approach with other approaches we rewrite Eq. (19) as

$$\hat{\underline{\mu}}_w = \sum_1^M \underline{g}_i \bar{\underline{w}}_i = \sum_1^{\underline{N}} \underline{w}_j / \underline{N} \tag{20}$$

where $\bar{\underline{w}}_i$ denotes the average waiting time in cycle i and $\underline{w}_j$ denotes waiting time of customer j $(j = 1, \ldots, \underline{N} = \Sigma_1^M \underline{n}_i)$ in the whole simulated period. So like Kabak they use a weighted average of the cycle averages. As in the traditional approaches, the estimator $\hat{\underline{\mu}}_w$ is the average of all simulated customers, so this estimator has the same efficiency as the traditional one; also see Crane and Iglehart (1972a, pp. 30-33). However, the analysis of $\hat{\underline{\mu}}_w$ is based on creating cycles. Crane and Iglehart (1972b) derived the following $(1 - \alpha)$ confidence interval for the ratio of two means, say $\eta = \mu_1/\mu_2$ $(\mu_2 \neq 0)$, assuming a large number of observations (M) so that the central limit theorem applies:

$$\frac{(\bar{\underline{x}}_1\bar{\underline{x}}_2 - k\underline{s}_{12}) - \underline{D}^{1/2}}{(\bar{\underline{x}}_2)^2 - k\underline{s}_{22}} \leq \eta \leq \frac{(\bar{\underline{x}}_1\bar{\underline{x}}_2 - k\underline{s}_{12}) + \underline{D}^{1/2}}{(\bar{\underline{x}}_2)^2 - k\underline{s}_{22}} \tag{21}$$

where

$$\bar{\underline{x}}_1 = \sum_{i=1}^M \underline{x}_{1i}/M \qquad \bar{\underline{x}}_2 = \sum_{i=1}^M \underline{x}_{2i}/M$$

$$\underline{s}_{12} = \underline{c\hat{o}v}(\underline{x}_1, \underline{x}_2) \qquad \underline{s}_{22} = \underline{v\hat{a}r}(\underline{x}_2)$$

$$k = (z^{\alpha/2})^2/M \qquad (z^{\alpha/2}\text{: upper } \alpha/2 \text{ point of standard normal})$$

$$\underline{D} = (\bar{\underline{x}}_1\bar{\underline{x}}_2 - k\underline{s}_{12})^2 - [(\bar{\underline{x}}_2)^2 - k\underline{s}_{22}][(\bar{\underline{x}}_1)^2 - k\underline{s}_{11}] \tag{22}$$

and the $\underline{x}_{1i}$ (and $\underline{x}_{2i}$) are independent but $\underline{x}_{1i}$ is correlated with $\underline{x}_{2i}$. When applying Eq. (21) to Eq. (19) we have $\underline{x}_{1i} = \underline{y}_i$ and $\underline{x}_{2i} = \underline{n}_i$. Note that the midpoint of the confidence interval given by Eq. (21) is

$$\frac{\bar{\underline{x}}_1 - k\underline{s}_{12}/\bar{\underline{x}}_2}{\bar{\underline{x}}_2 - k\underline{s}_{22}/\bar{\underline{x}}_2} \neq \frac{\bar{\underline{x}}_1}{\bar{\underline{x}}_2} \tag{23}$$

where the right-hand side of Eq. (23) is the naive estimator of $\eta = \mu_1/\mu_2$; if M approaches infinity then the left-hand side of Eq. (23) approaches $\bar{\underline{x}}_1/\bar{\underline{x}}_2$, since k becomes zero. So the point estimators in Eqs. (19) and (17) are only asymptotically unbiased. Fishman (1972b) derived an estimator for μ_1/μ_2 free of bias of order N [and a confidence interval equal to Eq. (21)]; also see Fishman (1972a). For simplicity's sake we might use the point estimator in the left-hand side of Eq. (23); this was obviously done by Crane and Iglehart (1972b, pp. 6-7). Note that Fishman (1972b) actually gives (more general) point estimators and confidence intervals for linear combinations of dependent variables, say $\underline{\eta} = \vec{a}'\underline{\vec{x}}/\vec{b}'\underline{\vec{x}}$, where $\vec{a}$ and $\vec{b}$ are vectors of known coefficients (e.g. $\vec{a}' = (1,0)$ above), and $\underline{\vec{x}}$ is a vector of correlated variables. He also presents simultaneous confidence intervals for the responses. (More about simultaneously valid intervals will be said in Part B of this chapter.)

The above approach requires that the system does indeed return to its empty state a number of times, so that several independent observations can be collected. If the utilization (or traffic intensity) increases then longer cycles will result. Hence, it takes more time to collect M cycles; however, the "subrun" and "autocorrelation" approaches also require more observations since high utilization implies strong serial correlation. As long as the system has a steady-state, it does return to its empty state; see Crane and Iglehart (1972a, pp. 10-11). (Otherwise we are interested in transient responses and replicated runs are required; see Sec. II.8.) Fishman (1972a and b; 1973) further investigated the estimation of a variety of responses in queuing systems (and showed that a number of responses are linear transformations of a few basic responses so that confidence intervals for the basic responses imply confidence intervals for the transformed responses). Note that confidence intervals for the responses of two systems separately, imply a joint confidence interval for the difference between the two responses. If

$$P(\underline{x}_1 < \mu_x < \underline{x}_2) = 1 - \alpha \tag{24}$$

and

$$P(\underline{y}_1 < \mu_y < \underline{y}_2) = 1 - \alpha \tag{25}$$

then

$$P(\underline{x}_1 - \underline{y}_2 < \mu_x - \mu_y < \underline{x}_2 - \underline{y}_1) \geq 1 - 2\alpha \tag{26}$$

as it follows from the Bonferroni inequality; see Chap. V, Part B, Eq. (10). Observe that Eq. (21) implies that we cannot apply $\text{var}(\underline{\hat{\mu}}_x - \underline{\mu}_y) = \text{var}(\underline{\hat{\mu}}_x) + \text{var}(\underline{\hat{\mu}}_y)$ since we do not estimate $\text{var}(\underline{\hat{\mu}})$. [Kabak, however, does use an estimate for $\text{var}(\underline{\hat{\mu}})$.]

(iii) In a subsequent paper Crane and Iglehart (1972c) generalized their approach to any Markov chain, e.g. an (s,S) inventory system or a repairmen system. Again the system is started in some particular state and cycles are started after the return to that state. For instance, start a cycle after the inventory ($\underline{x}$) reaches its maximum, S. How we arrived at $\underline{x} = S$ (i.e. the past history) is unimportant for the next history. All cycles (starting at $\underline{x} = S$) yield independent and identically distributed observations. We could have taken any other starting state because of the Markov chain properties. The efficiency of the estimator does not depend on the starting state selected. (A state may create long cycles but then each cycle contains more information.) For a further discussion of the renewal property of systems we refer to the original papers and their references.

Let us try to compare the various techniques for the estimation of the variance of the average of a continued run. Regarding Approach 2 we observe that Hauser et al. (1966) used Eq. (10). We are in favor of Eq. (12) since Fishman (1967) took account of the bias in the estimators of the covariances. Concerning Approach 1, we remark that Hauser et al. (1966, p. 83) and other authors used a fixed subrun length. We find Mechanic and McKay's procedure more attractive since it determines the length of the subruns in an iterative way. Hauser et al. (1966, p. 83) report that in their case-study the estimated variance in the variant based on subruns, agreed closely with that in the variant based on the individual observations. However, as Mechanic and McKay (1966, p. 5) state, the estimation of the individual correlations "can be extremely unwieldy" both in computer time and memory space; also compare Derriks (1971, pp. 5, 22). Indeed Hauser et al. found that their subrun procedure was twice as fast as their procedure based on the individual observations. Their publication is the only case-study comparing the procedures based on subruns and individual observations. Obviously we have to keep in mind that compared with Hauser et al. Fishman's procedure needs some extra computer time to find an acceptable value for k, the number of lags incorporated in Eq. (13). From Fishman (1967, p. 18) it follows that the calculations of $\hat{m}$ are repeated until $\hat{m}$ stabilizes. But also Mechanic and McKay's procedure requires some extra computer time since it is an iterative procedure for finding the appropriate subrun length. The relative merits of both approaches are also briefly discussed by Emshoff and Sisson (1971, pp. 201-202).[4)]

For relatively uncomplicated systems we recommend Approach 3 (based on the renewal property). For very complicated systems the experimenter may have problems discovering the Markovian structure of his simulated system so that he may decide to use, e.g. Mechanic and McKay's iterative subrun procedure. (We repeat that many simulations of real-life systems are actually transient-state simulations so that replicated runs should be used; see Sec. II.4. Some simulations, however, study relatively uncomplicated systems like multi-server systems in order to test and develop analytical results for such systems.)

We observe that the variance of the average $\bar{x}$ increases as the (positive) autocorrelation coefficients increase; compare Eq. (5). It is well-known that in simple queuing systems these autocorrelations increase as the utilization (or "traffic") of the system grows; compare the pictures in, e.g. Blomqvist (1971, p. 221) or Kosten (1968). The tables in Blomqvist (1967, pp. 165-166) and

(1969, p. 132) show that very large samples are required (because of the high variance) in queuing systems with high utilization. Note that in the subrun approach longer subruns are necessary if the autocorrelation is strong. Above we have already pointed out that increased traffic implies longer cycles in Approach 3.

To conclude this section we shall briefly discuss some procedures for the estimation of $\mathrm{var}(\bar{\underline{x}})$ that have only limited applicability in the simulation of complex systems. Gebhart (1963) and Reynolds (1972) derived expressions for $\mathrm{var}(\bar{\underline{x}})$ in a single-server Poisson queueing system. Gebhart's formula was applied by Healy (1964, pp. 12-15). Poisson queueing systems with identical parallel service stations were studied by Gürtler (1969, pp. 66-80). In such systems the autocorrelation decreases geometrically, i.e. $\rho_s = \rho_1^s$ so that $\Sigma_1^{\infty}\rho_s = \rho_1/(1 - \rho_1)$. [He describes how this geometric character can be tested; compare Tintner (1960, p. 296).] All of these approaches, however, seem to fail in complicated systems simulation. Baraldi (1969a and b) derived a sample size formula for certain queuing systems based on approximations (viz. neglecting the dependence between stochastic variables) and upper bounds for some expressions. He does not estimate $\mathrm{var}(\bar{\underline{x}})$ from the sample but gives an apriori value for the sample size. His formula may very well be invalid (because of approximations) and is too conservative (because of upper bounds). A conservative estimate is undesirable since simulation runs are expensive and, because of the positive autocorrelation, $\mathrm{var}(\bar{\underline{x}})$ is high so that many runs are required to obtain a reliable estimate. Finally Fishman (1967, pp. 2-3) and Fishman and Kiviat (1967, p. 27) stress the spectral analysis of the stationary process. But as Eqs. (12) through (14) above demonstrate, their estimation of the variance of the process does not use the spectrum.

V.A.3. FIXED SAMPLE SIZE AND ONE POPULATION

In this section we shall present some well-known results of mathematical statistics for determining confidence intervals and performing tests for the mean of one population or for the difference between the means of two populations. In the latter case we are not interested in the two means themselves, so that we are not confronted with problems of simultaneous inference making (see Part B). We shall also discuss the estimation of a binomial probability and quantiles. The results of this section will be used in the next section on sample size determination.

Let us suppose we are studying a <u>normal</u> population $N(\mu, \sigma^2)$. Based on n independent drawings $\underline{x}_i$ $(i = 1, 2, \ldots, n)$ from that population we want to make inferences about the population mean, μ. From the sample we determine the average

$$\bar{\underline{x}} = \sum_{i=1}^{n} \underline{x}_i / n \tag{27}$$

and the sample variance of $\underline{x}$ (denoted as $\underline{s}^2(\underline{x})$ or briefly $\underline{s}_x^2$)

$$\underline{s}^2(\underline{x}) = \underline{s}_x^2 = \sum_{i=1}^{n} (\underline{x}_i - \bar{\underline{x}})^2/(n - 1) \tag{28}$$

This yields the t-statistic based on $(n - 1)$ degrees of freedom

$$\underline{t}_{(n-1)} = \frac{\bar{\underline{x}} - \mu}{\underline{s}(\underline{x})/\sqrt{n}} = \frac{\bar{\underline{x}} - \mu}{s(\bar{\underline{x}})} \tag{29}$$

Define $t_{n-1}^{\alpha/2}$ as the upper $\alpha/2$ point of the (symmetric) distribution of $\underline{t}_{(n-1)}$, i.e.

$$P(\underline{t}_{(n-1)} \geq t_{n-1}^{\alpha/2}) = \alpha/2 \tag{30}$$

Then Eqs. (29) and (30) result in

$$P[\bar{\underline{x}} - t_{n-1}^{\alpha/2}\, \underline{s}(\bar{\underline{x}}) \leq \mu \leq \bar{\underline{x}} + t_{n-1}^{\alpha/2}\, \underline{s}(\bar{\underline{x}})] = 1 - \alpha \tag{31}$$

Hence, the point estimator of μ is $\bar{\underline{x}}$ and the associated $(1 - \alpha)$ <u>confidence interval</u> is given by

$$[\underline{\bar{x}} - t_{n-1}^{\alpha/2}\, \underline{s}(\underline{\bar{x}})], \qquad [\underline{\bar{x}} + t_{n-1}^{\alpha/2}\, \underline{s}(\underline{\bar{x}})] \tag{32}$$

In addition to estimation purposes the result in Eq. (31) can also serve testing purposes. Suppose that we want to test the hypothesis that $E(\underline{x})$ has the value μ_0, i.e. our "null hypothesis" is given by

$$H_0 : \mu = \mu_0 \tag{33}$$

If μ_0 is not contained in the interval, Eq. (32), then we reject H_0. This procedure results in two error sources. The Type I error or α-error is the probability of erroneously rejecting H_0, or

$$P(H_0 \text{ rejected} \mid \mu = \mu_0) = \alpha \tag{34}$$

The type II error or β-error is the probability of erroneously accepting H_0, i.e.

$$P(H_0 \text{ accepted} \mid \mu = \mu_1) = \beta \tag{35}$$

We can fix the value of α in Eq. (34). Then a specified value of μ_1 in the alternative hypothesis, $H_1 : \mu = \mu_1$, yields a particular value of β. Considering β as a function of μ_1 in Eq. (35) we obtain the operating characteristic of the test. The complement of β, i.e. $(1 - \beta)$, is called the power of the test. Also compare Eqs. (88) through (91) further in this section.

Next we shall suppose that we are studying two normal populations, $N(\mu_1, \sigma_1^2)$ and $N(\mu_2, \sigma_2^2)$ and that we are interested in the difference of their means, say δ. We can take the independent observations $\underline{x}_{1i}$ $(i = 1, 2, \ldots, n_1)$ and $\underline{x}_{2j}$ $(j = 1, 2, \ldots, n_2)$ from population 1 and 2, respectively. There are several ways to compare their means.

(i) The observations $\underline{x}_1$ and $\underline{x}_2$ may be "paired", i.e. $\underline{x}_{1i}$ and $\underline{x}_{2i}$ are dependent (while independence remains among the $\underline{x}_1$ and among the $\underline{x}_2$). For example, there may be n plots of land, two varieties being planted per plot so that plot i yields $\underline{x}_{1i}$ and $\underline{x}_{2i}$ (i = 1, ..., n). (Notice that in paired observations $n_1 = n_2 = n$.) In simulation and Monte Carlo studies this situation occurs when we use the same sequence of random numbers for two systems, so that sequence i gives response $\underline{x}_{1i}$ for system 1 and response $\underline{x}_{2i}$ for system 2. From the paired observations we can determine the <u>difference</u> $\underline{d}$, i.e.

$$\underline{d}_i = \underline{x}_{1i} - \underline{x}_{2i} \qquad (i = 1, 2, \ldots, n) \tag{36}$$

and the problem reduces to studying a single population with mean $\delta = \mu_1 - \mu_2$. Testing and estimating can be done using Eq. (31) with $\bar{\underline{x}}$ replaced by $\bar{\underline{d}}$. We refer to Fisz (1967, p. 432) for a discussion of the possibility that a hypothesis H_0 is rejected when applying a test on $\underline{d}$ while H_0 is not rejected when applying a test based on $\underline{x}_1$ and $\underline{x}_2$ separately as in (ii) to which we proceed now.

(ii) If both populations have the <u>same variance</u>, $\sigma_1^2 = \sigma_2^2 = \sigma^2$, then we can still use the familiar t-statistic. From, e.g. Fisz (1967, p. 353) it follows that the expression in Eq. (37) is distributed as $\underline{t}$ with $(n_1 + n_2 - 2)$ degrees of freedom.

$$\underline{t}_{(n_1+n_2-2)} = \frac{(\bar{\underline{x}}_1 - \bar{\underline{x}}_2) - (\mu_1 - \mu_2)}{[(n_1-1)\underline{s}_1^2 + (n_2-1)\underline{s}_2^2]^{1/2}} \left[\frac{n_1 n_2}{n_1 + n_2} (n_1 + n_2 - 2)\right]^{1/2} \tag{37}$$

where $\bar{\underline{x}}_1$ is the sample average for population 1, $\underline{s}_1^2$ is the sample variance for population 1, etc. The t-statistic has $(n_1 + n_2 - 2)$ degrees of freedom since we can pool the individual sums of squares.

(iii) If the populations may have <u>different</u> variances (so-called Behrens-Fisher problem) then we may consider a statistic analogous to Eq. (29), namely

$$\underline{t}' = \frac{\bar{\underline{d}} - \delta}{s(\bar{\underline{d}})} = \frac{(\bar{\underline{x}}_1 - \bar{\underline{x}}_2) - (\mu_1 - \mu_2)}{[(\underline{s}_1^2/n_1) + (\underline{s}_2^2/n_2)]^{1/2}} \tag{38}$$

This statistic is not exactly t-distributed, so we cannot take the $\alpha/2$ percentile of the t-distribution with $(n_1 + n_2 - 2)$ degrees of freedom; compare Eq. (37). Scheffé (1970, p. 1502) proposes as a possible solution taking the $\alpha/2$ percentile of the t-distribution with only (min - 1) degrees of freedom, where "min" denotes the minimum of n_1 and n_2. This procedure is conservative, i.e. too long confidence intervals may result or, equivalently, the type I error may be smaller than the nominal α value. Cochran and Cox (1957, pp. 100-101) suggest treating $\underline{t}'$ in Eq. (38) as an approximately t-distributed variable with the upper $\alpha/2$ point determined as a weighted average from the t-distribution with $(n_1 - 1)$ and $(n_2 - 1)$ degrees of freedom and (stochastic) weights depending on the estimated variances. Or

$$\underline{t}'^{\alpha/2} = \frac{\underline{w}_1\, t^{\alpha/2}_{n_1-1} + \underline{w}_2\, t^{\alpha/2}_{n_2-1}}{\underline{w}_1 + \underline{w}_2} \tag{39}$$

with

$$\underline{w}_1 = \frac{\underline{s}_1^2}{n_1} \quad \text{and} \quad \underline{w}_2 = \frac{\underline{s}_2^2}{n_2} \tag{40}$$

Observe that for equal sample sizes Eq. (39) reduces to Scheffé's procedure. Wang (1971) showed that the following variant of (iii), based on the results of Welch, gives satisfactory results. Put $\underline{t}'$ in Eq. (38) equal to a Student t-variable with degrees of freedom equal to $\underline{f}_w$, where

$$\underline{f}_w = \frac{[\underline{r}/(1 + \underline{r})]^2}{(n_1 - 1)} + \frac{[1/(1 + \underline{r})]^2}{(n_2 - 1)} \tag{41}$$

and

$$\underline{r} = \frac{\underline{w}_1}{\underline{w}_2} = \frac{\underline{s}_1^2\, n_2}{\underline{s}_2^2\, n_1} \tag{42}$$

Slightly more complicated variants of (iii) are examined by Mehta and Srinavasan (1970).

(iv) Scheffé also derived an exact statistic [in 1943; his statistic is presented in, e.g. Kendall and Stuart (1961, p. 144)]. Define

$$\underline{u}_i = \underline{x}_{1i} - (n_1/n_2)^{1/2}\, \underline{x}_{2i} \qquad (i = 1, 2, \ldots, n_1) \tag{43}$$

$$\bar{\underline{u}} = \sum_{i=1}^{n_1} \underline{u}_i/n_1 \tag{44}$$

assuming $n_1 \leq n_2$. Then the expression in Eq. (45) is exactly t-distributed with $(n_1 - 1)$ degrees of freedom.

$$\underline{t}_{(n_1-1)} = (\bar{\underline{x}}_1 - \bar{\underline{x}}_2 - \delta)\left[\frac{n_1(n_1 - 1)}{\Sigma(\underline{u}_i - \bar{\underline{u}})^2}\right]^{1/2} \tag{45}$$

When determining $\bar{\underline{x}}_2$ all n_2 values of $\underline{x}_2$ are used, but when determining $\underline{u}_i$ only n_1 values of $\underline{x}_2$ are randomly selected from the n_2 available values of $\underline{x}_2$. (Because of this randomization Scheffé (1970, p. 1503) does not recommend this approach.) Since in simulation and Monte Carlo experiments all experimental conditions are controlled, we may simply take the first n_1 observations on $\underline{x}_2$ instead of randomizing. Note that for equal sample sizes Eq. (43) reduces to Eq. (36) and approach (i) and (iv) become the same.

Surveys and extensions to other procedures for comparing two population means are given by Csörgö and Seshadri (1971), Press (1966), Scheffé (1970), Thomasse (1972), and Ying Yao (1963). Scheffé also shows how the power of the various tests can be evaluated numerically. Let us next consider the sensitivity of the t-statistic to the underlying assumptions.

Scheffé (1964, pp. 331-369) gives a survey of the effects of violations of independence, normality and common variance. Dependence among observations seriously affects the t-statistic. Nonnormality is unimportant for large samples; for small samples two populations can still be compared if we take equal sample sizes, $n_1 = n_2 = n$. Using Eq. (37) when actually the variances differ still gives valid results if we have equal sample sizes, even if these samples are small. We can also use distribution-free tests, e.g. the sign test or Wilcoxon's rank test for paired observations and the Mann-Whitney rank test for comparing means; see Keeping (1962, pp. 260-265) or the lucid exposé on nonparametric methods by Conover (1971). A sophisticated discussion of nonparametric methods (including extensions to $k > 2$ populations and multiple responses) is presented by Puri and Sen (1971). An application of a nonparametric test in a simulation context is shown by Meier et al. (1969, pp. 309-311). The robustness of point estimators of location (not the corresponding confidence intervals and tests) was extensively studied by Andrews et al. (1972). They examined about seventy different location estimators, and found that the sample mean is very sensitive to outliers. Therefore, they propose rejecting suspected outliers or, for heavy-tailed distributions, using the median; see Andrews et al. (1972, pp. 237-248). They also propose an alternative to the variance estimator that is less sensitive to extreme observations; see Andrews et al. (1972, pp. 81, 160). We shall not further discuss nonparametric tests since the t-statistic is robust and, moreover, this statistic gives simple explicit formulas for sample-size determination as we shall see in Sec. V.A.4. The sequentialized variants of distribution-free tests used for sample-size determination, are much more cumbersome; compare Geertsema (1970). (In Parts B and C we shall return to distribution-free tests, since in multiple comparison and ranking situations statistics may be used that are sensitive to nonnormality.) Fishman (1971, p. 36) suggests two other distribution-free approaches. The first one is the familiar Chebyshev inequality [proved in, e.g. Fisz (1967, p. 74)]; the other one is a Godwin inequality assuming a unimodal distribution. Unfortunately these two approaches are conservative and moreover assume a known variance. Let us see how well the assumptions of normality etc. are satisfied in simulation.

In simulation the assumption of normality may be well satisfied if the response of the run is an average. For the limit theorem formulated in Sec. V.A.2 shows that even the average of dependent observations may be approximately normally distributed. In any case, the normality assumption is not very crucial, as we have just seen. More important is the assumption of independence. As we have mentioned more often n independent observations can be obtained by replicating runs using n different sequences of random numbers. For a prolonged run (approximately) independent observations are created by splitting the run into n subruns. The subrun averages replace the observations $\underline{x}_i$ (i = 1, 2, ..., n) in the above procedures. Mechanic and McKay (1966, pp. 5,28) applied such an approach. However, instead of the t-distribution they used the normal distribution which is a good approximation if $n \geq 30$; see, e.g. Fisz (1967, p. 350). We may also look at the simulation as a long sequence of serially correlated observations, $\underline{x}_i$. Then the assumption of independence crucial for the t-statistic, is severely violated. Therefore we follow another approach. As we have just mentioned serially correlated observations $\underline{x}_i$ can still yield a sample average $\bar{\underline{x}}$ which is (approximately) normally distributed with mean μ and variance

$$\sigma^2(\bar{\underline{x}}) = \frac{\sigma^2(\underline{x})}{n} \left[1 + 2 \sum_{s=1}^{n} \left(1 - \frac{s}{n}\right) \rho_s \right] \tag{46}$$

where ρ_s denotes the autocorrelation with lag s. Hence, we replace Eq. (31) by

$$P[\bar{\underline{x}} - z^{\alpha/2}\, \sigma(\bar{\underline{x}}) \leq \mu \leq \bar{\underline{x}} + z^{\alpha/2}\, \sigma(\bar{\underline{x}})] = 1 - \alpha \tag{47}$$

where $z^{\alpha/2}$ is the upper $\alpha/2$ point of N(0,1). Unfortunately we do not know $\sigma(\bar{\underline{x}})$. Therefore we estimated $\sigma(\bar{\underline{x}})$ in Eq. (31) by $\underline{s}(\bar{\underline{x}}) = \underline{s}(\underline{x})/\sqrt{n}$ and corrected for this estimation replacing $z^{\alpha/2}$ by $t^{\alpha/2}_{(n-1)}$. In the same way we now replace $\sigma(\bar{\underline{x}})$ in Eq. (47) by its estimator but we continue using $z^{\alpha/2}$ as an approximation for $t^{\alpha/2}$. We point out that in Eq. (47) $\sigma(\bar{\underline{x}})$ is no longer equal to $\sigma(\underline{x})/\sqrt{n}$ but is now given by Eq. (46), where the serial correlation coefficients, ρ_s, can be estimated using the methods of Sec. V.A.2. As we shall see in the next section the above approaches, based on subruns or Eq. (46), were found to give correct confidence intervals.

For the third approach, based on truly independent cycles, we have already shown the confidence interval for the ratio of two means, this ratio being the unbiased estimator of the mean response of the simulated system; see Eq. (21) above. This confidence interval uses $z^{\alpha/2}$ because of the central limit theorem. When comparing two systems we do not use the t-statistic for differences as in (i) through (iv) of this section, but apply the confidence interval Eq. (26) based on the confidence intervals Eqs. (24) and (25) for the mean responses of the individual systems. The empirical results in Crane and Iglehart (1972b, pp. 6-7) and (1972c, pp. 16-21) show that Eq. (21) indeed gives correct results.

In some simulation and Monte Carlo studies we know that a binomial distribution will hold. We may, for instance, want to estimate the probability that some event occurs. For example, in an equipment failure model we may estimate the probability that the life of the equipment is longer than c units of time; such events have a constant probability from run to run and are independent. Hence, we can use the binomial model

$$P(\underline{x}_i = 1) = p$$

$$P(\underline{x}_i = 0) = 1 - p \qquad (i = 1, 2, \ldots, n) \tag{48}$$

where we score $\underline{x}_i = 1$ if the event occurs in run i $(i = 1, 2, \ldots, n)$, and $\underline{x}_i = 0$ if the event does not occur. Obviously p is estimated by

$$\hat{\underline{p}} = \frac{\Sigma_{i=1}^{n} \underline{x}_i}{n} \tag{49}$$

with

$$\mathrm{var}(\hat{\underline{p}}) = p(1 - p)/n \tag{50}$$

Confidence intervals for $\hat{\underline{p}}$ can be constructed from the table for the binomial distribution or from the normal or Poisson approximation; see, e.g. Keeping (1962, pp. 59, 64-68). Van der Waerden (1965, p. 29) points out that an unbiased estimator of $\mathrm{var}(\hat{\underline{p}})$ is obtained if we replace n in Eq. (50) by (n - 1), i.e.

$$\hat{\text{var}}(\hat{\underline{p}}) = \hat{\underline{p}}(1 - \hat{\underline{p}})/(n - 1) \tag{51}$$

Obviously for large sample sizes this refinement is negligible. When comparing two binomial probabilities, say p_1 and p_2, we can estimate these probabilities analogous to Eqs. (48) and (49). For independent runs we have

$$\text{var}(\hat{\underline{p}}_1 - \hat{\underline{p}}_2) = \text{var}(\hat{\underline{p}}_1) + \text{var}(\hat{\underline{p}}_2) \tag{52}$$

Confidence intervals may be constructed using the normal approximation and Eq. (52). [The normal approximation holds even better for the difference between the estimated probabilities than for each individual probability; see, e.g. Scheffé (1964, p. 332).] Van der Waerden (1965, p. 41) observes that when testing the hypothesis $H_0: p_1 = p_2 = p$ then Eq. (52) becomes

$$\text{var}(\hat{\underline{p}}_1 - \hat{\underline{p}}_2) = p(1 - p)(n_1 + n_2)/(n_1 n_2) \tag{53}$$

In Eq. (53) p can be estimated from the $n_1 + n_2$ observations on both populations, i.e. $\hat{\underline{p}}$ is the fraction of "successes" in the combined number of observations. Additional comments on binomial probabilities can be found in Conover (1971, pp. 95-104), Van der Waerden (1965, pp. 22-51), and Wehrli (1970, pp. 99-105).

A binomial model was used by Kurlat and Springer (1960, p. 476) in their "simulation" study of the reliability of an antitank-mine. (They built a physical "simulation" device instead of an abstract model.) In the distribution sampling example of Appendix I.2 where we estimated $p = P(\underline{x} < a)$ the binomial model is also valid. Flagle (1960, pp. 435-439) applied this model in his single-channel Poisson queueing system where the probability of a customer arriving in an empty system is estimated. However, in Flagle's example successive events are not independent so the binomial model is not correct.[5)]

Related to the percentile p, say $P(\underline{x} < c) = p$ with fixed c, is the quantile, say x_p:

$$p(\underline{x} < x_p) = p \tag{54}$$

For a discrete variable the quantile x_p is defined in Eq. (55):

$$P(\underline{x} < x_p) \leq p \ \ \underline{\text{and}} \ \ P(\underline{x} > x_p) \leq 1 - p \tag{55}$$

Note that Eq. (55) reduces to Eq. (54) for continuous variables. We shall restrict our attention to the continuous case; for the discrete case we refer to, e.g. Conover (1971, p. 31). An example of a quantile is the median, i.e. x_p for $p = 0.50$. Other quantiles of interest may be the upper or lower 5% point of a distribution of responses $\underline{x}$. In order to estimate x_p we have to arrange the observations in increasing order, say $x_{(1)} \leq x_{(2)} \leq \cdots \leq x_{(n)}$. These ordered observations $x_{(i)}$ are called the order statistics. The estimated or sample quantile is then $x_{(pn+1)}$, i.e. among the n observations a fraction $pn/n = p$ is smaller than this $x_{(pn+1)}$; compare Eq. (54). Actually pn does not need to be an integer, so pn is rounded downwards. [However, for the sample median we take $\underline{x}_{(i*)}$ with $i* = p(n + 1)$ for n is odd, and $[x_{(pn)} + \underline{x}_{(pn+1)}]/2$ for n is even; see David (1971, p. 4).] This rule corresponds with forming the empirical, or estimated distribution function

$$\hat{F}(x_{(i)}) = P(\underline{x} < x_{(i)}) = \frac{i - 1}{n} \qquad i = 1, \ldots, n \tag{56}$$

and determining which value of $x_{(i)}$ corresponds with p; see Exercise 6. In this way we have obtained an asymptotically unbiased (point) estimator of the quantile. [For an unbiased estimator, using the "jackknife" procedure we refer to Goodman et al. (1973).]

To test $H_0 : x_p = a$ we use the same approach as for the binomial case. So we score $\underline{x}_i = 1$ if $x < a$, apply Eqs. (49) and (50), and test if the resulting $\hat{\underline{p}}$ agrees with p; also see Conover (1971, pp. 104-110). This author further shows why the following confidence interval for the quantile x_p holds.

$$P(x_{(r)} \leq x_p \leq x_{(s)}) = 1 - \alpha \tag{57}$$

with

$$r = np - z^{\alpha/2} [np(1 - p)]^{1/2} \tag{58}$$

$$s = np + z^{\alpha/2} [np(1 - p)]^{1/2} \tag{59}$$

r and s being rounded upward to the next integer. For one-sided intervals we replace $\alpha/2$ by α in Eqs. (58) or (59) and delete $x_{(s)}$ or $x_{(r)}$ in Eq. (57). This confidence interval holds for any distribution, provided $n > 20$ so that the central limit theorem applies; otherwise binomial tables must be used instead of z^{α}; also see Conover (1971, pp. 110-115) and David (1971, pp. 13-15).

The estimation of quantiles is expensive in terms of _computer time_ _and_ _memory_ _space_, as we need to store and sort all n observations. Sorting is a standard subject of computer science; for a survey of sorting routines we refer to Martin (1971). To solve these computer problems a different approach is proposed by Goodman et al. (1973) and Lewis (1972, pp. 9-10); the reader may also check Andrews et al. (1972, pp. 55-56). Note that in simulation and Monte Carlo studies quantiles may indeed be of interest. In Monte Carlo studies we may be interested in the 5% critical value of a statistic $\underline{x}$. In simulation we may want to estimate the waiting times exceeded by no more than 5% of the customers; see, e.g. Van Frankenuysen and Schuringa (1971, p. 2). Note that Conover (1971, p. 299) and Holme (1972, p. 122) examined the construction of confidence bands for the _complete_ distribution. Once such bands are constructed simultaneously valid confidence bounds for all quantiles follow.

V.A.4. SAMPLE SIZE DETERMINATION FOR ONE POPULATION

Let us first consider the estimation of the mean μ of the normal population $N(\mu,\sigma^2)$, assuming a known σ^2. From the population we sample n independent observations $\underline{x}_1, \ldots, \underline{x}_n$ yielding the sample average $\bar{\underline{x}}$. Suppose we want our estimate to be less than

c units wrong. Because of sampling fluctuations we never can be 100% certain of achieving this goal. Therefore we further specify our reliability requirement as follows. We want to be $100(1 - \alpha)\%$ (e.g. 95%) certain that our estimator $\underline{\bar{x}}$ is not more than c units in error. Or

$$P(|\underline{\bar{x}} - \mu| \leq c) = 1 - \alpha \tag{60}$$

We know that the average $\underline{\bar{x}}$ of n independent normal variables $\underline{x}_i$ satisfies

$$P(|\underline{\bar{x}} - \mu| \leq z^{\alpha/2} \sigma(\underline{x})/\sqrt{n}) = 1 - \alpha \tag{61}$$

Hence if Eq. (60) should also be met then Eq. (62) must hold.

$$c = z^{\alpha/2} \sigma(\underline{x})/\sqrt{n} \tag{62}$$

or the sample size n should satisfy Eq. (63).

$$n = \left(\frac{z^{\alpha/2}}{c}\right)^2 \sigma^2(\underline{x}) \tag{63}$$

In simulation (and in most other applications) $\sigma^2(\underline{x})$ will be <u>unknown</u>. Therefore, we may decide to replace $\sigma^2(\underline{x})$ by its <u>estimator</u> and correct for this estimation by the use of Student's t-statistic, i.e. Eq. (61) is replaced by Eq. (64).

$$P(|\underline{\bar{x}} - \mu| \leq t^{\alpha/2}_{(n-1)} \underline{s}(\underline{x})/\sqrt{n}) = 1 - \alpha \tag{64}$$

This would yield

$$\underline{n} = \left(\frac{t^{\alpha/2}_{(n-1)}}{c}\right)^2 \underline{s}^2(\underline{x}) \tag{65}$$

Our convention of underlining stochastic variables clearly shows the default of this approach. Equation (64) is based on a <u>fixed</u> sample size n, whereas Eq. (65) shows that in this approach the sample size is actually stochastic since it depends on the estimator of the variance. Or following Tocher (1963, pp. 113-114), if we update $\underline{s}^2(\underline{x})$ in Eq. (65) after each additional observation, then $\underline{n} = n$ implies the following set of events:

$$2 < (t_1^{\alpha/2}/c)^2 \, \underline{s}_2^2, \quad 3 < (t_2^{\alpha/2}/c)^2 \, \underline{s}_3^2, \; \ldots, \; n \geq (t_{n-1}^{\alpha/2}/c)^2 \, \underline{s}_n^2 \tag{66}$$

where $\underline{s}_i^2$ $(i = 2, 3, \ldots, n)$ is the sequential estimator of σ^2 after i observations have been taken. These estimators are not independent so that it is not simple to evaluate the joint probability of Eq. (66), but it is clear that this probability does not need to be $1 - \alpha$.[6] Let us consider several alternatives for coping with this problem.

Anscombe (1953, pp. 7-8) proved that for large sample sizes the effect of estimating the unknown σ^2 sequentially, as it is done in Eq. (66), can be corrected by a simple adjustment. The adjustments, however, are so small that we would propose neglecting them. Later on it was proven by several authors that it is indeed asymptotically correct to replace the unknown variance (or variances when comparing two means) in the sample size formula based on a known variance, by a sequentially updated estimator; see Chow and Robbins (1965), Starr (1966a), Robbins et al. (1967), and Srivastava (1970). They proved that this approach is asymptotic "consistent" (i.e. the confidence interval contains μ with probability $1 - \alpha$) and asymptotic "efficient" (i.e. the expected sample size equals the sample size for known variance). Starr (1966a) also studied the small sample behavior of this approach and found that the deviations from the nominal $(1 - \alpha)$ coverage probability are very small. [For example, a slightly modified version for sequential confidence intervals for the mean resulted in a minimum coverage probability of 93% for a nominal $1 - \alpha = 95\%$ when varying σ/c between 0.5 and 6.75; see Starr (1966a, p. 44).] The small sample efficiency is also high. Robbins et al. (1967) give a Monte Carlo study of the small-sample behavior when comparing two means. Again only minor deviations from $(1 - \alpha)$ were found. [Starr (1966b) studied the efficiency of the sequential approach when a particular loss function is assumed; loss functions were also used by Starr and Woodroofe (1970) for estimating the variance or for exponential variates.]

Another alternative does not estimate $\sigma^2(\underline{x})$ sequentially but uses a double-sampling scheme, i.e. a pilot sample of n_0 observations is taken and this yields an estimator of σ^2 based on $(n_0 - 1)$ degrees of freedom, say $\underline{s}^2_{n_0}$. Analogous to Eq. (63) we have

$$\underline{n} = \left(\frac{t^{\alpha/2}_{n_0-1}}{c} \right)^2 \underline{s}^2_{n_0} \tag{67}$$

so that a second final sample is taken of $\underline{n} - n_0$ observations ($\underline{n} - n_0$ is rounded upwards to the next integer; obviously if $\underline{n} \leq n_0$, then no additional observations are required). The confidence interval is based on the over-all average of the $\underline{n}$ observations but as Eq. (67) shows $\underline{s}^2$ is based on only n_0 observations. This is the well-known approach devised by Stein; see, e.g. Kendall and Stuart (1961, p. 618). A variant on this approach can be found in Dudewicz (1972), where the over-all average is not used, but a particular linear combination of the averages of the first and second stage [also compare Part C, Eq. (38)]. Stein's method gives a higher confidence coefficient than Dudewicz's method in the case of inferences about the mean of a single population. In the case of inferences about the difference of two population means the situation is different as we shall see below. But first we shall compare the sequential and the double-sampling procedure.

A serious problem in the application of Stein's method is the choice of the pilot sample size. A large value for n_0 reduces $t^{\alpha/2}_{n_0-1}$ in Eq. (67) but may result in wasted observations (viz. when $n_0 > \underline{n}$). Since observations in simulation are expensive we shall usually prefer a more efficient procedure. Starr (1966a, pp. 44-47) compared Stein's procedure with the sequential procedure for a single population and found that the sequential approach is more efficient (but may yield a confidence level slightly less than the nominal value $1 - \alpha$ for particular σ/c values).

Next we shall consider the comparison of the means of two populations in more detail. Robbins et al. (1967) proposed a sequential procedure completely analogous to the sequential approach presented above for a single population; see Eq. (66). So, if the variances σ_1^2 and σ_2^2 were known then Eq. (62) would become

$$c = z^{\alpha/2}\,\sigma(\underline{\bar{x}}_1 - \underline{\bar{x}}_2) = z^{\alpha/2}\left(\frac{\sigma_1^2}{n_1} + \frac{\sigma_2^2}{n_2}\right)^{1/2} \tag{68}$$

Now it is well-known that the variance of $(\underline{\bar{x}}_1 - \underline{\bar{x}}_2)$ is minimized for a fixed total sample size n $(= n_1 + n_2)$ by the following allocation:

$$\frac{n_1}{n_2} = \frac{\sigma_1}{\sigma_2} \tag{69}$$

[See, e.g. Dudewicz and Dalal (1971, p. 2.1) or Tocher (1963, p. 107 with $p_i = 1$).] So the reliability requirement ($\delta = \mu_1 - \mu_2$ is contained within $(\underline{\bar{x}}_1 - \underline{\bar{x}}_2) \pm c$ with probability at least $1 - \alpha$) is satisfied with minimum total sample size by taking

$$n_1 = \left(\frac{z^{\alpha/2}}{c}\right)^2 \sigma_1(\sigma_1 + \sigma_2) \tag{70}$$

$$n_2 = \left(\frac{z^{\alpha/2}}{c}\right)^2 \sigma_2(\sigma_1 + \sigma_2) \tag{71}$$

so that

$$n = n_1 + n_2 = \left(\frac{z^{\alpha/2}}{c}\right)^2 (\sigma_1 + \sigma_2)^2 \tag{72}$$

In passing, we observe that for paired observations from the two populations we can use Eq. (63) with $\sigma^2(\underline{x}) = \sigma_1^2 + \sigma_2^2$. It is simple to prove that paired observations require a smaller total sample size if positive correlation between $\underline{x}_1$ and $\underline{x}_2$ indeed exists and both variances are equal [so that the optimum allocation in Eq. (69) yields $n_1 = n_2$ as it is the case in paired observations].

The sequential approach for unknown variances proposed by Robbins et al. (1967, p. 1385) proceeds as follows.

(a) In each stage calculate the usual estimators of σ_1^2 and σ_2^2 based on all observations available at that stage, i.e. if n_1 and n_2 observations have been taken calculate

$$s_1^2 = \sum_{i=1}^{n_1} (x_{1i} - \bar{x}_1)^2/(n_1 - 1) \tag{73}$$

where

$$\bar{x}_1 = \sum_{1}^{n_1} x_{1i}/n_1 \tag{74}$$

and s_2^2 is defined analogous to Eq. (73). Because of Eq. (69), take the next observation from population 1 if

$$\frac{n_1}{n_2} \le \frac{s_1}{s_2} \tag{75}$$

If (75) does not hold then the next observation comes from population 2.

(b) Let a_n be a sequence of positive constants such that

$$a_n \longrightarrow z^{\alpha/2} \quad \text{as} \quad n \longrightarrow \infty \tag{76}$$

We would propose replacing a_n by the $\alpha/2$ percentile of the t-statistic with degrees of freedom determined as discussed in Eq. (38). Notice, however, that these degrees of freedom may depend on the estimators s_1^2 and s_2^2 [see Eqs.(39) and (41)] and then we are replacing a_n by a sequence of positive stochastic variables that approach $z^{\alpha/2}$ for $n \longrightarrow \infty$. We conjecture that this approximation does not invalidate the procedure. It has the advantage of following the sequential procedure for a single population and the fixed-sample procedure for two populations (the latter one being discussed in the previous section). Robbins et al. (1967, p. 1387) use $a_n = z^{\alpha/2}[(n + 4)/(n - 4)]^{1/2}$. The authors propose three different stopping rules based on Eq. (68) through (72): Replace $z^{\alpha/2}$ by

a_n and σ_1 and σ_2 by $\underline{s}_1$ and $\underline{s}_2$ and stop sampling as soon as (i) Eq. (72) holds (with the indicated replacements and with $=$ replaced by $\geq$), (ii) Eq. (68) holds, and (iii) Eqs. (70) and (71) hold. These rules are such that the total sample size n is smallest for rule (i) and largest for rule (iii). The authors proved that these three rules are asymptotic consistent and efficient. The small-sample behavior, studied in a Monte Carlo experiment, was also found to be satisfactory.

Dudewicz (1972) presents the following double-sampling procedure for comparing two means.

(a) Take an initial sample of size n_0 (≥ 2) from each population and calculate the traditional estimators of the mean and variance.

(b) Take a second sample of size $(\underline{n}_i - n_0)$ from population i $(i = 1, 2)$, where

$$\underline{n}_i = \max(n_0 + 1, w^2\underline{s}_i^2) \qquad (77)$$

($w^2\underline{s}_i^2$ being rounded upwards) and

$$w = \frac{d_{n_0}^{\alpha/2}}{c} \qquad (78)$$

the critical constant $d_n^{\alpha/2}$ tabulated in Dudewicz (1972). Because of the limited availability of the original publication we have reproduced his table in our Table 1. (Note that the table is entered with $1-\alpha$ not $\alpha/2$.)

(c) Construct the desired interval

$$(\bar{\underline{x}}_1 - \bar{\underline{x}}_2) \pm c \qquad (79)$$

where $\bar{\underline{x}}_i$ is the average of all $\underline{n}_i$ observations. (A more exact interval is constructed from the weighted averages of the sample mean of stage 1 and 2; see Dudewicz [1972, Eq. (1.4)].

TABLE 1

Critical Constant $d_{n_0}^{\alpha/2}$ in Double-Sampling Procedure

For Comparing Two Means

$1-\alpha$ \ n_0	2	3	4	5	6	7	8	9
.75	2.00	1.37	1.21	1.14	1.10	1.07	1.05	1.04
.80	2.75	1.76	1.54	1.44	1.38	1.34	1.32	1.30
.85	3.93	2.27	1.94	1.80	1.72	1.68	1.64	1.62
.90	6.16	3.04	2.50	2.29	2.18	2.11	2.06	2.02
.95	12.63	4.57	3.50	3.11	2.91	2.79	2.71	2.66
.975	25.42	6.54	4.59	3.94	3.63	3.44	3.33	3.24
.99	63.68	10.28	6.31	5.14	4.60	4.30	4.11	3.98
.995	127.4	14.42	7.92	6.15	5.39	4.97	4.71	4.53
.999	624	31.9	13.3	9.13	7.48	6.65	6.16	5.84

$1-\alpha$ \ n_0	10	11	12	13	14	15	20	25	30
.75	1.03	1.02	1.02	1.01	1.01	1.00	.99	..98	.98
.80	1.29	1.28	1.27	1.26	1.26	1.25	1.24	1.23	1.22
.85	1.60	1.59	1.57	1.56	1.56	1.55	1.53	1.51	1.51
.90	2.00	1.98	1.96	1.95	1.94	1.93	1.90	1.88	1.87
.95	2.61	2.58	2.56	2.53	2.52	2.50	2.45	2.42	2.41
.975	3.18	3.13	3.09	3.06	3.04	3.02	2.95	2.91	2.88
.99	3.89	3.82	3.76	3.71	3.67	3.64	3.54	3.48	3.45
.995	4.41	4.31	4.24	4.18	4.13	4.09	3.96	3.89	3.85
.999	5.61	5.45	5.32	5.22	5.14	5.07	4.86	4.74	4.67

Before proceeding we make some general remarks on the sequential approach.

(i) For <u>updating</u> the estimator of the <u>variance</u> of a variable $\underline{x}$ we may use the familiar formula

$$\underline{s}^2(\underline{x}) = \sum_{i=1}^{n} \frac{(\underline{x}_i - \bar{\underline{x}})^2}{(n - 1)} = \sum_{i=1}^{n} \frac{\underline{x}_i^2}{(n - 1)} - n \frac{\bar{\underline{x}}^2}{(n - 1)} \tag{80}$$

or we may utilize the Helmert transformation, see, e.g. Tocher (1963, p. 114), i.e.

$$\underline{s}^2(\underline{x}) = \sum_{j=1}^{n-1} \frac{\underline{u}_j}{(n - 1)} \tag{81}$$

where

$$\underline{u}_j = \frac{(\Sigma_{i=1}^{j} \underline{x}_i - j\, \underline{x}_{j+1})^2}{j(j + 1)} \tag{82}$$

As we remarked in Sec. II.10 we may subtract a suitable constant to reduce loss of significant digits; the choice of this constant may be based on prior knowledge, debugging runs, etc.

(ii) We may start with any <u>initial sample size</u> n_0 (provided $n_0 \geq 2$). Large pilot samples tend to decrease the efficiency and to increase the consistency.

(iii) Instead of increasing the sample with one observation before we update $\underline{s}^2(\underline{x})$ etc. we may use a <u>multistage</u> approach where in each stage several observations are taken. This will increase the consistency and decrease the statistical efficiency. The overall efficiency may be increased since the stopping rule calculations are performed less often. In simulation the observations themselves take much computer time and therefore the number of observations per stage should be small. If we are comparing two systems then it may be cumbersome to generate observations alternatively for systems 1 and 2. In the few applications of stopping rules in simulation it is customary to use a multistage approach, where the number of observations per stage is fixed or is taken as in Eq. (83).

$$\underline{m}_{s+1} = (\underline{n}_s - n\underline{a}_s)f \qquad (0 < f < 1) \tag{83}$$

where $\underline{m}_{s+1}$ is the number of observations in stage $s+1$, $\underline{n}_s$ is the required number of observations as calculated after stage s, $\underline{na}_s$ is the actual number of observations available after stage s and f is a fraction. A procedure as in Eq. (83) was applied to simulation by Angers et al. (1970, pp. 5-6) with $f = 1/2$ and by Fishman (1971, p. 28) with $f = 1/2$ and $1/3$. We shall return to simulation applications of stopping rules in a moment.

(iv) The sequential approach for one or two populations is also valid for nonnormal distributions and large sample sizes (compare the central limit theorem); see Chow and Robbins (1965, p. 457), Robbins et al. (1967, p. 1391), and Srivastava (1970, p. 144). In simulation applications the populations are not necessary normal, so that the results for such applications provide additional information on the, possibly small-sample, behavior for nonnormal-populations. (As we shall see the results in simulation seem quite good.) Geertsema (1970) derived distribution-free procedures for a fixed-width confidence interval for the median of one symmetric distribution. The procedure sequentializes the confidence interval based on the sign test or Wilcoxon's one-sample test, exactly as Chow and Robbins (1965) did. His procedures are asymptotically correct. (Their asymptotic efficiency equals the efficiency in the usual fixed sample size comparisons between distribution-free tests and the parametric t-test.) His procedures are quite cumbersome since they require reordering of the ordered observations as more observations come available.

Let us consider the application of stopping rules in simulation and Monte Carlo experiments. Such applications are rare. If a stopping rule is applied the formulas for fixed sample sizes of Sec. V.A.3 are used and the unknown parameters are estimated (sequentially, multistage or two-stage) without even mentioning the error caused by the stochastic nature of the sample sizes in stopping rules. Examples are Angers et al. (1970), Bruzelius (1972), Crane and Iglehart (1972a, pp. 12-18), Fishman (1971), Flagle (1960), Geisler (1964a), and Prins (1962). Only Hauser et al. (1966, p. 82) explicitly discuss the deviations from the nominal $1 - \alpha$ caused by the sequential approach. Fortunately these deviations are small, as we have seen above. Other error sources in simulation can be the non-normality of the distribution and the biased estimation of the variance from a prolonged simulation run. As we saw in Sec. V.A.2 the variance σ^2 may be estimated from subruns, from the individual autocorrelation coefficients, or from cycles. The first approach was used by Mechanic and McKay (1966, p.34) who found that $(\bar{\underline{x}}-\mu)/\hat{\underline{\sigma}}$ was approximately normal in their simulated queueing systems. Fishman (1971) estimated σ^2 not from subruns but from the moving average representation and, unlike Mechanic and McKay, he did not fix the sample size but determined it in a multistage approach as in Eq. (83). He also used a relative reliability criterion, i.e. the length c of the confidence interval in Eq. (60) is replaced by $\gamma\mu$. For instance, if μ is to be estimated within 10% of its true value then $\gamma = 0.1$. Consequently in the sample size formula $\gamma\mu$ should be substituted for c so that besides σ we have to replace μ by its estimator. This creates an additional error source. "Relative" confidence intervals are not investigated in the statistical literature on sequential fixed-width confidence intervals. Fishman (1971, p. 35) gives results of the applications of the multistage procedure Eq. (83), for relative confidence intervals (with $\gamma = 0.3$) to a simple queuing system, using three different initial sample sizes n_0 (viz. 250, 500, and 1000). His results show that the confidence intervals more often miss the true mean than the expected α times. (Depending on n_0 the actual $1 - \alpha$ is equal to 0.73, 0.79, and 0.88 of the nominal $1 - \alpha$.) He suspects that the discrepancies from the nominal $1 - \alpha$ are caused partially by the relative criterion $\gamma\mu$. For additional discussion we refer to the original paper by Fishman (1971). Geisler (1964a) estimated σ^2 from the individual autocorrelation coefficients in his inventory simulations. He used $\alpha = 0.05$[7] and $\gamma = 1$ (i.e. the true mean is estimated within 100%) and estimated the required sample sizes from a pilot run of 500 observations.[8] This estimated sample size was taken for 25 separate inventory systems, each system being replicated 1000 times; see Geisler (1964b, p. 710). The resulting fraction of the

1000 replications yielding an estimate within 100% of the true mean, was indeed very close to the nominal $1 - \alpha$. Crane and Iglehart (1972a) applied a two-stage approach to the independent cycles of some multiserver simulations. [Actually they then used a conservative confidence interval; later on they derived the asymptotic exact interval (21).] Angers et al. (1970) also use a multistage stopping rule for independent simulation runs. They state (without proof) that their "stopping rule subroutine has already been used extensively in simulation. It has worked well;" see Angers et al. (1970, p. 7). We ourselves have also applied this type of approach to simulation and Monte Carlo experiments. Since the aim of these experiments was not to test the stopping rule, we did not obtain many replicates (but we did obtain results for a number of different systems). Let us consider these applications in some detail.

We applied a multistage approach (with 50 observations per stage, $\alpha = 0.05$, $\gamma = 0.1$ and critical constant $z^{\alpha/2}$) to the Monte Carlo estimation of the integral

$$\xi(\lambda, v) = \int_v^\infty \frac{1}{x} \lambda e^{-\lambda x}\, dx \qquad (\lambda,\ v > 0) \tag{84}$$

We refer to Chap. I for the Monte Carlo estimation of Eq. (84). We applied the stopping rule for several combinations of λ and v in Eq. (84) and for several Monte Carlo estimators (viz. crude sampling and three importance sampling estimators; see Chap. III). Figure 1 illustrates the fluctuations of $\underline{n}$ for three sets of (λ, v) and the crude Monte Carlo estimator. Figure 2 shows the fluctuations of the estimate $\hat{\xi}$ as the actual number of observations increases (until the required number is reached); in this figure horizontal lines indicate ξ and 90% and 110% of ξ, the true value ξ being calculated analytically. For other (λ, v) combinations and replications similar pictures were obtained. At first sight it might look strange that Fig. 2 does not look like Fig. 3. However, all previous observations are used to estimate ξ in stage j. Hence, $\underline{\hat{\xi}}_j$, the estimator in stage j is dependent on the estimators in the previous steps $(\underline{\hat{\xi}}_{j-1}, \underline{\hat{\xi}}_{j-2}, \ldots, \underline{\hat{\xi}}_2, \underline{\hat{\xi}}_1)$. So the probability of fluctuations of the estimates around the line $\hat{\xi} = \xi$ is small. (Also compare exercise 10.) We point out that

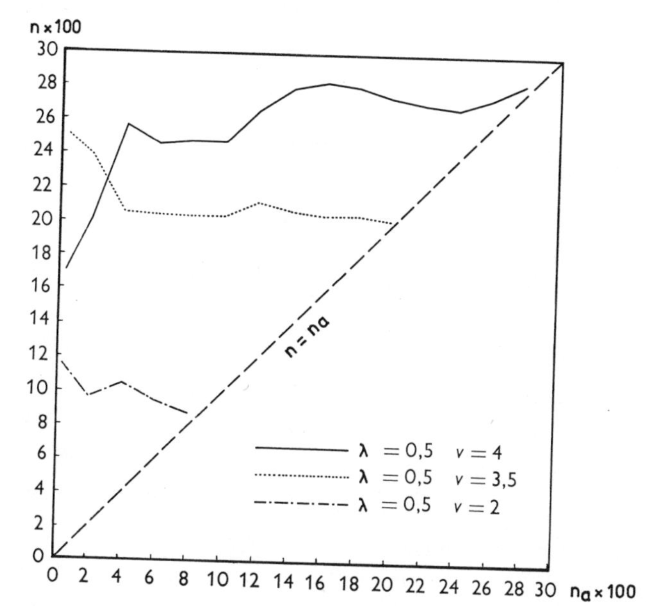

Fig. 1. Fluctuations of the required sample size n as the actual sample size n_a increases.

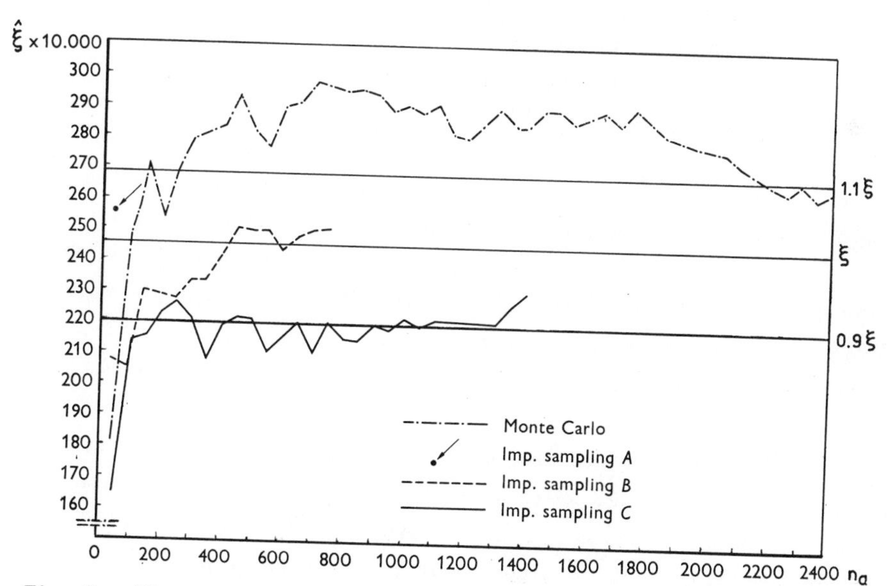

Fig. 2. Fluctuations of the estimate ξ as the actual sample size n_a increases.

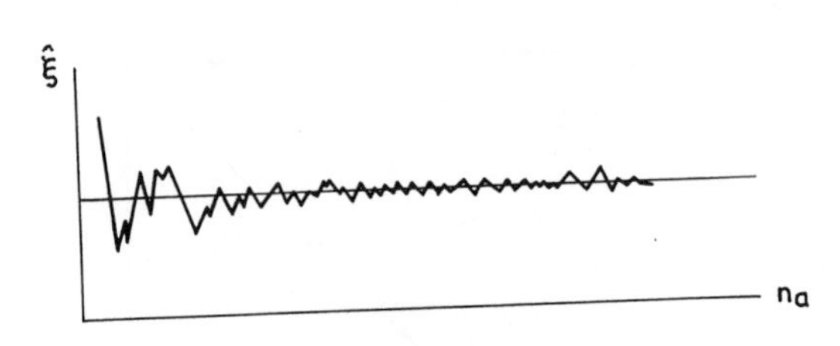

Fig. 3. Oscillating behavior.

Fig. 2 demonstrates that it seems unwise to continue sampling until the cumulative estimate $\hat{\xi}$ has "stabilized," for this practice may lead to unnecessarily large samples. Figure 2 shows further that in this example each estimate was indeed less than 10% wrong. Nevertheless, estimation errors larger than 10% are possible. As we have seen above the required sample size is based on multistage estimators of the variance and mean. Even if our formula for $\underline{n}$ is (approximately) correct there remains a probability α of estimation errors being larger than 10%. Since the estimation of $\xi(\lambda, v)$ was primarily aimed at the investigation of the effectiveness of importance sampling, we examined the correctness of our formula for $\underline{n}$ very briefly. So we took only ten independent replications of our formula. We found that one estimate was more than 10% in error. Our hypothesis is that p, the probability of such errors, is not more than α, i.e. 5% in our example. From the binomial distribution it follows that the probability of one or more defects in a sample of ten with $p = 0.05$ is 0.4013 so that we do not have to reject our null hypothesis. We further applied our stopping rule to the maintenance example described in Appendix I.3 (with $\alpha = 0.05$ and $\gamma = 0.1$) and we indeed estimated the difference between the mean number of bus trips within 10% of the true difference.

We also used our procedure in the distribution sampling example of Appendix I.2, where we estimate p, the probability that the life of a product consisting of two parts is smaller than a fixed quantity a. Our discussion of Eq. (48) shows that a binomial model holds in this situation. We used the normal approximation for the binomial distribution to find

$$n = \left(\frac{z_{\alpha/2}}{\gamma} \right)^2 \frac{(1 - p)}{p} \tag{85}$$

Observe that Eq. (85) implies that the sample size n increases as smaller values of p are estimated. We started with a sample of 150 observations and took 100 observations per step. This resulted in an estimate less than $\gamma = 10\%$ wrong. Wehrli (1970, pp. 99-101) discusses an approach such as Eq. (85) but also a different approach, where the estimator of p is only asymptotically unbiased but requires smaller sample sizes. He proceeds as follows.

(i) Calculate the asymptotically unbiased estimator

$$\hat{\underline{p}}' = \frac{\hat{\underline{p}} + 0.5(z^{\alpha/2})^2/n}{1 + (z^{\alpha/2})^2/n} \tag{86}$$

where $\hat{\underline{p}}$ denotes the usual estimator.

(ii) Calculate the length of the usual confidence interval around p using the normal approximation, i.e.

$$\underline{c} = \frac{z^{\alpha/2}[\hat{\underline{p}}(1 - \hat{\underline{p}})n + (z^{\alpha/2})^2/4]^{1/2}}{[n + (z^{\alpha/2})^2]} \tag{87}$$

As soon as this length $\underline{c}$ is small enough stop sampling. Wehrli's table shows that the required sample size is reduced with 4% to 25% for p ranging from 0.6 to 0.9; $\alpha = 0.05$ and $c = 0.05$. He does not mention the possible effect of making the sample size stochastic when estimating p.

To determine the sample size when estimating the pth quantile of the distribution of $\underline{x}$, say x_p, we may sequentialize the fixed-sample results for the confidence interval on x_p shown in Eq. (57) above. So we continue sampling until the interval $[x_{(r)}, x_{(s)}]$ has decreased to the prespecified length c. This nonparametric sequential procedure holds for large n $(n > 20)$ since Eqs. (58) and (59) are asymptotic; Farrell (1966) proved that sequential variants of fixed-sample rules for <u>finite</u> n do give the desired confidence interval. A <u>two-stage</u> rule was derived by Weiss (1960). A disadvantage of a two-stage approach is that it may overshoot the required final sample size; an advantage is that it does not require resorting all observations after each additional observation; a practical rule may be based on Eq. (83) or on a multi-stage procedure with a fixed number of additional observations per stage.

We observe that a single-stage procedure was proposed by Anderson and Thorburn (1972, pp. 4-5, 17-18). Their rule, however, is only approximative and requires knowledge of the distribution function of $\underline{x}$. Often the problem of sample size determination for quantiles is "solved" by continuing sampling until, e.g. the third digit after the decimal point does not change; see, e.g. Heuts (1971, p. 18) and Heuts and Rens (1972, p. 7). Of course such "numerical accuracy" does not guarantee statistically sound results. Confidence intervals for the difference between two quantiles $(x_p - x_q$ or $x_p - y_p)$ may be simply derived from Eq. (26); see David (1971, p. 15) for an alternative approach.

The sample size for comparing <u>two</u> binomial probabilities may be determined using the normal approximation. Al-Bayyati (1971) gives a different approach for <u>testing</u> the difference between two binomial probabilities but his procedure is conservative and consequently less attractive for expensive simulation runs. Let us have a closer look at testing in general (as opposed to estimation).

Instead of <u>estimating</u> the mean μ of the distribution of $\underline{x}$ we may <u>test</u> the null hypothesis H_0 against an alternative hypothesis H_1, where

$$H_0:\mu = \mu_0, \quad H_1:\mu = \mu_1 \tag{88}$$

It is well-known that for specified values of μ_0, μ_1, and α we can determine n such that a specified value of β is not exceeded, provided that the variance of $\underline{x}$ is known. For, assuming $\mu_1 > \mu_0$ and a one-sided test, we have

$$P(H_0 \text{ accepted}|H_1) = P\left(\frac{\underline{\bar{x}} - \mu_0}{\sigma/\sqrt{n}} < z^{\alpha} | \mu = \mu_1\right)$$

$$= P\left[\frac{(\underline{\bar{x}} - \mu_0) - (\mu_1 - \mu_0)}{\sigma/\sqrt{n}} < z^{\alpha} - \frac{(\mu_1 - \mu_0)}{\sigma/\sqrt{n}} | \mu = \mu_1\right]$$

$$= P\left[\underline{z} < z^{\alpha} - \frac{(\mu_1 - \mu_0)}{\sigma/\sqrt{n}}\right] = \beta \tag{89}$$

which yields

$$z^{\alpha} - \frac{(\mu_1 - \mu_0)}{\sigma}\sqrt{n} = -z^{\beta} \tag{90}$$

or

$$n = \frac{(z^{\alpha} + z^{\beta})^2}{(\mu_1 - \mu_0)^2}\sigma^2 \tag{91}$$

For the two-sided test we refer to, e.g. Robbins and Starr (1965, p. 1).

For an <u>unknown variance</u> σ^2 we might replace σ^2 by its estimator and z^{α}, z^{β} by t_v^{α} and t_v^{β}, v being the degrees of freedom of the estimator of σ^2. For the resulting sample size, H_0 is rejected if

$$\frac{\underline{\bar{x}} - \mu_0}{\underline{s}_x/\sqrt{n}} > t^{\alpha} \tag{92}$$

or using Eq. (91)

$$\underline{\bar{x}} > \mu_0 + \frac{t_v^{\alpha}\,\underline{s}_x}{\sqrt{n}} = \mu_0 + \frac{t_v^{\alpha}(\mu_1 - \mu_0)}{(t_v^{\alpha} + t_v^{\beta})} \tag{93}$$

In deriving Eq. (93) we neglected the stochastic character of the sample size. An exact solution is provided by Stein's <u>double-sampling</u> procedure; compare Dudewicz (1972, p. 4), i.e. estimate σ^2 from the first stage of n_0 observations, take a second sample of

($\underline{n} - n_0$) observations where $\underline{n}$ follows from Eq. (91) substituting $\underline{s}^2$, $t^{\alpha}_{n_0-1}$ and $t^{\beta}_{n_0-1}$, and finally reject H_0 if Eq. (93) holds (with $v = n_0 - 1$). Robbins and Starr (1965, p. 2) show that the double-sampling procedure is not asymptotically efficient; for a poorly chosen initial sample size the procedure may be quite inefficient. They studied the sequentialized analogue of the above testing procedure. (Actually they examined the two-sided test.) The authors found that their procedure is asymptotically correct (i.e. the α and β errors are satisfied) and asymptotically efficient.

For two populations with paired observations $\underline{d} = \underline{x}_1 - \underline{x}_2$ the procedure runs exactly as above. For independent observations Dudewicz (1972, p. 5) gives a double-sampling procedure which requires use of Table 1 above and a trial-and-error lookup in one more table [given in Dudewicz (1972) and Dudewicz and Dalal (1971)]. Chapman (1950) gives a double-sampling procedure for testing the ratio of two means. [His Table II is expanded by Dudewicz (1972, Table II).] However, the problem of testing the difference between two means can be approached in a different way. We shall usually compare two means in order to select the best one, say the highest one. This selection problem might be explicitly formulated as a testing problem. We want to discriminate among the following three hypotheses.

$$H_{-1}: \mu_1 < \mu_2, \quad H_0: \mu_1 = \mu_2, \quad H_1: \mu_1 > \mu_2 \tag{94}$$

The sample size could be determined specifying an α- and β-error; compare also the discussion on sequential tests for selecting one of three hypotheses in Wetherill (1966, pp. 30-40). However, we may also follow an approach that corresponds with the general formulation used when selecting the best mean from among k (≥ 2) populations. So we estimate μ_h by the sample mean

$$\bar{\underline{x}}_h = \sum_{i=1}^{n_h} \underline{x}_{hi}/n_n \qquad (h = 1, 2) \tag{95}$$

We shall select as the population with the highest mean the population yielding the highest sample mean. We want to realize a certain minimum probability, P^* or $(1 - \alpha)$, of making a correct selection (abbreviated to CS), i.e.

$$P(CS) = P[(\bar{x}_1 > \bar{x}_2 | \mu_1 > \mu_2) \text{ or } (\bar{x}_1 < \bar{x}_2 | \mu_1 < \mu_2)] \geq P^* = 1 - \alpha \tag{96}$$

Now consider Fig. 4, where $\mu_1 - \mu_2 = \delta$ is supposed to be positive; $\bar{d}$ denotes $\bar{x}_1 - \bar{x}_2$. The probability of a wrong selection is equal to the shaded area. This probability decreases with increasing sample sizes. But it is also clear that if δ differs only slightly from zero then the probability of a wrong selection is still considerable. So small differences between μ_1 and μ_2 are difficult to detect and require very large sample sizes. But in the case of a small difference we may not care to select the wrong population and therefore, it would be undesirable to use large samples for discovering such a small difference. Therefore, we assume that we do not care when we reach a wrong conclusion and $|\delta|$ is smaller than, say, δ^* units $(\delta^* > 0)$. A difference larger than or equal to δ^* we do want to detect with probability at least P^* $(= 1 - \alpha)$. Or

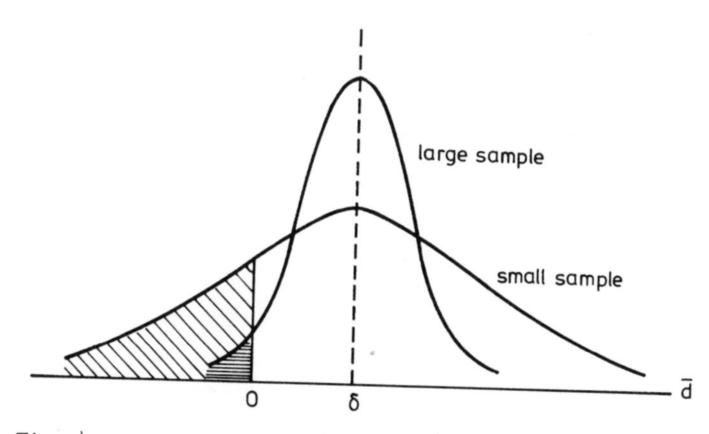

Fig. 4. Relationship Among P(CS), Sample Size and δ.

$$P(CS) \geq P^* \quad \text{if} \quad |\mu_1 - \mu_2| \geq \delta^* \tag{97}$$

where P^* and δ^* are specified by the experimenter. Equation (97) is called the indifference zone approach. To derive the sample sizes we formulate Eq. (97) as Eqs. (98) and (99).

$$P(\underline{\bar{d}} > 0 | \delta \geq \delta^*) \geq 1 - \alpha \tag{98}$$

$$P(\underline{\bar{d}} < 0 | \delta \leq -\delta^*) \geq 1 - \alpha \tag{99}$$

To compare two means, assuming normality but not necessarily equal variances, we may use, e.g. the Cochran and Cox statistic in Eq. (39) or Scheffé's statistic in Eq. (45). From Eqs. (38) and (39) it follows that Eq. (100) holds approximately.

$$P\left[\frac{\underline{\bar{d}} - \delta}{\underline{s}(\underline{\bar{d}})} > -\underline{t}'^{\alpha}\right] = 1 - \alpha \tag{100}$$

So. Eq. (100) implies

$$P(\underline{\bar{d}} > 0) = 1 - \alpha \tag{101}$$

provided

$$-\underline{t}'^{\alpha} \cdot \underline{s}(\underline{\bar{d}}) + \delta = 0 \tag{102}$$

It is simple to derive that Eq. (102) is satisfied for

$$\underline{n}_h = \frac{2\underline{s}_h^2(0.5t^{\alpha}_{n_1-1} + 0.5t^{\alpha}_{n_2-1})^2}{\delta^2} \tag{103}$$

where we used Eqs. (38) and (39) and the sample allocation rule

$$\frac{\underline{n}_1}{\underline{n}_2} = \frac{\underline{s}_1^2}{\underline{s}_2^2} \tag{104}$$

The relation in Eq. (104) means that we take more observations from the population with the higher variablility. The ratio in Eq. (104) was inspired by the well-known fact that for a fixed total sample size the variance of $(\bar{x}_1 - \bar{x}_2)$ is minimized if $n_1/n_2 = \sigma_1/\sigma_2$; taking $n_1/n_2 = s_1^2/s_2^2$ instead of $n_1/n_2 = s_1/s_2$ simplifies the derivation of the required sample sizes.[9] From Eq. (103) it follows that for $\delta = \delta^*$ the required sample sizes are given by Eq. (105).

$$n_h = \frac{2s_h^2(0.5t^{\alpha}_{n_1-1} + 0.5t^{\alpha}_{n_2-1})^2}{(\delta^*)^2} \tag{105}$$

If actually $\delta > \delta^*$ then Eq. (105) yields "overprotection," i.e.

$$P(\bar{d} > 0) > 1 - \alpha \tag{106}$$

Consequently taking n_h as in Eq. (105) satisfies Eq. (98). Analogous to Eqs. (100) through (106) we find that Eq. (105) also meets Eq. (99). Hence Eq. (97) is satisfied (neglecting of course the stochastic character of n_1 and n_2).

In the same way we can derive from Scheffé's statistic, Eq. (45), that the smallest sample size, n_1, should satisfy

$$n_1 = \frac{(t^{\alpha}_{n_1-1})^2 \cdot s^2(u)}{(\delta^*)^2} \tag{107}$$

where

$$u_i = x_{1i} - (n_1/n_2)^{1/2}\, x_{2i} \qquad (i = 1, 2, \ldots, n_1) \tag{108}$$

Notice that we would have to recalculate u_i (old and new values) when (n_1/n_2) changes; to do so we would need all individual values x_{1i} and x_{2i} and this might require much computer storage. It is simple to derive an alternative expression for the numerator of $s^2(u)$, namely

$$\sum_{i=1}^{n_1} (\underline{u}_i - \bar{\underline{u}})^2$$

$$= \sum_i (\underline{x}_{1i}^2) + \frac{n_1}{n_2} \sum_i (\underline{x}_{2i}^2) - 2\left(\frac{n_1}{n_2}\right)^{1/2} \sum_i (\underline{x}_{1i}\underline{x}_{2i})$$

$$- \frac{1}{n_1} (\sum_i \underline{x}_{1i})^2 + \frac{2}{(n_1 n_2)^{1/2}} (\sum \underline{x}_{1i})(\sum \underline{x}_{2i}) - \frac{1}{n_2} (\sum_i \underline{x}_{2i})^2 \qquad (109)$$

This result looks complicated but it permits the recalculation of $\underline{s}^2(\underline{u})$ by updating only sum terms so that no previous individual values of $\underline{x}_{1i}$ and $\underline{x}_{2i}$ are needed; also compare our comments on Eq. (80). The sequential application of Eq. (107) runs as follows. If the actual sample size is smaller than $\underline{n}_1$ then one additional value is generated from the population with the smallest actual sample size. If both populations have the same actual sample size then the condition $n_1 \leq n_2$ in Scheffé s statistic, implies that from both populations one observation is obtained.

We have not proven that the above sequential selection approaches do satisfy the probability requirement in Eq. (97). It is a heuristic procedure inspired by the knowledge that sequentialized versions of estimation procedures do work; see the discussion of Eq. (65). We have some further experimental data resulting from the application of the sample size formulas Eqs. (103) and (107). Before we present these data we remark that the experimenter who finds a selection formulation as in Eq. (97) more adequate than a testing or confidence interval formulation, may also use the multiple ranking procedures (for $k \geq 2$ populations) presented in Part C.

We applied Eqs. (103) and (107) to the simulation of the well-known newspaper-boy problem formulated, e.g. in Naylor et al. (1967, pp. 177-178) as follows. "A newsboy buys papers for 4¢ each and sells them for 10¢ each. At the end of each day the newspaper publisher will pay him 2¢ each for his unsold papers. Daily demand

(D) for papers has (a given discrete) probability function." We used simulation to estimate the difference between the profits of the newsboy when ordering two different amounts of newspapers. These simulation results are then used to select from the two possible numbers of newspapers the one giving the highest profit.[10)] In our application of the stopping rules based on Cochran-Cox and Scheffé we tried to satisfy the underlying assumption of normality by defining as one observation the average of ten individual observations. Because of the central limit theorem we expect that this average better satisfies the normality assumption. (Alternatively we could take one observation instead of ten observations and rely on the robustness of the t-statistic; the Cochran and Cox and the Scheffé statistic, however, are defined differently from the usual t-statistic.) We refer to Kleijnen (1967, pp. 11-18) for the various flow charts for this case study. The results for $P^* = 0.95$ and $\delta^* = 4$ are given in Table 2 where the expected differences are calculated from the analytical solution for the expected profit. In Table 2 "min" denotes the number of observations (i.e. averages of ten individual observations) that are generated before the formula for the sample size is applied for the first time. The part of the table between the dashed horizontal lines concerns situations where $|\delta|<\delta^*$. In situations where $|\delta| \geq \delta^*$ Scheffé's statistic always yielded a correction selection while the Cochran and Cox statistic gave one wrong selection (the circled entry in Table 2). The table for the actual and desired sample sizes in Kleijnen (1967, p. 20) shows that this wrong selection was reached when the simulation was stopped immediately after the minimum number of observations (i.e. two observations) was taken; when this minimum was raised to five a correct selection resulted (see column 7). That the Cochran and Cox statistic performs less well than Scheffé's statistic may be explained by the approximate nature of their statistic. We observe that the Cochran and Cox statistic required higher sample sizes but took less time given the way we wrote our computer program; for details see Kleijnen (1967, pp. 21-22). When $|\delta| < \delta^*$ wrong selections are very well possible; compare the signs of the entries between the dashed lines of Table 2.

TABLE 2

Experimental Results for the Stopping Rule Based on the Cochran and Cox and the Scheffé Statistic

Number of newspapers compared	Expected difference (δ)	Estimated differences ($\bar{d}$)				
		Scheffé		Cochran and Cox		
		min = 2		min = 2		min = 5
		Replicate 1	Replicate 2	Replicate 1	Replicate 2	Replicate 1
(1)	(2)	(3)	(4)	(5)	(6)	(7)
75- 80	26.80	22.00	29.12	25.33	26.00	25.33
80- 85	24.00	26.19	20.50	27.46	18.00	24.07
85- 90	23.20	26.29	27.17	19.20	29.81	23.33
90- 95	15.20	13.74	14.90	38.20	16.26	13.66
95-100	11.60	11.67	13.55		10.27	13.50
100-105	4.00	0.36	6.13	4.23	6.96	3.34
105-110	-0.80	2.55	-7.85	0.04	-3.14	2.23
110-115	-2.00	a)	-2.81	-3.71	-2.09	-3.25
115-120	-7.60	a)	a)	-5.23	-4.41	-6.88

[a] Not calculated because of long computing time required.

So far we have seen that when we are not interested in a confidence interval for the mean then we may specify the β error and the alternative hypothesis (i.e. μ_1) and estimate the unknown variance from a pilot sample; (see Eq. (91). Instead of this double-sampling procedure of Stein, we may use a selection procedure with an indifference zone; see Eq. (96). A third alternative consists of a fully sequential testing approach, not neglecting the stochastic character of the sample size. This approach yields useful procedures provided no nuisance parameters (like σ^2) exist. It is based on Wald's "sequential probability ratio test" (or SPRT) for deciding between two simple hypotheses, $H_0: \Theta = \Theta_0$ and $H_1: \Theta = \Theta_1$. The SPRT runs as follows. If we take a sample of n observations $\underline{x}_i$ (i = 1, 2, ..., n) from the density function $f(x)$, then the "probability" of obtaining a particular set of values x_i is[11)]

$$P = \prod_{i=1}^{n} f(x_i | \Theta = \Theta_h) \tag{110}$$

given that the parameter Θ of $f(x)$ has the value Θ_h. If we consider the observations of the sampling process, $\underline{x}_i$, as stochastic then the probability in Eq. (110) becomes a stochastic variable, $\underline{P}$. The probability ratio or likelihood ratio $\underline{\ell}$ is defined as the ratio of the probability of the sampled observations under hypothesis $H_1: \Theta = \Theta_1$ and $H_0: \Theta = \Theta_0$, respectively, i.e.

$$\underline{\ell} = \frac{\underline{P}_1}{\underline{P}_0} \tag{111}$$

An obvious decision rule is to continue sampling as long as this ratio is close to one; to accept H_1 if this ratio is high and to accept H_0 if the ratio is small. We want to realize

$$P(H_0 \text{ rejected} | \Theta_0) = \alpha \tag{112}$$

$$P(H_0 \text{ accepted} | \Theta_1) = \beta \tag{113}$$

Taking into account the stochastic character of both the observations $\underline{x}_i$ and the sample size $\underline{n}$, Wald derived that Eqs. (112) and (113) are approximately realized if we continue sampling as long as

$$\frac{\beta}{1-\alpha} < \underline{\ell} < \frac{1-\beta}{\alpha} \tag{114}$$

and accept H_0 as soon as

$$\underline{\ell} \leq \frac{\beta}{1-\alpha} \tag{115}$$

or accept H_1 as soon as

$$\underline{\ell} \geq \frac{1-\beta}{\alpha} \tag{116}$$

For a proof of Eqs. (114) through (116) we refer to, e.g. Ghosh and Freeman (1961, pp. 38-43) and Wetherill (1966, pp. 14-16). The <u>sample size</u> of such a procedure is smaller than in fixed sample plans. (We do not know any studies on the efficiency of the SPRT compared with the approximate sequential selection procedures neglecting the stochastic character of the sample size that we applied above. Both approaches, however, use different problem formulations; compare the parameters P^* and δ^* versus α, β, μ_0, μ_1.)

It is simple to derive a SPRT for the mean μ of a normal population provided σ^2 is known, for the standard deviation σ of a normal population (μ known or not), the mean λ of a Poisson population or the binomial parameter p; see, e.g. Ghosh and Freeman (1961, pp. 44-48). We think that the <u>binomial</u> case is important for simulation and Monte Carlo experiments. Using the formulation in Wetherill (1966, p. 17) the SPRT for this case states that we should continue sampling as long as Eq. (117) holds.

$$\ell n\left(\frac{\beta}{1-\alpha}\right) < \underline{r}\ \ell n\left(\frac{p_1}{p_0}\right) + \underline{s}\ \ell n\left(\frac{1-p_1}{1-p_0}\right) < \ell n\left(\frac{1-\beta}{\alpha}\right) \tag{117}$$

where $\underline{r}$ and $\underline{s}$ denote the number of defectives and effectives, respectively, and p is the probability of a defective (ℓn denotes the logarithm). Corneliussen and Ladd (1970) investigated the exact statistical properties of the binomial sequential test. Most applications of Eq. (117) have been in the area of quality control; see Wetherill (1966, p. 27). The test was also applied by Kurlat and Springer (1960, p. 477) in their "physical simulation" of an anti-tank-mine. We do not know any applications in simulation and Monte Carlo experiments though the binomial distribution may be a realistic problem formulation in such experiments.

Another realistic problem formulation requires testing the mean μ of a normal population $N(\mu, \sigma^2)$ while σ^2 is unknown. Unfortunately this problem is not solved satisfactorily by the exact sequential theory. For one approach is to solve the reformulated problem

$$H_0: \mu = \mu_0, \qquad H_1: |\mu - \mu_0|/\sigma > \delta_1 \tag{118}$$

i.e. the distance of μ relative to μ_0 is specified as a multiple of the unknown standard deviation σ. Then a test can indeed be devised; see Wetherill (1966, p. 46). Unfortunately the formulation of Eq. (118) is unrealistic for simulation and Monte Carlo studies. The other approach is to estimate σ^2 from a pilot-sample of n_0 observations. Next observations are taken one at a time until the SPRT touches one of the two boundaries. Such a procedure has been developed by Paulson (1964, pp. 1048-1052). In his approach $\underline{\bar{x}}$ is based on all $\underline{n}$ observations, but $\underline{s}^2$ uses only the first n_0 observations! His procedure was applied by Sasser et al. (1970, pp. 290, 292-294) in their simulation of an inventory system. We further remark that in sequential ANOVA the original hypothesis, $H_0: \mu_1 = \mu_2 = \cdots = \mu_k$, is also reformulated, resulting in an unrealistic problem formulation; compare Ghosh and Freeman (1961, p. 72) and Wetherill (1966, p. 71). So it is not surprising that SPRT's are not much applied either in the field of simulation and Monte Carlo or outside this field. For an extensive study of the SPRT we refer to Ghosh (1970). We also mention Fu (1970); though the title of his book may not suggest it, this book nevertheless gives a thorough treatment of the SPRT, modified boundaries (to limit the maximum number of observations), nonparametric variants, choosing among more than two hypotheses, etc.

Summarizing, for testing purposes SPRT's are exact and are more efficient than fixed sample size plans, but usually they are not suitable for testing hypotheses in simulation and Monte Carlo studies; a noteworthy exception is the binomial SPRT. It would be interesting to investigate the SPRT for testing μ in $N(\mu, \sigma^2)$ when σ^2 is reestimated after each step and to compare such an approximate SPRT with a double-sampling and an indifference zone approach. In estimation problems the situation is different. Several authors derived that for not too small sample sizes an approximate sequential procedure derived from the fixed sample size formula gives the correct confidence level and approaches the exact sequential procedure; also compare Johnson (1957).

Finally, a different type of procedure is available if we are willing to specify a particular loss function, possibly together with a prior distribution for the parameters of interest. Then we can try to develop a decision theoretic solution. Maurice (1957), e.g. derived a "minimax" procedure for choosing between two populations using sequential sampling while assuming a common known variance. Grundy et al (1954, p. 318) minimize the "average risk" applying a double-sample plan, assuming a known variance. Hayes (1969) describes a fixed-sample Bayesian approach for choosing between two alternatives assuming known variances; however, this approach does not result in a simple procedure. Details on the decision theoretic approach in sequential sampling can be found in Whetherill (1966, pp. 11-12, 85-110, 134-141, 187-189). For the simulation of inventory systems Brenner (1965) and (1966) studied economic sample sizes which, however, seem too specific for general use in simulation. Naylor (1971) gives a brief survey of various stopping rules for simulation; Dutton and Starbuck (1971, pp. 592-593) give some more references.

APPENDIX V.A.1. CORRELATION BETWEEN CONSECUTIVE RUNS

In this appendix we shall derive the correlation coefficient for the averages of two consecutive subruns, or briefly runs. Denote the average of these two runs by $\bar{\underline{x}}_1$ and $\bar{\underline{x}}_2$, respectively, where each run comprises a individual observations, say $\underline{x}_{1i}$ (i = 1, 2, ..., a) and $\underline{x}_{2j}$ (j = 1, 2, ..., a), respectively. So from the definition of the covariance it follows that

$$\text{cov}(\bar{\underline{x}}_1, \bar{\underline{x}}_2)$$

$$= E\left\{\left[\frac{\Sigma_{i=1}^{a} \underline{x}_{1i}}{a} - \frac{E(\Sigma_{i=1}^{a} \underline{x}_{1i})}{a}\right]\left[\frac{\Sigma_{j=1}^{a} \underline{x}_{2j}}{a} - \frac{E(\Sigma_{j=1}^{a} \underline{x}_{2j})}{a}\right]\right\}$$

$$= E\left(\left\{\frac{\Sigma_{i=1}^{a} [\underline{x}_{1i} - E(\underline{x}_{1i})]}{a}\right\}\left\{\frac{\Sigma_{j=1}^{a} [\underline{x}_{2j} - E(\underline{x}_{2j})]}{a}\right\}\right)$$

$$= \frac{1}{a^2} E\left\{\sum_{i=1}^{a}\sum_{j=1}^{a} [\underline{x}_{1i} - E(\underline{x}_{1i})]\,[\underline{x}_{2j} - E(\underline{x}_{2j})]\right\}$$

$$= \frac{1}{a^2} \sum_{i=1}^{a}\sum_{j=1}^{a} E\left\{[\underline{x}_{1i} - E(\underline{x}_{1i}) \quad \underline{x}_{2j} - E(\underline{x}_{2j})]\right\}$$

$$= \frac{1}{a^2} \sum_{i=1}^{a}\sum_{j=1}^{a} \text{cov}(\underline{x}_{1i}, \underline{x}_{2j}) \tag{1.1}$$

We assume that $\underline{x}_{1i}$ and $\underline{x}_{2j}$ form <u>stationary</u> stochastic sequences. This stationarity implies that the covariances do not vary over time but vary only with the "lag," i.e. the number of periods between the two stochastic variables, as we saw in Eq. (3) above. Since $\underline{x}_{21}$, $\underline{x}_{22}$, is the first, second, ... observation after $\underline{x}_{1a}$ we have:

$$\text{cov}(\underline{x}_{1i}, \underline{x}_{2j}) = c_{(a+j-i)} \tag{1.2}$$

This yields the following table.

TABLE 1.1

Value of $(a + j - i)$ as i and j Range From 1 Through a

		i						
j	a+j	1	2	3	···	(a-2)	(a-1)	a
1	a+1	a	a-1	a-2		3	2	1
2	a+2	a+1	a	a-1		4	3	2
3	a+3	a+2	a+1	a		5	4	3
⋮	⋮							
(a-2)	2a-2	2a-3	2a-4	2a-5		a	a-1	a-2
(a-1)	2a-1	2a-2	2a-3	2a-4		a+1	a	a-1
a	2a	2a-1	2a-2	2a-3		a+2	a+1	a

This table together with the above relations gives

$$\begin{aligned}\operatorname{cov}(\bar{\underline{x}}_1,\bar{\underline{x}}_2) &= \frac{1}{a^2}[c_1 + 2c_2 + 3c_3 + \cdots + (a-1)c_{a-1} + ac_a + (a-1)c_{a+1} \\ &\qquad + \cdots + 3c_{2a-3} + 2c_{2a-2} + c_{2a-1}] \\ &= \frac{1}{a^2}\left(\sum_{s=1}^{a} sc_s + \sum_{s=1}^{a-1} sc_{2a-s}\right) \qquad (1.3)\end{aligned}$$

In order to determine the correlation instead of the covariance we need to know the variance of $\bar{\underline{x}}_1$. Using Eq. (5) we obtain

$$\operatorname{var}(\bar{\underline{x}}_1) = \frac{1}{a}[\sigma^2 + \frac{2}{a}\sum_{s=1}^{a}(a-s)c_s] \qquad (1.4)$$

Since the standard deviations of $\bar{\underline{x}}_1$ and $\bar{\underline{x}}_2$ are the same, the correlation between $\bar{\underline{x}}_1$ and $\bar{\underline{x}}_2$ is given by Eq. (1.5).

$$\rho(\bar{\underline{x}}_1, \bar{\underline{x}}_2) = \frac{a^{-2}\,(\Sigma_{s=1}^{a}\ sc_s + \Sigma_{s=1}^{a-1}\ sc_{2a-s})}{a^{-1}[\sigma^2 + 2\ \Sigma_{s=1}^{a}\ c_s(a-s)/a]}$$

$$= \frac{\Sigma_{s=1}^{a}\ \frac{s}{a}\,\rho_s + \Sigma_{s=1}^{a-1}\ \frac{s}{a}\ \rho_{(2a-s)}}{1 + 2\ \Sigma_{s=1}^{a}\ \frac{a-s}{a}\ \rho_s} \qquad (1.5)$$

APPENDIX V.A.2. MECHANIC AND McKAY'S SUBRUN PROCEDURE

Mechanic and McKay (1966, pp. 18-23) give the following algorithm: Let the total sample comprise N individual observations ($\underline{x}_t$, t = 1, ..., N). For subruns or batches of size a, b = 4a, c = 4b, ..., s. There should be at least three different batchsizes, and the largest batchsize should yield at least 25 batchsize means, i.e. $N/s \geq 25$.

For each batchsize (a, b, ..., s) calculate the sample batchmeans ($\bar{\underline{x}}_{ai}$, $\bar{\underline{x}}_{bj}$, ... with i = 1, ..., N/a, etc.). From these batchmeans calculate the (biased) estimators of the variance of the overall average $\bar{\underline{x}}$, e.g.

$$\hat{\underline{V}}_a^2 = \frac{\hat{\underline{\sigma}}_a^2}{\alpha} \qquad (\alpha = N/a) \qquad (2.1)$$

where

$$\hat{\underline{\sigma}}_a^2 = \sum_{i=1}^{\alpha} \frac{(\bar{\underline{x}}_{ai} - \bar{\underline{x}})^2}{\alpha - 1} \qquad (2.2)$$

(Observe that $\hat{\underline{V}}_a^2$ becomes a less biased estimator as the batchsize a increases so that the batchmeans become more independent.)

Estimate the "<u>weighted average</u> $\bar{\rho}_{ab}$ of the autocorrelation of batchmean data over <u>the first (K-1) lags</u>" (K = b/a = 4); see Mechanic and McKay (1966, p. 11). Or

$$\bar{\rho}_{ab} = 2 \sum_{k=1}^{K-1} \frac{w_k \rho_a(k)}{K - 1} \qquad (2.3)$$

where the weights are

$$w_k = \frac{K - k}{K} \qquad (2.4)$$

and $\rho_a(k)$ denotes the serial correlation lag k between the averages of batches of size a. Obviously if the batchaverages are independent we have $\rho_a(k) = 0$ even for $k = 1$, and $\bar{\rho}$ becomes zero. Mechanic and McKay (1966, p. 12) derived the following expression for estimating $\bar{\rho}_{ab}$ containing no autocorrelation coefficients!

$$\hat{\underline{\rho}}_{ab} = \frac{\underline{V}_b^2}{3\underline{V}_a^2} - \frac{1}{3} \qquad (2.5)$$

Calculate such estimators for all batchsizes except the largest one, i.e. calculate $\hat{\underline{\rho}}_{ab}, \hat{\underline{\rho}}_{bc}, \ldots, \hat{\underline{\rho}}_{qr}$ $(q = r/4 = s/16)$. (Since the $\hat{\underline{\rho}}$ are sensitive to estimation errors they are calculated from at least 100 batches. The confidence interval, however, is calculated from an estimated variance based on only 25 batches.)

Accept $\hat{\underline{\rho}}_{\ell,4\ell}$ (briefly $\hat{\underline{\rho}}$) as being small enough

(i) if it is in the range $.05 \leq \hat{\underline{\rho}} < .50$ and
  (1) it is not the first $\hat{\underline{\rho}}$ in the sequence
  (2) it is less than its predecessor
  (3) all the following $\hat{\underline{\rho}}$, if any, are a monotonically decreasing sequence (with the exception of a last unbroken chain of $\hat{\underline{\rho}}$ which may oscillate as long as they are $\leq .05$)

(ii) if it is $\leq .05$ and all the remaining $\hat{\underline{\rho}}$ are $\leq .05$. If the sequence of $\hat{\underline{\rho}}$ does not satisfy criterion (i) or (ii) then the total sample size N should be increased. Finally calculate the variance from the next largest batchsize, i.e.

$$\hat{\underline{\sigma}}_{\bar{x}}^2 = \hat{\underline{V}}_{16\ell}^2 \qquad (2.6)$$

EXERCISES

1. Prove that the variance of $\underline{\bar{x}} = \Sigma_{t=1}^{n} \underline{x}_t/N$ with $\underline{x}_t$ being r-dependent stationary, is given by formula (5).

2. Prove that Eq. (11) is equivalent to the following formula [taken from Fishman (1967)]:

$$\text{var}(\underline{\bar{x}}) = \frac{\sigma^2}{N} \sum_{s=-\infty}^{\infty} \rho_s \qquad (N \longrightarrow \infty)$$

3. If $P(\underline{x}_1 < \mu_x < \underline{x}_2) = 1 - \alpha$ and $P(\underline{y}_1 < \mu_y < \underline{y}_2) = 1 - \alpha$ then derive a (conservative) confidence interval for μ_1/μ_2.

4. Assume that Θ is a linear function of λ, say $\Theta(\lambda) = a + b\lambda$, and the following confidence intervals hold: $P[\underline{\Theta}_1 < \Theta(\lambda_1) < \underline{\Theta}_2] \geq 1 - \alpha_1$, $P[\underline{\Theta}_3 < \Theta(\lambda_2) < \underline{\Theta}_4] \geq 1 - \alpha_2$. Prove the following result given by Crane and Iglehart (1972a, p. 4):

$$P\left[\underline{\Theta}_1 + \frac{(\lambda - \lambda_1)}{(\lambda_2 - \lambda_1)}(\underline{\Theta}_3 - \underline{\Theta}_1) \leq \Theta(\lambda) \leq \underline{\Theta}_2 + \frac{(\lambda - \lambda_1)}{(\lambda_2 - \lambda_1)}(\underline{\Theta}_4 - \underline{\Theta}_2) \quad \text{for all } \lambda_1 \leq \lambda \leq \lambda_2\right] \geq 1 - \alpha_1 - \alpha_2$$

   Do you know an alternative approach to testing the effect of λ on Θ, i.e. the sensitivity of λ?

5. Will in general a binomial model hold in terminating or non-terminating systems?

6. A sample of $n = 5$ observations is taken; the observations in increasing order are $x_{(1)}, x_{(2)}, \ldots, x_{(5)}$. Estimate the median from the order statistics and from the empirical distribution function; the same for the lower quartile $x_{.25}$.

7. Consider a Monte Carlo experiment with a particular multiple ranking procedure. Each replication shows whether the procedure did or did not "work" (i.e. selected the best population). The null-hypothesis is that the procedure does work, i.e. $E(\hat{\underline{p}}) \geq P^*$. Derive the number of replications that guarantees a type-I error of no more than 1% and a type-II error not exceeding 10% for the alternative hypothesis $H_1: p = .85$ (while $P^* = .90$).

8. Prove that $\mathrm{var}(\bar{\underline{x}}_1 - \bar{\underline{x}}_2)$ is minimized taking $n_1/n_2 = \sigma_1/\sigma_2$ subject to the restriction of a fixed total sample size, $n = n_1 + n_2$, and assuming independent observations.

9. Paired observations are taken from two populations. Prove that the required sample size in unpaired observations as given by Eq. (72) is larger when $\sigma_1 = \sigma_2$ (and the paired observations show positive covariance).

10. Consider the following type of estimator. Sample $\underline{x}$ 50 times independently and take its average $\hat{\underline{\xi}}_1$. Next take 100 more independent observations $\underline{x}$ and calculate its average $\hat{\underline{\xi}}_2$. Proceed taking larger samples of size 150, 200, etc. Do you expect a behavior as in Fig. 2 or 3? What is the probability that all $\hat{\underline{\xi}}_j$ $(j = 1, 2, \ldots, m)$ are, say, higher than ξ (assuming that the central limit theorem applies)?

11. Derive the required sample sizes in Eq. (103) using the Cochran and Cox statistic for testing the difference of two means.

12. Derive the alternative expression Eq. (109) for Scheffé's statistic.

13. Derive the formula for the expected profit of the newspaper boy when he buys c newspapers; compare column (2) in Table 2.

14. Simulate a simple queuing problem.
    (a) Estimate the variance of the run average using the subrun approach.
    (b) Estimate this variance from the individual autocorrelation coefficients.

(c) Estimate this variance from independent cycles.

(d) Determine the sample size required to estimate the steady-state mean μ within c units, with probability $(1 - \alpha)$.

(e) As (d) but within $\gamma\mu$ with probability $(1 - \alpha)$.

(f) Investigate experimentally whether the procedure in (d) is consistent.

Note. Schmidt and Taylor (1970, pp. 558-574) gave 43 more problems requiring simulation and the procedures of Part A.

NOTES

1. Blomqvist (1967, p. 162; 1968, p. 187) studied the single-server FIFO system with exponential interarrival and service-times analytically. He found that the (positive) covariances of the waiting times decrease exponentially to zero. Blomqvist (1970, p. 121) proved that in the transient phase the covariances also decrease monotonically to zero.

2. Blomqvist (1968, pp. 188-189; 1970, p. 125) proved that Eq. (11) also gives the asymptotic mean square error (i.e. transient state bias may exist) for a simple queuing system started in the empty state.

3. This variant replaces the infinite number of autocorrelations ρ_s in Eq. (11) by a finite number (p) of parameters b_s [b_s being related to a_s in Eq. (15)]. Unfortunately we have to estimate the adequate value of p.

4. A brief discussion of the various techniques for determining the reliability of the average of a simulation run can also be found in Dear (1961, pp. 21-22), Emshoff and Sisson (1971, pp. 193-195, 198-202), Hillier and Lieberman (1968, pp. 463-465) and Fishman and Kiviat (1967, p. 27).

5. Put $\underline{x}_i = 1$ if customer i $(i = 1, 2, \ldots, n)$ arrives into an empty system. Let $\underline{v}_i$ denote the time between the arrival of customer $(i - 1)$ and i; $\underline{s}_i$ the service time for customer i, and $\underline{w}_i$ his waiting time. Then $P(\underline{x}_i = 1 | \underline{x}_{i-1} = 1) = P(\underline{s}_{i-1} < \underline{v}_i)$ while $P(\underline{x}_i = 1 | \underline{x}_{i-1} = 0) = P(\underline{s}_{i-1} + \underline{w}_{i-1} < \underline{v}_i)$.

6. In Tocher (1963, p. 114) a printing error occurs: The t-statistic should be squared in the expressions

$$s_i^2 \geq i(\ell/2)^2/t_\alpha^{(i)} \qquad (i = 1, 2, \ldots) .$$

7. Geisler (1964a, p. 263) states that his approach is based on the "Central Limit Theorem" and on "Tchebycheff's inequality." Actually this inequality would imply that Geisler's factor (1.96) for 95% confidence should be $(0.05)^{-1/2} = \sqrt{20}$. His factor 1.96 is still correct because of the central limit theorem for dependent observations sampled from a stationary process.

8. It would be misleading to call his procedure a two-stage approach with $n_0 = 500$ for the 500 observations are serially correlated. [500 observations were taken since they seem to yield a good estimate of σ^2; see Geisler (1964a, p. 269).] Moreover he replicated his calculations ten times. The average of these ten replications were tabulated. These averages were used in Geisler (1964b).

9. See Dudewicz and Dalal (1971, Secs. 2 and 3). They proved that the total required sample size for the allocation (104) as compared with the optimal allocation is, e.g. 3% higher for $\sigma_1^2/\sigma_2^2 = 2$ and 27% higher for $\sigma_1^2/\sigma_2^2 = 10$.

10. Actually the problem in Naylor et al. is to determine the optimum number of newspapers in the range of all possible numbers of papers. However, that problem can better be solved using the Kiefer-Wolfowitz procedure or a one-dimensional variant of response surface methodology; compare Wetherill (1966, pp. 154-157).

11. For a continuous stochastic variable P in (110) is not a probability but is a joint density function; cf. for continuous $\underline{x}$ the probability of $\underline{x}$ being exactly x is zero. The SPRT holds for both continuous and discrete variables.

REFERENCES

Adhikari, A.K. (1967). "Simulation of queuing problems," Opsearch, 4, 49-60.

Al-Bayyati, H.A. (1971). "A rule of thumb for determining a sample size in comparing two populations," Technometrics, 13, 675-677.

Anderson, H. and D. Thorburn (1972). "Determining the daily need of cash for a savings bank; simulation versus analytical solution," in Working Papers, Volume 2, Symposium Computer Simulation versus Analytical Solutions for Business and Economic Models, Graduate School of Business Administration, Gothenburg (Sweden).

Andréasson, I.J. (1971). Antithetic and Control Variate Methods for the Estimation of Probabilities in Simulations, Report NA 71.41, Department of Information Processing, The Royal Institute of Technology, Stockholm (Sweden).

Andrews, D.F., P.J. Bickel, F.R. Hampel, P.J. Huber, W.H. Rogers, and J.W. Tukey (1972). Robust Estimates of Location, Princeton University Press, Princeton, New Jersey.

Angers, C., C. Grenier, and F.A. Hurtubise (1970). A Stopping Rule Subroutine for Digitial Computer Simulation, DREV R-629/70, Defence Research Establishment Valcartier, Quebec (Canada).

Anscombe, F.J. (1953). "Sequential estimation," J. Roy. Stat. Soc., Series B, 15, 1-29.

Baraldi, S. (1969a). "Previsione e limitazione degli errori statistici nelle simulazioni numeriche su calcolatore," (Prediction and limitation of statistical errors in numerical simulations on a computer.) Revista di Ingegneria, 224-227.

Baraldi, S. (1969b). "Previsioni teoriche e misure sperimentali degli errori statistici in certe simulazioni numeriche," (Theoretical predictions and experimental measurements of statistical errors in particular numerical simulations), Rivista di Ingegneria, 395-401.

Blomqvist, N. (1967). "The covariance function of the M/G/1 queuing system," Skandinavisk Aktuarietidskrift, 50, 157-174.

Blomqvist, N. (1968). "Estimation of waiting-time parameters in GI/G/1 queuing system, Part I: general results," Skandinavisk Aktuarietidskrift, 51, 178-197.

Blomqvist, N. (1969). "Estimation of waiting-time parameters in GI/G/1 queuing system, part II: heavy traffic approximations," Scandinavisk Aktuarietidskrift, 52, 125-136.

Blomqvist, N. (1970). "On the transient behaviour of the GI/G/1 waiting-times," Skandinavisk Aktuarietidskrift, 53, 118-129.

Blomqvist, N. (1971). "Seriestorlek vid systemsimulering -ett exempel" (Runlengths in systems simulation--an example), Särtryck ur Statistik Tidskrift, 3, 220-225, 244.

Brenner, M.E. (1965). "A relation between decision making penalty and simulation sample size for inventory systems," Operations Res. 13, 433-443.

Brenner, M.E. (1966). "A cost model for determining the sample size in the simulation of inventory systems," J. Ind. Eng., 17, 141-144.

Bruzelius, L.H. (1972). "Estimating endogeneous parameters in a dynamic simulation model," in Working Papers, Volume 1, Symposium Computer Simulation versus Analytical Solutions for Business and Economic Models, Graduate School of Business Administration, Gothenburg (Sweden).

Chapman, D.G. (1950). "Some two sample tests," Ann. Math. Stat., 21, 601-606.

Chow, Y.S. and H. Robbins (1965). "On the asymptotic theory of fixed-width sequential confidence intervals for the mean," Ann. Math. Stat., 36, 457-462.

Clark, S.R., T.A. Rourke, and J.M. Wren (1972). "A note on the reproducibility of discrete-event simulation studies," Infor, 10, 194-200.

Cochran, W.G. and G.M. Cox (1957). Experimental Designs, 2nd edition, Wiley, New York.

Conover, W.J. (1971). Practical Nonparametric Statistics, Wiley, New York.

Conway, R.W. (1963). "Some tactical problems in digital simulation," Management Sci., 10, 47-61.

Corneliussen, A. and D.W. Ladd (1970). "On sequential tests of the binomial distribution," Technometrics, 12, 635-646.

Crane, M.A. and D.L. Iglehart (1972a). A New Approach to Simulating Stable Stochastic Systems: I-General Multi-Server Queues, Technical Report No. 86-1, Control Analysis Corporation, Palo Alto, California.

Crane, M.A. and D.L. Iglehart (1972b). Confidence Intervals for the Ratio of Two Means with Application to Simulations, Technical Report No. 86-2, Control Analysis Corporation, Palo Alto, California.

Crane, M.A. and D.L. Iglehart (1972c). Simulating Stable Stochastic Systems, II: Markov Chains. Technical Report No. 86-3, Control Analysis Corporation, Palo Alto, California.

Csörgö, M. and V. Seshadri (1971). "Characterizations of the Behrens-Fisher and related problems (a goodness of fit point of view)," Theory Prob. Applications, 16, 23-35.

David, H.A. (1971). Order Statistics, Wiley, New York.

Dear, R.E. (1961). Multivariate Analyses of Variance and Covariance for Simulation Studies Involving Normal Time Series, Field note 5644, System Development Corporation, Santa Monica, California.

Derriks, J.C. (1971). Onderzoek t.a.v. de Implementatie van de Methode van Mechanic en McKay in Simulatie-Technieken (Investigation with respect to the implementation of Mechanic and McKay's method in simulation techniques), Graduate thesis, tweede studierichting, Technische Hogeschool, Delft (The Netherlands).

Dudewicz, E.J. (1972). Statistical Inference with Unknown and Unequal Variances, Department of Statistics, The University of Rochester, Rochester, New York.

Dudewicz, E.J. and S.R. Dalal (1971). Allocation of Observations in Ranking and Selection with Unequal Variances, The University of Rochester, Rochester, New York.

Dutton, J.M. and W.H. Starbuck (1971). Computer Simulation of Human Behavior, Wiley, New York.

Emshoff, J.R. and R.L. Sisson (1971). Design and Use of Computer Simulation Models, second printing, MacMillan, New York.

Farrell, R.H. (1966). "Bounded length confidence intervals for the p-point of a distribution function III," Ann. Math. Stat., 37, 586-592.

Fishman, G.S. (1967). Digital Computer Simulation: The Allocation of Computer Time in Comparing Simulation Experiments, RM-5288-1-PR, The Rand Corporation, Santa Monica, California; Operations Res., 16, 1968, 280-295, 1087.

Fishman, G.S. (1968). Spectral Methods in Econometrics, R-453-PR, The Rand Corporation, Santa Monica, California.

Fishman, G.S. (1971). "Estimating sample size in computing simulation experiments," Management Sci., 18, 21-38.

Fishman, G.S. (1972a). Output Analysis for Queueing Simulations, Technical Report No. 56, Department of Administrative Sciences, Yale University, New Haven, Connecticut.

Fishman, G.S. (1972b). Estimation in Multiserver Queueing Simulations, Technical Report No. 58, Department of Administrative Sciences, Yale University, New Haven, Connecticut.

Fishman, G.S. (1973). Statistical Analysis of Multiserver Queueing Simulations, Technical Report No. 64, Department of Administrative Sciences, Yale University, New Haven, Connecticut.

Fishman, G.S. and P.J. Kiviat (1967). Digital Computer Simulation: Statistical Considerations, RM-5387-PR, The Rand Corporation, Santa Monica, California.

Fisz, M. (1967). Probability Theory and Mathematical Statistics, 3rd edition, Wiley, New York.

Flagle, C.D. (1960). "Simulation techniques," in Operations Research and Systems Engineering, (C.D. Flagle, W.H. Huggins, and R.H. Roy, eds.), The John Hopkins Press, Baltimore.

Fraser, D.A.S. (1957). Nonparametric Methods in Statistics, Wiley, New York.

Fu, K.S. (1970). Sequential Methods in Pattern Recognition and Machine Learning, second printing, Academic, New York.

Gebhart, R.F. (1963). "A limiting distribution of an estimate of mean queue length," Operations Res., 11, 1000-1003.

Geertsema, J.C. (1970). "Sequential confidence intervals based on rank tests," Ann. Math. Stat., 41, 1016-1026.

Geisler, M.A. (1964a). "The sizes of simulation samples required to compute certain inventory characteristics with stated precision and confidence," Management Sci., 10, 261-286.

Geisler, M.A. (1964b). "A test of a statistical method for computing selected inventory model characteristics by simulation," Management Sci., 10, 709-715.

Ghosh, B.K. (1970). Sequential Tests of Statistical Hypotheses, Addison-Wesley, Reading, Massachusetts.

Ghosh, B.K. and H. Freeman (1961). Investigation of Sequential Methods in Design and Analysis of Experiments (Introduction to Sequential Experimentation; Sequential Analysis of Variance), Technical Report R-11, FEA MRS 60-7j, Field Evaluation Agency, Fort Lee, Virginia.

Goodman, A.S., P.A.W. Lewis, and H.E. Robbins (1973). "Simultaneous estimation of large numbers of extreme quantiles in simulation experiments," Commun. Stat. In press.

Grundy, P.M., D.H. Rees, and M.J.R. Healy (1954). "Decision between two alternatives--how many experiments?" Biometrics, 10, 317-323.

Gürtler, H. (1969). Quantitative Modelle zur Optimierung des Schalterverkehrs in Einem Postamt. (Quantitative models for optimizing counter traffic in a postoffice), Doctoral dissertation, Wilhelms-Universität, Münster (Germany).

Hauser, N., N.N. Barish, and S. Ehrenfeld (1966). "Design problems in a process control simulation," J. Ind. Eng., 17, 79-86.

Hayes, R.H. (1969). "The value of sample information," in The Design of Computer Simulation Experiments (T.H. Naylor, ed.), Duke University Press, Durham, North Carolina.

Healy, T.L. (1964). On the Solution of Queueing Problems by Computer Simulation, Operations Research Memorandum Op 8-15, Operations Evaluation, Advanced Development Division, National Cash Register Company, Dayton, Ohio.

Heuts, R.M.J. (1971). Parameter Estimation in the Exponential Distribution, Confidence Intervals and a Monte Carlo Study for Some Goodness of Fit Tests, EIT 22, Tilburg Institute of Economics, Katholieke Hogeschool, Tilburg (The Netherlands).

Heuts, R.M.J. and P.J. Rens (1972). A Monte Carlo Study on the Kuyper Test Statistic for Testing Exponentiality (Two Different Approaches), RC-Report No. 13, Rekencentrum Katholieke Hogeschool, Tilburg (The Netherlands).

Hillier, F.S. and G.J. Lieberman (1968). Introduction to Operations Research, second printing, Holden-Day, San Francisco, California.

Holme, I. (1972). "On the Construction of Confidence Bands for Distribution Functions," in Working Papers, Volume 1. Symposium Computer Simulation versus Analytical Solutions for Business and Economic Models, Graduate School of Business Administration, Gothenburg (Sweden).

Huisman, F. (1969). Statistische Aspekten van Simulatie (Statistical aspects of simulation), Report No. 2, Afdeling Werktuigbouwkunde, Technische Hogeschool Twente, Enschede (The Netherlands).

Huisman, F. (1970). Bepaling van de Deelrungrootte bij Simulatie (Determination of the Subrun-Length in Simulation), Report 005, Afdeling Werktuigbouwkunde, Technische Hogeschool Twente, Enschede (The Netherlands).

Johnson, N.L. (1957). "Sequentially determined confidence intervals," Biometrika, 44, 279-281.

Kabak, I.W. (1968). "Stopping rules for queueing simulations," Operations Res., 16, 431-437.

Keeping, E.S. (1962). Introduction to Statistical Inference, Van Nostrand, Princeton, New Jersey.

Kendall, M.G. and A. Stuart (1961). The Advanced Theory of Statistics, Volume 2, Griffin, London.

Kendall, M.G. and A. Stuart (1963). The Advanced Theory of Statistics, Volume 1, second edition, Griffin, London.

Kleijnen, J.P. (1967). Reliability and Simulation (Obtainable at Katholieke Hogeschool, Tilburg, The Netherlands).

Kosten, L. (1968). Statistische Aspecten van Simulatie (Statistical Aspects of Simulation), Afdeling Algemene Wetenschappen, Technische Hogeschool, Delft (The Netherlands).

Kurlat, S. and M. Springer (1960). "Sequential analysis of the reliability of an antitank-mine simulation system," Operations Res., 8, 473-486.

Lewis, P.A.W. (1972). Large-Scale Computer-Aided Statistical Mathematics, Naval Postgraduate School, Monterey, California. (To appear in Proceedings Computer Science and Statistics: 6th Annual Symposium on the Interface, Western Periodical, Hollywood, California.)

Martin, W.A. (1971). "Sorting," Computing Surveys, 3, 147-174.

Maurice, R. (1957). "A minimax procedure for choosing between two populations using sequential sampling," J. Roy. Stat. Soc., Series B, 19, 225-261.

Mechanic, H. and W. McKay (1966). Confidence Intervals for Averages of Dependent Data in Simulations II, Technical Report 17-202, IBM Advanced Systems Development Division, Yorktown Heights, New York.

Mehta, J.S. and R. Srinavasan (1970). "On the Behrens-Fisher problem," Biometrika, 57, 649-655.

Meier, R.C., W.T. Newell, and H.L. Pazer (1969). Simulation in Business and Economics, Prentice-Hall, Englewood Cliffs, New Jersey.

Mihram, G.A. (1972). Simulation: Statistical Foundations and Methodology, Academic, New York.

Naylor, T.H. (1971). Computer Simulation Experiments with Models of Economic Systems, Wiley, New York.

Naylor, T.H., J.L. Balintfy, D.S. Burdick, and K. Chu (1967). Computer Simulation Techniques, 2nd printing, Wiley, New York.

Paulson, E. (1964). "Sequential estimation and closed sequential decision procedures," Ann. Math. Stat., 35, 1048-1058.

Press, S.J. (1966). "A confidence interval comparison of two test procedures proposed for the Behrens-Fisher problem," J. Amer. Stat. Assoc., 61, 454-466.

Prins, H.J. (1962). Een Monte-Carlo-Methode om een Integraal met Vooraf Gegeven Nauwkeurigheid te Schatten, Waarbij het Oppervlak van het Gebied, Waarover Geintegreerd Wordt, Mede Geschat Moet Worden (A Monte Carlo Method to estimate an integral with predetermined accuracy where the integrating surface must also be estimated), Report No. 9/62, Philips' Gloeilampenfabrieken N.V., Nat. Lab. Groep Statistiek, Eindhoven (The Netherlands).

Puri, M.L. and P.K. Sen (1971). Nonparametric Methods in Multivariate Analysis, Wiley, New York.

Reynolds, J.F. (1972). "Asymptotic properties of mean length estimators for finite Markov queues," Operations Res., 20, 52-57.

Robbins, H., G. Simons, and N. Starr (1967). "A sequential analogue of the Behrens-Fisher problem," Ann. Math. Stat., 38, 1384-1391.

Robbins, H. and N. Starr (1965). Remarks on Sequential Hypothesis Testing, Technical Report No. 68, Department of Statistics, University of Minnesota, Minneapolis.

Sasser, W.E., D.S. Burdick, D.A. Graham, and T.H. Naylor (1970). "The application of sequential sampling to simulation: an example inventory model," Communications ACM, 13, 287-296.

Scheffé, H. (1964). The Analysis of Variance, 4th printing, Wiley, New York.

Scheffé, H. (1970). "Practical solutions of the Behrens-Fisher problem," J. Amer. Stat. Assoc., 65, 1501-1508.

Schmidt, J.W. and R.E. Taylor (1970). Simulation and Analysis of Industrial Systems, Richard D. Irwin, Inc., Homewood, Illinois.

Srivastava, M.S. (1970). "On a sequential analogue of the Behrens-Fisher problem," J. Roy. Stat. Soc., Series B, 32, 144-148.

Starr, N. (1966a). "The performance of a sequential procedure for the fixed-width interval estimation of the mean," Ann. Math. Stat., 37, 36-50.

Starr, N. (1966b). "On the asymptotic efficiency of a sequential procedure for estimating the mean," Ann. Math. Stat., 17, 1173-1185.

Starr, N. and M. Woodroofe (1970). Further Remarks on Sequential Estimation: The Exponential Case, Technical Report No. 7, Department of Statistics and Statistical Research Laboratory, The University of Michigan, Ann Arbor.

Thomasse, A.H. (1972). Het Behrens-Fisher Probleem (The Behrens-Fisher problem), Scriptie o.l.v. prof. dr. J. Hemelrijk, Mathematisch Instituut, Universiteit van Amsterdam, Amsterdam.

Tintner, G. (1960). Handbuch der Ökonometrie (Handbook of Econometrics), Springer Verlag, Berlin.

Tocher, K.D. (1963). The Art of Simulation, The English Universities Press Ltd., London.

Van der Waerden, B.L. (1965). Mathematische Statistik (Mathematical Statistics), second printing, Springer Verlag, Berlin.

Van Frankenhuysen, J.H. and A.W. Schuringa (1971). Wachttijden bij Lifttransport in een Hoog Kantoorgebouw (Waiting times at elevators in a high-rise office building), Report BW 6/71, Afdeling Mathematische Besliskunde, Stichting Mathematisch Centrum, Amsterdam.

Wang, Y.Y. (1971). "Probabilities of the type I errors of the Welch tests for the Behrens-Fisher problem," J. Amer. Stat. Assoc., 66, 605-608.

Wehrli, M. (1970). "Zur Stichprobenreduktion bei Monte Carlo Simulationen." (On reduction of sample sizes in Monte Carlo simulations), Unternehmensforschung, 14, 97-108.

Weiss, L. (1960). "Confidence intervals of preassigned length for quantiles of unimodal populations," Naval Res. Logistics Quart., 7, 251-256.

Wetherill, G.B. (1966). Sequential Methods in Statistics, Methuen London and Wiley, New York.

Ying Yao (1963). On the Comparison of the Means of Two Populations with Unknown Variance, ARL 63-106, Aerospace Research Laboratories, Wright-Patterson Air Force Base, Ohio.

V.B. MULTIPLE COMPARISON PROCEDURES

V.B.1. INTRODUCTION AND SUMMARY

In Sec. V.A we considered the problem of determining the reliability of statements on the mean of one population, or the difference between the means of two populations, the sample size being fixed. Now we shall consider the general case of k (≥ 2) populations. Unlike Sec. V.A.4 (and later on in Sec. V.C), the number of observations is not to be chosen in such a way that the best mean is selected with predetermined reliability. Instead there is a fixed sample size of n_i observations from population π_i $(i = 1,2,\ldots,k)$. Such a situation arises in simulation and Monte Carlo studies that are in a rather exploratory phase. Then we are not yet looking for the best system but are trying to determine if and how a factor influences the system response, i.e. we are gaining insight into the problem. Notice that the factor or factors are assumed to be qualitative; for quantitative factors a regression analysis approach would be more adequate.[1)]

In Sec. 2 we shall discuss several types of error rates that may be distinguished in situations with multiple (i.e. more than one) statements. The experimentwise error rate will be selected as an appropriate rate. In Sec. 3 parametric and nonparametric multiple comparison procedures (MCP) will be presented for various experimental objectives and situations. The experimental objectives considered are: confidence bounds for comparisons with a control $(\mu_i - \mu_0)$, all pairwise comparisons $(\mu_i - \mu_{i'})$, linear contrasts in the means ($\Sigma\, c_i\mu_i$ with $\Sigma\, c_i = 0$), linear functions $(\Sigma\, c_i\mu_i)$, means themselves (μ_i); selection of a subset containing the best population or a subset containing all populations better than a standard population. In addition to experimental situations with only one factor, we examine situations with more factors. In Sec. 4 the efficiency and robustness of MCP are discussed. Recommendations for the use of particular MCP in simulation and Monte Carlo experiments are made. In Sec. 5 more MCP for the above experimental situations are briefly discussed. Further MCP for other objectives and situations are mentioned. The reader looking for a procedure for his particular problem, may skip the detailed exposé in Sec. 3 and consult Sec. 4.

V.B.2. ERROR RATES

Let us first consider the traditional error rate α in confidence interval and significance statements on only one population. Such a confidence interval statement may be "μ is contained in the interval $\bar{x} \pm t_{(n-1)}^{\alpha/2} s(\bar{x})$." In Sec. V.A.3 we saw that this confidence statement can be easily converted into a significance statement on the hypothesis $H_0: \mu = \mu_0$ (for any μ_0). If the confidence interval does not contain μ_0, then we reject H_0. Since the confidence interval approach is more general than the significance testing approach, we shall usually phrase our discussion in confidence interval (estimation) terms. Thus, we are following Tukey who emphasized the advantage of using confidence intervals; see, e.g. Kurtz et al. (1965, pp. 148-149) or Tukey (1953, pp. 247-256).[2)]

Let S denote a statement, such as a confidence statement on μ or a significance statement on $H_0: \mu = \mu_0$. Using the error rate α means that the probability is α that the stated confidence interval does not cover the true value of μ, and that if H_0 holds then the probability is α that the significance statement is false (i.e. H_0 is erroneously rejected). Observe that in the testing approach we have to add "if H_0 holds." In the following discussion the qualification "under the nullhypothesis" is always assumed when S is interpreted as a significance statement instead of a confidence interval statement. A statistician makes many statements during his lifetime. Each statement can be seen as a trial with probability of success $(1 - \alpha)$ and probability of failure α. For N experiments with one statement per experiment the expected number of false statements is αN and hence, the expected fraction of wrongly evaluated experiments is $\alpha N/N = \alpha$.

Now consider the example of an experiment where the means of k populations are examined, n independent observations being taken from each population. When we compare population 1 with 2 we might use the familiar t-statistic to derive a first statement S_1 like "$\mu_1 - \mu_2$ is contained in the interval $(\bar{\underline{x}}_1 - \bar{\underline{x}}_2) \pm t^{\alpha/2}_{(2n-2)}\underline{s}(\bar{\underline{x}}_1 - \bar{\underline{x}}_2)$" as we saw in Sec. V.A.3. In the same way we can derive a statement S_2 on $\mu_3 - \mu_4$. The (marginal) probability that S_1 is false, is α; the probability that S_2 is false is also α. However, we may also consider the two statements together, for they refer to a single experiment. Let $\underline{S}_i = 0$ denote that statement i is false and $\underline{S}_i = 1$ that statement i is correct $(i = 1, 2, \ldots)$. Then the joint probability that both (independent) statements are correct is

$$P(\underline{S}_1 = 1 \wedge \underline{S}_2 = 1) = P(\underline{S}_1 = 1)\, P(\underline{S}_2 = 1)$$
$$= (1 - \alpha)^2 < (1 - \alpha) \qquad (1)$$

More generally, if we make m independent statements, each with confidence level $1 - \alpha$, then the (joint) probability that all m statements are correct, is

$$P(\underline{S}_1 = 1 \wedge \underline{S}_2 = 1 \wedge \cdots \wedge \underline{S}_m = 1) = \prod_{i=1}^{m} P(\underline{S}_i = 1)$$
$$= (1 - \alpha)^m \qquad (2)$$

where $(1 - \alpha)^m$ is much smaller than $(1 - \alpha)$, if m is large. Consequently if we are making many statements in an experiment, each individual statement with a confidence level $(1 - \alpha)$, then "the" conclusion on the experiment may show a high proportion of false statements!

The value of α used in each individual statement is called the "per comparison error rate." Instead of controlling this error rate we may control the probability that "the" conclusion on the experiment is correct, i.e. the probability that all statements concerning the experiment are correct, or

$$P(\underline{S}_1 = 1 \wedge \underline{S}_2 = 1 \wedge \dots \wedge \underline{S}_m = 1) \tag{3}$$

The experimentwise error rate, say α_E, is the probability that not all statements are correct or equivalently the probability that one or more statements on the experiment are false:

$$\alpha_E = P(\underline{S}_1 = 0 \vee \underline{S}_2 = 0 \vee \dots \vee \underline{S}_m) = 0 \tag{4}$$

(Observe that $A \vee B$ means A or B, or A and B.) An α_E of, e.g. 10% means that in 90 out of 100 experiments all statements made for a particular experiment are correct, or conversely, in 10 out of 100 experiments a set of populations is "misjudged," i.e. at least one statement on the set is false. The calculation of α_E is simple if all statements are independent for we can then use Eq. (2) above. Unfortunately the statements are often dependent since they concern, e.g. $(\bar{\underline{x}}_1 - \bar{\underline{x}}_2)$, $(\bar{\underline{x}}_1 - \bar{\underline{x}}_3)$, etc., where several statements have a common element $(\bar{\underline{x}}_1)$. There are special techniques for the determination of α_E in case of dependent statements. The crucial point in the concept of error rates, however, is not the dependence or independence of statements but the existence of more than one (related) statement. More statements means that the error rate is changed if the statements are taken together; the (in)dependence of the statements effects the degree of change in the resulting error rate [compare Ryan (1959, p. 34)].

Several other error rates can be defined. Miller (1966, pp. 5-12, 28-35, 84-85, 102-107) gives an excellent exposé on the error rate per comparison, experimentwise, per experiment (is not experimentwise) and on Duncan's p-level significance or "protection" levels. Verhagen (1963) proposes a "caution level" and Seeger (1966, pp. 134-141) describes Eklund's proportion of false significance statements.

There are many more references on the various error rates.[3] Which error rate should be controlled is a controversial matter. Each type of error rate has its advocates. Nevertheless, in our opinion the per comparison and the experimentwise error rates are the two fundamental types. For the first one fixes the marginal probability that the individual statement is correct; the second one fixes the joint probability that all statements constituting "the" conclusion on the experiment, are correct.[4] The per comparison rate is well-known from the traditional theory on testing and confidence intervals; an experimentwise error rate is realized in many MCP that have been developed since the 1950's. Remains the question how to choose between these two error rates.

When we make only one statement in an experiment since there are only one or two populations then the per comparison error rate and the experimentwise error rate coincide. If the experiment results in a conclusion implying more than one statement then an experimentwise error rate is justified in our opinion. An example of a conclusion implying more statements is, e.g. "$\mu_1 - \mu_2$ is contained in interval I_1 and $\mu_1 - \mu_3$ in interval I_2," or in significance terms "μ_1 differs from μ_2 but not from μ_3" (supposing that I_1 does not contain zero but I_2 does). Both statements are more meaningful when taken together. Therefore we used the formulation that the statements taken together give "the" conclusion of the experiment. If the statements based on a single experiment stand on their own, then it is justified to use the traditional per comparison error rate. Often, however, we have a sample of n_i observations from population i $(i = 1, \ldots, k)$ and we examine which of the k means are different or we look for the largest mean. In these situations the statements about the comparisons among the k means should obviously be considered together. Indeed statisticians agree that in such comparisons the per comparison rate would yield too many false statements. For if there are m statements the expected number of false statements woould be αm, where α is the per comparison error rate.[5] Even while statisticians agree that in such a situation a per comparison error rate would be undesirable there is no consensus on the alternative error rate as we have seen. Yet we think that the majority of statisticians[6)] favor an experimentwise error rate since it keeps the number of experiments with all statements being correct under control.

There is a price we pay for the control of the experimentwise error rate. For if we want all m statements on the experiment to be correct, then the lengths of the individual confidence intervals increase. For instance, in one technique the estimated standard deviation is not multiplied by $t_v^{\alpha/2}$ but by the higher value $t_v^{\alpha/(2m)}$. In testing terms the MCP imply that the test is less sensitive to deviations from $H_0:\mu_1 = \mu_2 = \cdots = \mu_k$, i.e., the procedure has less power than the nonsimultaneous procedure. Miller (1966, pp. 32-33) mentions three alternatives for remedying these long confidence intervals and small power.

(i) Increase the experimentwise error rate α_E. Dunn (1964, p. 248), for instance, suggests "a value of [α_E] considerably larger than the traditional .05" and she uses an α_E of 20% in her example. A complication may be that for some MCP we need special tables and these tables give values only for α_E is 5% or 1%; [see, e.g. the nonparametric procedure of Steel (1959a)].

(ii) Decrease the number of statements m. This number m depends on the number of factors and the number of levels per factor. If we have only one factor at many, say k, levels then we are in favor of realizing α_E, for all m statements on the k means. If, however, there are many statements because there are more factors then we may decide to preserve the power of the test by controlling α_E only for all statements on main effects, or all statements on one particular factor. In this way "families" of closely related statements are formed within the total set of m statements, and per family the error rate is controlled. These error rates are called the familywise error rates instead of the experimentwise error rate. In Sec. 3 we shall return to the formation of families. With Miller (1966, pp. 34-35) we agree that the choice of families is a subjective affair. (Anyhow the experimenter should be aware of the various error rates that are at stake.)

(iii) Increase the sample size. Given the sample size we can try to calculate the power of the test. This, however, is a difficult problem; see Miller (1966, pp. 102-107) and also Hartley (1955, pp. 51-52), Ryan (1959, pp. 36-37), Scheffé (1964, p. 71) and Tukey (1953). We may also try to choose the sample size in such a way that the probability of a correct ranking of the k means is guaranteed. Then we are in the realm of multiple ranking procedures which will be discussed in Sec. V.C.

V.B.3. MULTIPLE COMPARISON PROCEDURES

Bonferroni Procedure

Let there be m statements each with a per comparison error rate α_i $(i = 1, 2, \ldots, m)$. For independent statements we can apply Eqs. (2) and (4) to find Eq. (5) where the last equality in Eq. (5) holds if all statements use the same $\alpha_i = \alpha$.

$$\begin{aligned} \alpha_E &= 1 - P(\text{all statements correct}) \\ &= 1 - P(\underline{S}_1 = 1 \wedge \underline{S}_2 = 1 \wedge \ldots \wedge \underline{S}_m = 1) \\ &= 1 - \prod_{i=1}^{m} (1 - \alpha_i) = 1 - (1 - \alpha)^m \end{aligned} \tag{5}$$

Usually, however, the statements constituting the conclusion of the experiment, are dependent. Then we can still establish an upper limit on α_E. First consider the Venn diagram in Fig. 1. This figure shows that the probability of event A_1 and/or A_2 satisfies

$$\begin{aligned} P(A_1 \vee A_2) &= P(A_1) + P(A_2) - P(A_1 \wedge A_2) \\ &\leq P(A_1) + P(A_2) \end{aligned} \tag{6}$$

Hence

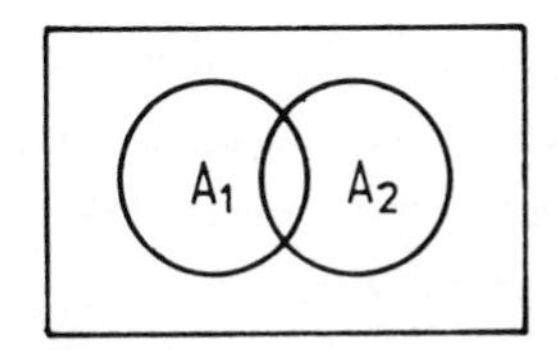

Fig. 1. Venn diagram.

$$\begin{aligned} P(A_1 \vee A_2 \vee A_3) &= P[(A_1 \vee A_2) \vee A_3] \\ &\leq P(A_1 \vee A_2) + P(A_3) \\ &\leq P(A_1) + P(A_2) + P(A_3) \end{aligned} \quad (7)$$

Or, in general

$$P(A_1 \vee A_2 \vee \dots \vee A_m) \leq \sum_{i=1}^{m} P(A_i) \quad (8)$$

Now let A_i denote the event of a false statement ($\underline{S}_i = 0$). Then Eq. (8) yields

$$P(\underline{S}_1 = 0 \vee \underline{S}_2 = 0 \vee \dots \vee \underline{S}_m = 0) \leq \sum_{i=1}^{m} P(\underline{S}_i = 0) \quad (9)$$

If each individual statement uses a per comparison error rate α_i, then Eq. (9) is equivalent to Eq. (10).

$$\alpha_E \leq \sum_{i=1}^{m} \alpha_i \quad (10)$$

This bound is called a Bonferroni inequality; see Miller (1966, p. 8). (Also compare Exercise 1.) So in a multiple comparison we can make m possibly dependent statements, each statement using the traditional per comparison error rate α_i ($i = 1, \dots, m$); then we can keep the experimentwise error rate α_E below some desirable level, say 10%, by using α_i such that $\Sigma_1^m \alpha_i = 0.10$. The application of this Bonferroni inequality in multiple comparisons has been advocated especially by Dunn (1961) and (1964); also see Miller (1966, pp. 15-16, 67-70), Scheffé (1964, p. 80) or Seeger (1966, p. 123). A simple example follows here.

Suppose we have n normally distributed independent observations from each of k populations and we want to compare all k means with each other. From V.A.3 we know that two means μ_i and $\mu_{i'}$ ($i \neq i'$) can be compared with each other using a t-statistic with 2n-2 degrees of freedom, viz.

$$\underline{t}_{2n-2} = \frac{(\bar{\underline{x}}_i - \bar{\underline{x}}_{i'}) - (\mu_i - \mu_{i'})}{(\underline{s}_i^2 + \underline{s}_{i'}^2)^{1/2}} n^{1/2} \tag{11}$$

There are $m = k(k-1)/2$ comparisons to be made. Hence the experimentwise error rate does not exceed, say, α' if per individual comparison we use an error rate α'/m. So the following m statements have a joint confidence level of at least $(1 - \alpha')$.[7]

$$(\mu_i - \mu_{i'}) \in (\bar{\underline{x}}_i - \bar{\underline{x}}_{i'}) \pm t_{2n-2}^{\alpha'/(2m)} (\underline{s}_i^2 + \underline{s}_{i'}^2)^{1/2}/n^{1/2} \tag{12}$$

We emphasize that the Bonferroni approach can also be applied to statements not based on the t-statistic. Dunn (1964), e.g. applied a nonparametric rank test per individual comparison; see (41) below.

Let us consider the advantages and disadvantages of the Bonferroni approach. (A deeper understanding of the relative merits of this approach is possible only after the other MCP have been presented; see Sec. 4.) Its limitations are: (i) We have to know before we examine the sampling results how many statements will be made (or specify a maximum for this number). (ii) The procedure is conservative as only an upper bound on α_E is known. The first disadvantage is unimportant if, e.g. we have k populations and we know beforehand that we want to compare all k populations with a standard population $(m = k)$ or with each other $(m = k(k-1)/2)$. (As we shall see Scheffé's MCP permits to look first at the data and to decide next which comparisons to make.) The second disadvantage may also be unimportant for Dunn (1961) has shown that under certain circumstances this procedure gives smaller confidence intervals than competing exact MCP. We shall return to the efficiency of various MCP in Sec. 4. Next consider the advantages of the Bonferroni technique.

(i) We can use statistics traditional in non-multiple situations that are familiar to most experimenters.

(ii) The procedure has a wider application range than the other MCP. For as we shall see most MCP assume normally distributed independent observations with a common variance, an equal number of observations per population and a one-factor layout. In the Bonferroni approach we need no normality assumption since we can apply nonparametric statistics (or rely on the known robustness of the t-statistic). Unequal variances are no problem in, e.g. the revised t-statistics we presented in Sec. V.A.3. The number of observations per population may vary in several traditional parametric and nonparametric tests. Higher-way layouts form no problem either as we shall see later on. Finally, we point out that as far as we know the Bonferroni approach is the only technique that permits <u>dependent</u> observations (created in simulation by the use of common random numbers)! For per individual comparison $\mu_i - \mu_{i'}$ we can apply a parametric or nonparametric statistic for paired observations ($\underline{d}_j = \underline{x}_{ij} - \underline{x}_{i'j}$ where $j = 1, \ldots, n$) with a (per comparison) error rate α'/m.

One final point concerns the <u>tables</u> needed for this approach. The available tables give critical values of the statistic for the traditional levels of 10%, 5%, 1%, etc. In a multiple comparisons situation, however, we need the levels α'/m, where α' is, say, 10%. A widely used statistic for individual comparisons is the t-statistic. In Scheffé (1964, p. 80) the following approximation for the upper α point of the t-statistic with v degrees of freedom is given.

$$t_v^\alpha \approx z^\alpha + [(z^\alpha)^3 + z^\alpha]/(4v) \tag{13}$$

where z^α is the upper α point of the standard normal variate $N(0,1)$. Dunn (1961, p. 55) gives a table for a two-sided t-test with an experimentwise error rate not higher than 5% of 1% and varying v and m; her table is partially reproduced in Miller (1966, p. 238). More sources for the determination of the α'/m point of the t-distribution are given by Miller (1966, p. 70). In nonparametric procedures we may use the normal distribution as the asymptotic distribution of the statistic. Then the determination of the α'/m point is usually no problem since detailed tables for $N(0,1)$

are available in many publications. We observe that other inequalities stricter than the Bonferroni one, have been derived by Khatri (1967) but they apply only to normal populations and are more cumbersome.

Experimental Situations

The Bonferroni approach is applicable to any situation where more than one statement is made. Specially devised MCP, however, apply only to special situations. In an experiment with k populations (e.g. one factor at k levels) we may have various experimental objectives.

(i) Comparisons between the means μ_i of k experimental populations and the mean μ_0 of a standard or control population: $\mu_i - \mu_0$ ($i = 1, \ldots, k$). In simulation such a standard system may be the existing system configuration, the priority rule presently in use, etc.

(ii) All pairwise comparisons: $\mu_i - \mu_{i'}$ ($i, i' = 1, \ldots, k$; $i < i'$). If there is no standard population then (ii) is an obvious extension of (i).

(iii) All "contrasts" among the k means. A contrast is a linear function in the k means, say $\Sigma_{i=1}^{k} c_i \mu_i$, where the known coefficients c_i satisfy $\Sigma_1^k c_i = 0$. A contrast is, e.g. $(\mu_1 + \mu_2)/2 - \mu_3$. The above comparisons, $\mu_i - \mu_0$ and $\mu_i - \mu_{i'}$, are also particular contrasts with coefficients $c_i = +1$, $c_0 = -1$, $c_{i'} = 0$ $(i' \neq i)$ and $c_i = +1$, $c_{i'} = -1$, $c_{i''} = 0$ $(i < i', i'' \neq i, i'' \neq i')$, respectively. As an example of a contrast in simulation we may consider a queuing system with three possible priority rules, say (1) FIFO (2) LIFO (3) random. We may want to examine the pairwise comparisons $\mu_1 - \mu_2$, $\mu_1 - \mu_3$ and $\mu_2 - \mu_3$ plus the average performance of the systematic priority rules vs. the random rule, i.e. $(\mu_1 + \mu_2)/2 - \mu_3$. Other examples outside the field of simulation are given by Crouse (1969, p. 38), Dunn (1964, p. 243) and Scheffé (1964, p. 66).

(iv) All linear functions of the means: $\Sigma_1^k c_i\mu_i$ where the known coefficients need not satisfy $\Sigma c_i = 0$. For example, if we also want to study the means themselves then, when considering the mean μ_p, we have $c_p = 1$ and $c_{p'} = 0$ $(p' \neq p)$.

Observe that the set of statements covered by (i) is a subset of (ii). Likewise (ii) is contained in (iii), while (iii) is contained in (iv). Obviously the MCP valid for a particular set of statements, are also valid for a subset of that set. Nevertheless, most times it is better to apply a procedure especially devised for a certain situation, since such a procedure may give shorter confidence intervals than the more general procedure that cannot make use of the particularities of a specific situation. There are sets of statements not identical to any one of the four sets listed above. For example, the set of statements on the means only is contained in (iv) but (iv) is more general. Hence, we can apply a general procedure (in the example a procedure for linear functions) but such an approach is conservative, i.e. it yields too long confidence intervals. We shall return to this problem in Sec. 4.

Often an experiment is aimed at finding the best population. There are special procedures for selecting a subset containing the best population with prescribed probability or for selecting a subset containing all populations better than a standard.

Besides various experimental objectives like the ones listed above, we may distinguish experimental designs with one factor or more factors. We have just discussed one-factor layouts. In a higher-way layout we may have experimental objectives similar to the ones given above. For example, pairwise comparisons among the k_j levels of factor j $(j = 1, 2, \ldots, J)$, but no pairwise comparisons among levels of factor j and levels of another factor, j' $(j \neq j')$. We shall now study several MCP for various experimental situations.

Comparisons Between Experimental Populations and a Standard

Suppose we have n_i observations from experimental population i $(i = 1, \ldots, k)$ and n_0 observations from the standard or control population. We like to compare the means μ_i of the experimental populations with the standard mean μ_0. We shall present Dunnett's MCP based on normal observations and Steel's nonparametric ranking procedure; some more distribution-free techniques will be briefly discussed.

Dunnett's Parametric Procedure

Dunnett (1955; 1964) uses the traditional ANOVA assumptions, viz. the observations are normally distributed and independent with a common unknown variance σ^2 and means μ_j $(j = 0, 1, \ldots, k)$. This common variance is estimated by the usual pooled estimator or the mean square for error:

$$\underline{s}_v^2 = \frac{\Sigma_{j=0}^{k} \Sigma_{g=1}^{n_j} (\underline{x}_{jg} - \bar{\underline{x}}_j)^2}{\Sigma_{j=0}^{k} (n_j - 1)} \tag{14}$$

where v denotes the degrees of freedom so that

$$v = \sum_{j=0}^{k} (n_j - 1) \tag{15}$$

Hence the marginal distribution of

$$\underline{t}_i = \frac{\bar{\underline{x}}_i - \bar{\underline{x}}_0}{\underline{s}_v \left(\frac{1}{n_i} + \frac{1}{n_0} \right)^{1/2}} \qquad (i = 1, 2, \ldots, k) \tag{16}$$

is a t-distribution with v degrees of freedom. The $\underline{t}_i$ are dependent since they all have the common terms $\bar{\underline{x}}_0$ and $\underline{s}_v$. By looking at the maximum[8] of the k statistics $\underline{t}_i$ Dunnett found the following results for equal sample sizes $n_i = n_0 = n$.

(i) One-sided confidence intervals. The k statements in Eq. (17) hold with experimentwise error rate $\alpha_E = \alpha$, the critical point $d^{\alpha}_{k,v}$ being tabulated by Dunnett (1955, pp. 1117-1118) and Krishnaiah and Armitage (1966, pp. 41, 51) for α 1% or 5% and varying k and v. Values of $d^{\alpha}_{k,v}(\sqrt{2})$ are also tabulated in Gupta and Sobel (1957).[17]

$$\mu_i - \mu_0 \geq (\bar{x}_i - \bar{x}_0) - d^{\alpha}_{k,v} s_v \frac{\sqrt{2}}{n} \qquad (i = 1, \ldots, k) \tag{17}$$

In Eq. (17) it is assumed that we are looking for experimental means larger than the control. If the right-hand side of Eq. (17) is positive then $\bar{x}_i$ is significantly higher. If a "better" mean is a mean smaller than the standard, then $\bar{x}_i$ is significantly smaller if

$$(\bar{x}_0 - \bar{x}_i) - d^{\alpha}_{k,v} s_v \frac{\sqrt{2}}{n} > 0 \tag{18}$$

If we are not looking for better means but for means different from the control then we should use the following two-sided confidence intervals.

(ii) Two-sided confidence intervals. The critical constant $d'^{\alpha}_{k,v}$ is tabulated in Dunnett (1964, pp. 488-489) for $k = 1(1)12$, 15, 20 and α and v as above.

$$(\mu_i - \mu_0) \in (\bar{x}_i - \bar{x}_0) \pm d'^{\alpha}_{k,v} s_v \frac{\sqrt{2}}{n} \tag{19}$$

Unequal variances and unequal sample sizes will be discussed in Sec. 4. A presentation of Dunnett's MCP is also given in Miller (1966, pp. 76-81) who reproduced Dunnett's tables.

Steel's Rank Sum Test

Steel (1959a) derived a nonparametric procedure yielding confidence intervals for a possible shift in location, i.e. all populations are assumed to have identical distributions except for possibly different means. (So variances and other moments do not vary over the populations.) He further assumes that all observations are independent while the critical constants are only tabulated for equal sample sizes. Following Miller (1966, pp. 144-145), we can describe Steel's procedure as follows.

From the control population we have the observations ($\underline{x}_{01}$, ..., $\underline{x}_{0n}$) and from the experimental population, i, the observations ($\underline{x}_{i1}$, ..., $\underline{x}_{in}$). For each experimental population combine its observations with the observations on the control. Arrange these 2n observations in ascending order, irrespective of the original population. Assign the ranks 1, 2, ..., 2n to these ordered observations, the smallest observation having rank 1, etc. Now consider only the observations from the experimental population. The ranks of these n observations are denoted as $\underline{r}_{i1}$, $\underline{r}_{i2}$, ..., $\underline{r}_{in}$. (The ranks $\underline{r}_{01}$, ..., $\underline{r}_{0n}$ of the standard population are not used.) Hence the rank sum of the experimental population i is $\underline{r}_i = \Sigma_{g=1}^{n} \underline{r}_{ig}$ and this is the familiar Wilcoxon statistic for comparing two populations; see, e.g. Conover (1971, p. 223) or Wilks (1963, p. 460). To obtain confidence intervals and tests that hold simultaneously for all k differences the largest statistic among the k two-sample statistics $\underline{r}_i$ has to be determined. (Compare our discussion of Dunnett.[8]) So determine $\underline{r} = \max_i \underline{r}_i$. The critical value of $\underline{r}$ guaranteeing an experimentwise error rate α, say $r_{k,n}^{\alpha}$, is tabulated by Miller (1966, p. 250) for k = 2(1) 10, α 1% or 5% and n between 6 and 100.[9] Any population with rank sum $\underline{r}_i \geq r_{k,n}^{\alpha}$ is inferred to be shifted positively away from the standard. (Consequently the single composite hypothesis $H_0: \mu_0 = \mu_1 = \cdots = \mu_k$ is rejected when $\max_i \underline{r}_i \geq r_{k,v}^{\alpha}$.) For a two-sided test both high and low rank sums may indicate significance. Therefore, the rank sum statistic $\underline{r}_i$ is replaced by $\underline{r}_i^* = \max[\underline{r}_i, n(2n+1) - \underline{r}_i]$, where $n(2n+1) - \underline{r}_i$ is the rank sum if the observations were assigned ranks in reverse order (rank 1 to the largest observation, etc.). The critical value $r_{k,n}^{*\alpha}$ of $\max_i \underline{r}_i^*$ is tabulated by Miller (1966, p. 251) for k = 2(1)9, n = 4(1)20, α = 1%, 5%.[10] An application of this nonparametric test is given in Steel (1959a, pp. 563-564).

If we want confidence intervals instead of significance tests for the differences $\mu_i - \mu_0$, then we may apply a graphical procedure described in Miller (1966, pp. 145-146). However, since in simulation and Monte Carlo experiments a computer is available it is simpler to apply a numerical technique also presented in Miller (1966, p. 149) and based on Lehmann (1963). Calculate the n^2 differences

$$\underline{d}_{gh} = \underline{x}_{ig} - \underline{x}_{0h} \qquad \begin{matrix}(g = 1, \ldots, n)\\(h = 1, \ldots, n)\end{matrix} \tag{20}$$

Order these differences in ascending order, i.e.

$$\underline{d}_{(1)} < \underline{d}_{(2)} < \cdots < \underline{d}_{(n^2)} \tag{21}$$

For a <u>one-sided</u> confidence interval calculate the critical value $a_{k,n}^{\alpha}$.

$$a_{k,n}^{\alpha} = n^2 + \frac{n(n+1)}{2} - r_{k,n}^{\alpha} \tag{22}$$

Then the confidence intervals with experimentwise error rate α are

$$\underline{d}_{(a_{k,n}^{\alpha}+1)} \leq \mu_i - \mu_0 < \infty \qquad (i = 1, \ldots, k) \tag{23}$$

For <u>two-sided</u> confidence intervals replace $r_{k,n}^{\alpha}$ by $r_{k,n}^{*\alpha}$ in Eq. (22) and calculate

$$\underline{d}_{(a_{k,n}^{*\alpha}+1)} \leq \mu_i - \mu_0 \leq \underline{d}_{(n^2-a_{k,n}^{*\alpha})} \tag{24}$$

(See also Exercise 3.)

We remark that as in other nonparametric MCP the critical values $r_{k,n}^{\alpha}$ are derived using the knowledge that under H_0 each possible ranking of the $(k+1)n$ observations is equally likely. Hence, the exact distribution of the test statistic requires the cumbersome evaluation of expressions involving factorials. This exact distribution is approximated applying a multivariate normal distribution. Further, theoretically no ties can occur since continuous distributions are assumed. In practical applications ties do occur. In simulation and Monte Carlo applications, however, the responses are quantitative variables measured in many digits so that the probability of ties is very small. If ties are nevertheless met in such applications, the quoted publications should be consulted. Remember that we pointed out in Part A of this chapter that the ranking of observations is expensive in terms of computer time and memory space since we need to store and sort all observations. For a survey of sorting routines we referred to Martin (1971); also see Lewis (1972, pp. 8-9). Examples of the computer time involved in estimators based on ranking (and other rules) are provided by Andrews et al. (1972, p. 105).

Other Procedures

There are several other distribution-free MCP that are either inappropriate or inferior compared with Steel's rank test. For example, the multiple sign tests developed by Steel (1959b) and Rhyne and Steel (1965) assume a block effect. In simulation and Monte Carlo experiments block effects do not exist since replicates of a particular population differ only because of random error and not because of a systematic block effect. Rhyne and Steel (1965) further point out that their nonparametric MCP assumes that the populations have a common variance.[11)] Miller (1966, pp. 165-172) discusses other MCP developed by Nemenyi and Dunn and based on the Kruskall-Wallis method of ranking in a one-way layout. [The Kruskall-Wallis method is discussed by Conover (1971, p. 256).] As Miller observes, these procedures have an unpleasant property since the comparisons between two populations depend on the observations from the other populations. Moreover it is difficult to transform these tests into confidence intervals. Gabriel and Sen (1968, p. 309), Peritz (1971), and Puri and Sen (1971, pp. 244-254) give alternative nonparametric ranking tests (that also apply to all pairwise comparisons) [also see Gupta et al. (1971), p. 10)].

Unfortunately they do not compare their tests with Steel's test. In Part C we shall discuss several more procedures for situations involving a control population, but in that part the sample size will be determined in a two-stage or a sequential way.

All Pairwise Comparisons

If there is no standard population then we may compare all populations with each other, i.e. we study $\mu_i - \mu_{i'}$ $(i < i')$ resulting in $k(k-1)/2$ pairwise comparisons. First we shall present Tukey's parametric procedure, next Steel's rank test, and some other MCP.

Tukey's Studentized Range Procedure

Tukey's procedure is based on the studentized range. This range is defined, e.g. by Scheffé (1964, p. 28) as follows. Let $\underline{z}_j$ $(j = 1, \ldots, w)$ be w independent observations from the normal distribution $N(\mu, \sigma_z^2)$. Then the range of these $\underline{z}_j$ is

$$\underline{R}_w = \max_j \underline{z}_j - \min_j \underline{z}_j \tag{25}$$

Let $\underline{s}_v^2$ be an independent estimator of σ_z^2 with v degrees of freedom, i.e. $v\underline{s}_v^2/\sigma_z^2$ is $\underline{\chi}_v^2$ and statistically independent of $\underline{R}_w$. Then the studentized range, say $\underline{q}_{w,v}$, is defined as

$$\underline{q}_{w,v} = \frac{\underline{R}_w}{\underline{s}_v} \tag{26}$$

Miller (1966, pp. 47-48) lists various publications containing tables of the studentized range. Harter (1960a, p. 671) points out that several publications give inaccurate critical points. More accurate tables based on Harter can be found in, e.g. Miller (1966, pp. 234-237) for w between 2 and 100, v between 1 and ∞, α 5% or 1%. Since in MCP a higher α level may be used, we mention that among the detailed tables in Harter (1960b) there is a table for α 10%.

Tukey's procedure assumes that there are n observations from each of the k populations. All observations are independent, normally distributed with common variance σ^2 and possibly different means μ_i. Then the following confidence intervals hold with an experimentwise error rate α.

$$(\mu_i - \mu_{i'}) \in (\bar{x}_i - \bar{x}_{i'}) \pm q^{\alpha}_{k,k(n-1)} \frac{s}{\sqrt{n}} \qquad (i, i' = 1, \ldots, k) \tag{27}$$

where $\underline{s}^2$ is the pooled estimator of σ^2 based on $k(n-1)$ degrees of freedom and $q^{\alpha}_{k,k(n-1)}$ is the upper α point of the studentized range with parameters k and $k(n-1)$.

Since the proof of Eq. (27) is so simple we shall give it here. Put $\underline{z}_i = \bar{\underline{x}}_i - \mu_i$ so that all $\underline{z}_i$ have the same mean zero. All $\underline{z}_i$ have a common variance σ^2/n. Hence, the $\underline{z}_i$ are independent observations from $N(0, \sigma^2/n)$. The pooled estimator of their variance, $var(\underline{z}) = \sigma^2/n$, with $v = k(n-1)$ degrees of freedom, is

$$\hat{\underline{\sigma}}^2_z = \frac{\underline{s}^2_v}{n} = \frac{1}{n} \sum_{i=1}^{k} \sum_{g=1}^{n} \frac{(\underline{x}_{ig} - \bar{\underline{x}}_i)^2}{k(n-1)} \tag{28}$$

Now it is well-known that $\underline{s}^2_i$ and $\bar{\underline{x}}_i$ are independent, see Fisz (1967, p. 346). Hence $\hat{\underline{\sigma}}^2_z$ is independent of $\bar{\underline{x}}_i - \mu_i = \underline{z}_i$ so that by definition

$$\frac{\max_i \underline{z}_i - \min_i \underline{z}_i}{\hat{\underline{\sigma}}_z} = q_{k,k(n-1)} \tag{29}$$

Or with probability $(1 - \alpha)$ we have

$$\max_i \underline{z}_i - \min_i \underline{z}_i \le q^{\alpha}_{k,k(n-1)} \frac{\underline{s}_v}{\sqrt{n}} \tag{30}$$

which is equivalent to

$$|\underline{z}_i - \underline{z}_{i'}| \leq q^{\alpha}_{k,k(n-1)} \frac{\underline{s}_v}{\sqrt{n}} \quad \text{for all } i, i' \tag{31}$$

which yields Eq. (27).

Tukey's method can be generalized to all contrasts and all linear combinations, to observations with a particular correlation pattern or with known (varying) variance ratios, and to higher-way layouts [see Miller (1966, pp. 39-42), Scheffé (1964, pp. 73-75), Sen (1969), and Tukey (1953)]. We shall return to these generalizations, but first we consider a distribution-free procedure for pairwise comparisons.

Steel's Rank Test

Let us have n observations from each population i ($i = 1, \ldots, k$). All observations are independent but not necessarily normal. The density functions of the populations are identical except for a possible shift of location. As in Steel's test for comparisons with a control, we calculate the rank $\underline{r}_{ii'}$ for population i when compared with population i'. (Now i' is not identical to 0 but runs from $i+1$ to k.) The two-sided test statistic is

$$\underline{r}^*_{ii'} = \max[\underline{r}_{ii'}, n(2n+1) - \underline{r}_{ii'}] \tag{32}$$

The rank statistic in a multiple comparison situation with experimentwise error rate is the maximum of these $k(k-1)/2$ statistics $\underline{r}^*_{ii'}$, i.e. it is

$$\underline{r}^* = \max_{i,i'} (\underline{r}^*_{ii'}) \tag{33}$$

The null-hypothesis of equal means is rejected if $\underline{r}^* \geq r^{**\alpha}_{k,n}$ (two asterisks distinguish this critical value from $r^{*\alpha}_{k,n}$ in the control population situation). Confidence intervals for $\mu_i - \mu_{i'}$ can be constructed applying Eqs. (20) through (24) substituting $r^{**\alpha}_{k,n}$. Miller (1966, pp. 157, 252) tabulated the critical values for $k = 2(1)10$, n between 6 and 100, and α 1% or 5%. An application of the procedure is given in Steel (1960, pp. 199-200). Later on Gabriel and Lachenbruch (1969) found that the tables for Steel's procedure are very conservative, especially for high k and low n.

Other Nonparametric Procedures

Miller (1966, pp. 138-143, 165-172) discusses a multiple sign test assuming blocks and a Kruskall-Wallis type test. As we have seen for the control population situation these procedures are inferior compared with the rank test. Alternative procedures are given by Peritz (1971) and Puri and Sen (1971, pp. 244-254, 328-331) as we remarked before. Tobach et al. (1967) derived a procedure that is a specialization of the method developed by Dunn (1964) to which we shall return later on.

Gabriel's Parametric Procedure for Testing Subsets of Means

The above parametric and nonparametric MCP can be used to test if $\mu_i = \mu_{i'}$. Gabriel (1964) derived a procedure for a related problem. He tests if subsets of the total set of k means are homogeneous. His test is closely related to the traditional ANOVA F-test for the hypothesis $H_0 : \mu_1 = \cdots = \mu_k$, i.e. for the homogeneity of the total set. If the F-test rejects this hypothesis at a confidence level $1 - \alpha$ then Gabriel's procedure can be applied to determine which subsets are significant and which are nonsignificant. This subsequent analysis does not effect the experimentwise error rate α, i.e. the probability that one or more statements declare a subset significant, whereas it actually contains homogeneous means, remains α; also compare our discussion of the next procedure, Scheffé's test. Unfortunately the procedure has the following disadvantage in our opinion (but maybe the reader disagrees). Gabriel (1964, p. 463) gives an example where the populations ranked in order of decreasing sample averages are: D C B F A E H G. Now D and G are declared nonsignificant, but the larger subset D C H G is declared significant. So the inclusion of the populations C and H which give sample averages between the extremes

D and G, makes the subset significant. In our opinion this is an undesirable property.[12)]

Linear Constrasts

Scheffé's Parametric Procedure

A linear contrast, say Ψ_j, was defined as

$$\Psi_j = \sum_{i=1}^{k} c_{ij}\mu_i \qquad (j = 1, 2, \ldots) \tag{34}$$

the known coefficients c_{ij} satisfying

$$\sum_{i=1}^{k} c_{ij} = 0 \qquad (j = 1, 2, \ldots) \tag{35}$$

Notice that Eqs. (34) and (35) define infinitely many contrasts. An unbiased estimator of Ψ_j is

$$\underline{\hat{\Psi}}_j = \sum_{i=1}^{k} c_{ij}\underline{\hat{\mu}}_i = \sum_{i=1}^{k} c_{ij}\underline{\bar{x}}_i \tag{36}$$

which has variance

$$\begin{aligned} \text{var}(\underline{\hat{\Psi}}_j) = \sigma_\Psi^2 &= \sum_i c_{ij}^2 \text{ var}(\underline{\bar{x}}_i) = \sum_i c_{ij}^2 \frac{\sigma^2}{n_i} \\ &= \sigma^2 \sum_i \frac{c_{ij}^2}{n_i} \end{aligned} \tag{37}$$

assuming that all individual observations $\underline{x}$ are independent with common variance σ^2; the number of observations per population may be different. The unknown σ^2 in Eq. (37) can be estimated by the traditional pooled estimator $\underline{s}_v^2$ which, on substitution into Eq. (37) yields $\underline{\hat{\sigma}}_\Psi^2$. If it is further assumed that the individual observations, $\underline{x}$, are normally distributed, then the probability is $(1 - \alpha)$ that all contrasts Ψ_j simultaneously satisfy the inequalities

$$|\Psi_j - \hat{\Psi}_j| \leq S\hat{\sigma}_{\Psi} \qquad (j = 1, 2, \ldots) \tag{38}$$

with

$$S = [(k - 1)\, F^{\alpha}_{k-1,v}]^{1/2} \tag{39}$$

v denoting the degrees of freedom of the estimator of σ^2, i.e. $v = \Sigma_i\,(n_i - 1)$ in a one-way layout. This result was derived by Scheffé; its proof can be found in Scheffé (1964, pp. 67-72) while a different derivation is given by Miller (1966, pp. 48-53, 63-66). The derivation hinges on the projection of a confidence ellipsoid (based on the F-statistic) onto the axes.

The Scheffé procedure, Eq. (38), can be applied as follows. Scheffé proved that the ANOVA F-test rejects $H_0 : \mu_1 = \cdots = \mu_k$ if and only if Eq. (38) yields at least one contrast Ψ significantly different from zero. So a first step in the analysis of the experiment may be the application of the F-test for H_0. If this F-test is not significant then we know that Eq. (38) will not yield any significant contrast. Hence it would be useless to calculate Eq. (38).[13)] If the F-test rejects H_0 then a logical next step is to determine which contrasts caused this rejection. However, since there are infinitely many contrasts we cannot calculate all contrasts. Instead we may study the experimental results and determine which contrasts may be of interest. These contrasts can be tested for significance by applying Eq. (38). It may happen that H_0 is rejected but that none of the calculated contrasts are significant [see Miller (1966, p. 51)]. The procedure developed by Gabriel (1964, pp. 469-470) may be used to hunt for significant contrasts. Observe that Scheffé's procedure permits examination of the data before a decision on the contrasts to be made, is reached; in the Bonferroni approach the total number of comparisons must be specified before the data are investigated. An application of Scheffé's procedure is given in Seeger (1966, pp. 120-122).

Miller (1966,p. 53) shows how Scheffé's technique can be generalized to observations having unequal variances σ_i^2 and/or being correlated. Unfortunately it must be supposed that the variance ratios $\sigma_i^2/\sigma_{i'}^2$, are known and/or that the correlation coefficients $\rho_{ii'}$ are known. For the adjustments on the confidence intervals, Eq. (38), we refer to Miller.

Dunn's Ranking Procedure

While there are several nonparametric procedures for pairwise comparisons there are not many distribution-free procedures for contrasts. We mention the approaches derived by Crouse (1969), Dunn (1964), and Puri and Sen (1971, pp. 236-240, 244-254). The former two MCP restrict the contrasts to comparisons of the averages of two subsets of the k populations, i.e. to

$$\varphi_j = \frac{\Sigma_{i'} \mu_{i'}}{k_1} - \frac{\Sigma_{i''} \mu_{i''}}{k_2} \qquad (j = 1, 2, \ldots) \tag{40}$$

where i' runs over k_1 populations and i'' over k_2 populations ($1 \leq k_1, k_2 < k$ and $k_1 + k_2 \leq k$). Nevertheless these contrasts seem to be the only contrasts of interest to an experimenter.[14] The MCP are similar insofar as they are based on ranking observations and on the asymptotic distribution of the test statistic. We shall restrict our discussion to Dunn's procedure which is quite simple as it relies on the Bonferroni inequality; it runs as follows.[15]

Decide which m contrasts will be examined. Rank the $N_j = \Sigma_{i'} n_{i'} + \Sigma_{i''} n_{i''}$ observaticns involved in contrast φ_j, from smallest to largest. Per population we calculate the rank sums, $\underline{r}_{i'j}$ and $\underline{r}_{i''j}$. (An index j is used with $\underline{r}_{i'j}$ and $\underline{r}_{i''j}$ since the ranks sums vary over the contrasts; for simplicity of presentation j is not used with $n_{i'}$ and $n_{i''}$.) Calculate the difference in mean ranks between the two subsets, i.e.

$$\underline{d}_j = \frac{\Sigma_{i'} \underline{r}_{i'j}}{\Sigma_{i'} n_{i'}} - \frac{\Sigma_{i''} \underline{r}_{i''j}}{\Sigma_{i''} n_{i''}} \qquad (j = 1, 2, \ldots) \tag{14}$$

The variance of $\underline{d}_j$ is

$$\text{var}(\underline{d}_j) = \sigma_j^2 = \left[\frac{N_j(N_j + 1)}{12}\right] \left(\frac{1}{\Sigma n_{i'}} + \frac{1}{\Sigma n_{i''}}\right) \qquad (42)$$

In the derivation of the distribution of the test statistic it is assumed that all k density functions are identical except for a possible shift in location. Under the null-hypothesis of equal means the $\underline{d}_j$ have an asymptotic multivariate normal distribution. Hence we can test each $\underline{d}_j$ individually using the two-sided critical values $z^{\alpha/(2m)}$ of the table for $N(0,1)$. So, if

(i) $-z^{\alpha/(2m)} < \underline{d}_j/\sigma_j < z^{\alpha/(2m)}$ then do not reject the hypothesis of no difference between the means of the two subsets containing the k_1 and k_2 populations of the particular contrast φ_j.

(ii) $\underline{d}_j/\sigma_j > z^{\alpha/(2m)}$ then reject H_0 in favor of the hypothesis that the mean of the k_1 populations is larger than that of the k_2 populations.

(iii) $\underline{d}_j/\sigma_j < -z^{\alpha/(2m)}$ then accept the hypothesis that the mean of the k_1 populations is smaller.

An application of this test is given in Dunn (1964, pp. 243-247). As we remarked Tobach et al. (1967) derived a variant of Dunn's procedure especially for paired comparisons.

Linear Functions

As we mentioned before linear functions of the means, say, $\Psi_j = \Sigma_i c_{ij}\mu_i$, become relevant if, besides contrasts in the means, we want to study the means themselves. Scheffé's S-technique is very versatile since it also applies to linear functions. All we have to do is to replace $(k - 1)$ by k in the definition of S in Eq. (39). The rationale is that the linear functions of the k means span a space of k dimensions, while linear contrasts are restricted by one side-condition, viz. $\Sigma_i c_{ij}\mu_i = 0$; see Scheffé (1964, p. 70).

Tukey's method for pairwise comparisons can be generalized to linear contrasts and linear functions of the means, but these generalized MCP most times give wider confidence intervals than Scheffé's technique [see Miller (1966, pp. 39-41, pp. 43-44), and Scheffé (1964, pp. 74-79)]. An obvious alternative is to apply parametric or nonparametric techniques per linear function and to follow the Bonferroni approach. If we are not interested in means plus contrasts but only in the means themselves then other techniques are available as we shall see next.

Means Only

Studentized Maximum Modulus

First we define the studentized maximum modulus. Let $\underline{z}_j$ $(j = 1, \ldots, w)$ be w independent observations from $N(0, \sigma_z^2)$ and let $\underline{s}_v^2$ be an unbiased estimator of σ_z^2 with v degrees of freedom and independent of each $\underline{z}_j$. Then the maximum modulus of the $\underline{z}_j$ is $\max_j |\underline{z}_j|$ and consequently the studentized maximum modulus is

$$\underline{m}_{w,v} = \frac{\max_j |\underline{z}_j|}{\underline{s}_v} \quad (j = 1, \ldots, w) \tag{43}$$

Only limited tables of $\underline{m}_{w,v}$ are available. Pillai and Ramachandran tabulated the upper 5% point for $w = 1(1)8$ and v between 5 and ∞; their table is reproduced in Miller (1966, p. 239). Some more critical values were calculated by Dunn and Massey (1965). Since in MCP high values of the experimentwise error rate may be used, we have reproduced those values which Dunn and Massey calculated for the upper 10% and 20%; see Table 1.[16)]

Next consider a one-way layout with k populations, each population yielding n_i normally distributed independent observations with a common variance σ^2 and mean μ_i $(i = 1, \ldots, k)$.

TABLE 1

Upper α Point of Studentized Maximum Modulus

	v							
	4		10		30		∞	
	α		α		α		α	
w	.10	.20	.10	.20	.10	.20	.10	.20
2	2.66	2.01	2.19	1.76	2.03	1.66	1.95	1.62
6	3.51	2.74	2.77	2.33	2.50	2.17	2.38	2.09
10	3.89	3.06	3.03	2.58	2.71	2.39	2.56	2.29
20	4.38	3.47	3.33	2.90	2.98	2.66	2.79	2.54

Hence the $(\bar{\underline{x}}_i - \mu_i)\sqrt{n_i}$ are independent observations from $N(0,\sigma^2)$. An unbiased estimator of σ^2 is the usual pooled estimator $\underline{s}_v^2$ with $v = \Sigma(n_i - 1)$ degrees of freedom which is independent of $(\bar{\underline{x}}_i - \mu_i)\sqrt{n_i}$. So

$$\frac{\max_i |\bar{\underline{x}}_i - \mu_i| \sqrt{n_i}}{\underline{s}_v} = \underline{m}_{k,v} \qquad (44)$$

and the following statements have a joint confidence level of $(1-\alpha)$.

$$|\mu_i - \bar{\underline{x}}_i| \leq m^{\alpha}_{k,v}\underline{s}_v/\sqrt{n_i} \qquad (i = 1, \ldots, k) \qquad (45)$$

Observe that the studentized maximum modulus can be applied to any set of k estimators, e.g. k estimated regression coefficients $\hat{\underline{\beta}}_i$, provided that these estimators are independent and an independent estimator of their common variance is available. We shall return to this type of application. An extension to unequal variances but known variance ratios is possible. The studentized maximum modulus can also be used for linear functions, but tends to give longer confidence intervals than other parameteric procedures [see Miller (1966, p. 73) and Scheffé (1964, p. 72)].

Independent t-Statistics

If the k statistics $\bar{\underline{x}}_i$ have unequal, unknown variances σ_i^2 we cannot apply the studentized maximum modulus. If the individual observations are independently and normally distributed then we can estimate σ_i^2 by

$$\underline{s}_i^2 = \frac{\Sigma_{g=1}^{n_i} (\underline{x}_{ig} - \bar{\underline{x}}_i)^2}{(n_i - 1)} \tag{46}$$

which has $v_i = n_i - 1$ degrees of freedom and is independent of $\bar{\underline{x}}_i$ (and $\bar{\underline{x}}_{i'}$ and $\underline{s}_{i'}^2$, $i' \neq i$). Hence we have the following k independent t-statistics, based on v_i degrees of freedom.

$$\underline{t}_{v_i} = \frac{\bar{\underline{x}}_i - \mu_i}{\underline{s}_i/\sqrt{n_i}} \quad (i = 1, \ldots, k) \tag{47}$$

So two-sided confidence bounds with a joint confidence level of $1-\alpha$ are realized if we select critical constants a_i in such a way that the individual t-statistics satisfy

$$P(|\underline{t}_{v_i}| \leq a_i) = (1 - \alpha)^{1/k} \tag{48}$$

a_i in Eq. (48) is identical to the upper $[\frac{1}{2} + \frac{1}{2}(1 - \alpha)^{1/k}]$ critical point in the t-table for v_i degrees of freedom. For equal degrees of freedom $v_i = v$ these critical points were tabulated by Dunn and Massey (1965) for some values of v and k. Their values are reproduced in Table 2 for α is 10% and 20%. (For α is 0.01, 0.025, 0.05, 0.30, 0.40, and 0.50 we refer to Dunn and Massey.) Other critical values for any v_i and k can be calculated from a detailed t-table possibly using the approximation for t_v^α in Eq. (13). Comparison of Tables 1 and 2 shows that Table 2 yields wider confidence intervals (the standard deviation cannot be based on a pooled estimator).

TABLE 2

Upper $[\frac{1}{2} + \frac{1}{2}(1 - \alpha)^{1/k}]$ Point of t-Distribution

	v							
	4		10		30		∞	
	α		α		α		α	
k	.10	.20	.10	.20	.10	.20	.10	.20
2	2.75	2.08	2.21	1.78	2.03	1.67	1.95	1.62
6	3.91	3.09	2.85	2.41	2.52	2.19	2.38	2.09
10	4.54	3.64	3.14	2.71	2.71	2.41	2.56	2.29
20	5.52	4.47	3.55	3.11	3.01	2.71	2.79	2.54

Subset Containing Best Population

Above we considered several procedures where means are compared with each other or with a standard. These MCP yield confidence bounds for $\mu_i - \mu_{i'}$ or $\mu_i - \mu_0$ with a prescribed joint confidence level $1 - \alpha$. In a testing approach we want to decide if the means are different or not, or in what direction they are different. Then the probability that one or more significances are false is fixed to be α, provided the null-hypothesis of equal means holds. The experimentwise error rate is not controlled under an alternative hypothesis, i.e. the power is not controlled (and often difficult to calculate) [see Miller (1966, pp. 102-107)]. Often what we like to find out is which system has the highest mean. The above MCP may be used to decide which populations are significantly better; compare, e.g. Eq. (18) and the other confidence intervals. (If the confidence interval does not contain zero then a significant difference results.) However, there are procedures explicitly aimed at selecting with prescribed probability a subset from the total set of k populations such that this subset contains the best population (or contains all populations not worse than the standard if there is a standard population). Or

$$P(CS) \geq P^* \tag{49}$$

where P^* is the prescribed probability and CS stands for "correct selection," i.e. the subset does contain the best population. Relation (49) holds for all configurations of the means; the equality sign holds if all means are equal (compare H_0 in the above MCP); otherwise the inequality sign holds. The subset has a stochastic size, say $\underline{s}$. The $\underline{s}$ populations in the subset are not ranked so that for the selection of the best population additional experimentation is required (or the selection must be based on imponderables). The advantage of the subset approach is that after the pilot phase, where n observations from each population are taken, $k - \underline{s}$ inferior populations can be eliminated. In Part C we shall present so-called multiple ranking procedures which prescribe how many observations should be taken from each population in order to select the best population with prescribed confidence; yet in the preliminary phase of an investigation it makes sense to take a fixed pilot number of observations from each population. For a discussion on the subset approach versus the above MCP and ranking procedures we also refer to Desu and Sobel (1968, p. 402), Gupta (1965, pp. 225-227), Gupta and Santner (1972), Gupta and Sobel (1958, pp. 235-236), and Miller (1966, pp. 226-229).

If the performance of a system is measured by the mean response then we shall look for the system with the highest mean (or lowest mean depending on the problem, compare, e.g. high profit or low cost). Another criterion may be the variance, the performance increasing as the variance decreases; see e.g. the simulation study on national income by Naylor et al. (1968), where an economic policy is to be selected that minimizes the fluctuations of national income. We shall return to this point later.

Strangely enough most textbooks do not cover the subset approach. (Even in the extensive survey by Miller it is only briefly mentioned.) Yet this approach seems very useful in pilot studies and we shall, therefore present some procedures of this type, viz. a procedure for selection of the best population and one for comparisons with a standard, assuming normality (parametric procedure) or not (distribution-free procedure).

Gupta's Parametric Procedure

Gupta (1965, pp. 235-236) gives a procedure for k normal populations with unknown means μ_i and a common unknown variance σ^2. From each population an equal number (n) of independent observations is available. This yields the sample averages $\bar{\underline{x}}_i$ $(i = 1, \ldots, k)$ and the pooled estimator $\underline{s}_v^2$ with $v = k(n-1)$ degrees of freedom. His procedure selects a subset which contains the population with the highest mean with probability at least P^*, P^* being specified by the experimenter (as α was specified for the above MCP). Population i is included in the subset if

$$\bar{\underline{x}}_i \geq \bar{\underline{x}}_{max} - D\underline{s}_v / \sqrt{n} \qquad (50)$$

where the constant D depending on k, n and P^* is tabulated in Gupta and Sobel (1957, pp. 962-964) while for other k and n values of $D/\sqrt{2}$ are given in Dunnett (1955, pp. 1117-1118) and Krishnaiah and Armitage (1966, pp. 41, 51).[17] If we want to find the smallest mean then we can put $\underline{x}' = -\underline{x}$ and apply the rule to $\underline{x}'$. The derivation of the procedure hinges on the fact that $P(CS)$ reaches its minimum value if all means are equal; for equal means $P(CS)$ is a rather simple integral which is set equal to P^* and is solved for D.

Gupta (1965, pp. 229-232) studied the properties of a procedure of this type and found the following.

(i) If all parameters (μ_i) are equal then $P(CS)$ attains its minimum value P^* and the expected subset-size, $E(\underline{s})$, its maximum P^*k.

(ii) If $\mu_i \geq \mu_{i'}$ then the probability that population i is contained in the subset is not smaller than that of population i'.

(iii) An (unattractive) procedure for guaranteeing the inclusion of the best population in the subset is putting all k populations in the "subset." Gupta's procedure selects a subset of size $\underline{s}$ where $\underline{s}$ may be smaller than k (since $1 \leq s \leq k$). The probability of including the best population in a subset of given size s (s not known beforehand) is maximized by Gupta's procedure.

There are also techniques for selecting a subset containing the smallest variance σ_i^2 of k variables with distributions $N(\mu_i, \sigma_i^2)$, or the smallest scale parameter of gamma distributions, or the smallest binomial parameter p_i of binomially distributed variables [see Gupta (1965), Gupta and Nagel (1971) and Gupta and Sobel (1960, 1962)]. Instead of these special procedures we may use Eq. (50) hoping that this rule is robust enough; first we may apply a (logarithmic or arcsin) transformation to the original data (see Table 4 in Chap. IV). Goel (1972) examines the subset approach for Poisson distributions. Recently Gupta and Panchapakesan (1972) gave an extensive survey on the subset approach for parametric (normal, gamma, binomial, Poisson, negative binomial), nonparametric, multivariate, decision-theoretic and Bayesian situations.

Desu and Sobel (1968) give another subset approach. Unlike Gupta's rule their technique yields a <u>subset-size</u> that is not stochastic but is <u>known</u> in advance. They assume a common and known variance σ^2 and use an "indifference zone," i.e. only if the best mean $\mu_{(k)}$ is δ^* units better than the other means, the probability of CS is guaranteed; if $\mu_{(k)}$ is less than δ^* units better we do not care. So

$$P(CS) \geq P^* \quad \text{if} \quad \mu_{(k)} - \mu_{(k-1)} \geq \delta^* \tag{51}$$

We prefer Gupta's procedure since Desu and Sobel's approach requires a known variance and specification of δ^*. We mention their procedure here since it <u>might</u> be used for the selection of the <u>sample size</u> n in Gupta's rule. For if n is small then $\underline{s}$ may have the value k, i.e. after the pilot phase no population can be eliminated. Therefore, we may specify δ^* (and P^*) and the guessed value of σ^2 and use Desu and Sobel's table to compute the value of n that makes s smaller than k [see Desu and Sobel (1968, pp. 404-407)]. Their study has been extended by Gupta and Santner (1972); the latter authors also show the relationships between the subset approach (with either $\underline{s} \leq k$ or $\underline{s} < k$) and the sample size selection approach discussed in part C of this chapter. Sobel (1969) also generalized the Gupta (1965) and Desu and Sobel (1968) procedures and derived a number of rules for selecting the t (≥ 1) best populations, with or without an indifference zone.

Rizvi and Sobel's Nonparametric Procedure

Rizvi and Sobel (1967) developed a nonparametric procedure for selecting with probability of at least P^*, a subset containing the population with the highest median. They assume only that there is an (odd) equal number (n) of independent observations from each of the k populations; the populations have continuous distributions which may have different forms, i.e. the populations are not required to differ only in location [also compare the companion paper by Sobel (1967, p. 1804)]. We derived from Rizvi and Sobel (1967, p. 1789) that we can apply their rule as follows.

(i) Per population arrange the observations in increasing order, i.e. obtain the order statistics $\underline{x}_{(1)i} < \underline{x}_{(2)i} < \cdots < \underline{x}_{(j)i} < \cdots < \underline{x}_{(n)i}$ $(i = 1, \ldots, k)$.

(ii) Find the value of the critical constant, say a,[18] in Table 3 in Rizvi and Sobel (1967, pp. 1801-1802).

(iii) For each population determine the ath order statistic, $\underline{x}_{(a)i}$ $(i = 1, \ldots, k)$, and find the maximum of these k values,

$$\underline{x}\ \max_{(a)} = \max_i [\underline{x}_{(a)i}] \tag{52}$$

(iv) Include population i in the subset if

$$\underline{x}_{(r)i} \geq \underline{x}\ \max_{(a)} \tag{53}$$

with $r = (n+1)/2$, i.e. $x_{(r)i}$ is the sample median.

If we are looking for the smallest median, then, as we derived from Rizvi and Sobel (1967, p. 1799), the rule becomes:[19] Include population i in the subset if

$$\underline{x}_{(r)i} \leq \underline{x}\ \min_{(n+1-a)} \tag{54}$$

with

$$\underline{x}\ \min_{(n+1-a)} = \min_{i}[\underline{x}_{(n+1-a)i}] \tag{55}$$

and as above $r = (n+1)/2$ and a is found in Rizvi and Sobel's Table 3.

For high values of P^* and small values of n it is known _a priori_ that the subset will consist of all k population.[20)] Rizvi and Sobel's Table 2 shows which minimum value n should have for a specified P^* to prevent such degeneracy of the procedure. Analogous to Gupta's parametric procedure, Rizvi and Sobel (1967, pp. 1790-1792, 1800) prove that the maximum expected value of $\underline{s}$ is reached when all k distributions are the same and this maximum is approximately equal to kP^*. A population with a higher median has a higher chance of being included in the subset containing the highest median. Rizvi and Sobel (1967, pp. 1792-1798) studied the _asymptotic_ efficiency of their rule and found that for normal populations with a common known variance their rule is (obviously) inefficient compared with Gupta's parametric procedure, but for exponential distributions their rule is much more efficient (by factor 2).

Other Procedures

Bartlett and Govindarajulu (1968) and Patterson (1965) give nonparametric procedures but they assume populations identical except for a possible location shift. Barlow and Gupta (1969) and Gupta and McDonald (1970; 1972) present distribution-free procedures assuming "stochastically increasing" distributions, i.e. $F_1(x) \geq F_2(x)$ for all x if the parameters Θ satisfy $\Theta_1 < \Theta_2$ (cf. normal populations with $\mu_1 < \mu_2$ and common variance). Barlow and Gupta (1969) compare the efficiency of their procedure with that of Rizvi and Sobel; no procedure is uniformly better for the class of distributions they study. A subset procedure for normal populations with unequal variances has been derived by Dudewicz (1972) and Dudewicz and Dalal (1971). Since it is a double-sampling approach we shall discuss it in Part C.

Subset Containing All Populations Better than the Standard

Gupta and Sobel's Parametric Procedure

Gupta and Sobel (1958) derived a procedure assuming n independent observations from each of $(k-1)$ experimental populations and one standard population; all populations have normal distributions with a common unknown variance σ^2. (The authors also give variants for known, possibly different variances σ_i^2 and for a known standard mean μ_0.) They include a population i in the subset if

$$\underline{\bar{x}}_i \geq \underline{\bar{x}}_0 - D\underline{s}_v/\sqrt{n} \tag{56}$$

where $\underline{s}_v^2$ is the pooled estimator of σ^2 with $v = k(n-1)$ degrees of freedom; D is the critical constant depending on P^*, k and v and is tabulated in Gupta and Sobel (1957, pp. 962-964), while $D/\sqrt{2}$ is also given by Dunnett (1955, pp. 1117-1118) and Krishnaiah and Armitage (1966, pp. 41, 51) for other values of k and v. [Read the value for D from the column corresponding with $p = k-1$; the critical constants in Eqs. (56) and (50) are found in the same tables, one parameter being the total number of populations including a possible standard population.] Observe that the subset-size may run from 0 to k-1.

Note that Eq. (17) shows that Dunnett (1955) would conclude that population i is better than the standard if

$$\underline{\bar{x}}_i \geq \underline{\bar{x}}_0 + D\underline{s}_v/\sqrt{n} \tag{57}$$

where the critical constant D is so selected that the control population is chosen if all populations have the same mean, i.e. if the null-hypothesis holds. Comparing Eq. (57) with Eq. (56) shows that in Dunnett's approach the hypothesis of all experimental means being <u>not</u> better than the standard, is rejected only if there is strong evidence against H_0. ($\bar{\underline{x}}_i$ must be better than $\bar{\underline{x}}_0$ <u>plus</u> a factor $D\underline{s}_v/\sqrt{n}$.) In Gupta and Sobel's approach there is no H_0 favoring the standard, so that more populations are included in the subset.

Rizvi, Sobel, and Woodworth's Nonparametric Procedure

Rizvi et al. (1968, pp. 2076-2077) give a procedure analogues to the Rizvi and Sobel (1967) procedure for subset selection in case of no control population. One assumption is more restrictive, viz. the assumption of stochastically ordered distributions. So they assume that distribution i is either stochastically better [$F_i(x) \leq F_0(x)$ for all x] or worse. (Observe that this restriction is less strict than the assumption of distributions differing only in location.) The procedure runs analogous to the Rizvi-Sobel procedures.

(i) Arrange the observations in increasing order per population, i.e. obtain the order statistics $\underline{x}_{j(i)}$ $(j = 1, ..., n)$.

(ii) Find the critical constant a (equal to r - c in their notation) in Rizvi et al. (1968, p. 2079) or Rizvi and Sobel (1967).

(iii) Put population i in the subset if $\underline{x}_{(r)i} \geq \underline{x}_{(a)0}$ where $r = (n + 1)/2$.

Observe that step (iii) is the "natural" alternative for steps (iii) and (iv) in the Rizvi-Sobel procedure. Again degeneracy of the procedure is possible, i.e. all experimental populations may be included in the subset. If population i is stochastically better than population j then it is more likely that i is selected in the subset. If we want to include in the subset medians <u>smaller</u> than the control median then analogous to Eq. (54) we include population i if $\underline{x}_{(r)i} \leq \underline{x}_{(n+1-a)0}$.

Other Procedures

Rizvi et al. (1968, p. 2086) also derive a procedure based on <u>Steel's ranking</u> statistic for comparisons between the experimental and the control means discussed earlier. This alternative procedure again assumes stochastically ordered distributions. For this class of distributions the procedure is less efficient but for a smaller class (e.g. location shift only) it may be more efficient [see Rizvi et al. (1968, p. 2087)]. They also derived an asymptotic procedure based on sample means, assuming stochastically ordered distributions (plus a technical condition concerning the variance and fourth central moment of the control population). This procedure works only for large sample sizes; it may be more or less efficient than the other nonparametric procedures.

Gupta and Sobel (1958, pp. 239-244) also give parametric procedures for variances σ_i^2 (or more generally for scale parameters of gamma distributions) and for binomial parameters p_i. Alternatively we may decide to use Eq. (56) after a logarithmic or arcsine transformation. The subset approach in case of a standard is also discussed by Gupta (1965, pp. 234-235, 243-244). Desu (1970) gives a procedure for selecting a subset not containing all the inferior populations but this seems a less attractive problem formulation. Gupta and Panchapakesan (1968; 1972) list literature on the subset approach in multivariate and other situations. Procedures for sample size determination in situations involving a control will be discussed in Part C.

Multi-Factor Designs

Introduction

Until now we have limited our discussion to the situation with a single factor at k levels, e.g. k different machines or k queuing rules. Next we shall examine the case of more than one factor, e.g. a simulation experiment where one factor is type of queuing discipline, another factor is type of service time distribution, etc. For, say, three factors we can study main effects, two-factor interactions, and three-factor interactions; if factor C is at c levels then there are c main effects of factor C, viz.

$\alpha_i^c = \eta_{..i} - \eta_{...}$ $(i = 1, \ldots, c)$ with $\Sigma \alpha_i^c = 0$, etc. (see Chap. IV). We may wish to realize a predetermined probability that all statements on the experiment are correct. The price for a fixed experimentwise error rate is wide confidence intervals, or low power for individual effects. Therefore we may decide to use familywise error rates instead of a single, overall, experimentwise error rate. For example, if there are three factors each at many levels, all statements on the main effects of one particular factor may be considered as constituting one family of statements and we want to be $(1 - \alpha)$ sure that all statements of that family are correct. If, however, the factors have only a small number of levels, then the statements on the main effects of all factors may be seen as one family. So the formation of families is subjective. Observe that the various familywise error rates for a single experiment need not have the same values. For we can take a higher error rate for the family of statements on main effects and a lower rate for the interactions if we are more anxious to discover main effects. For a discussion on families of statements we further refer to Miller (1966, pp. 10-12, 31-35) and also to Kurtz et al. (1965, p. 154), Tukey (1953), and Sec. 2 above. A case study involving the creation of families of statements will be presented in Chap. VI. We shall now discuss some techniques for the realization of familywise or even experimentwise error rates in an experiment with more factors.

General Procedure: Bonferroni

If we use an error rate α_i for statement i $(i = 1, \ldots, m)$ then the error rate for the family of m statements does not exceed $\Sigma \alpha_i$ as the Bonferroni inequality in Eq. (10) stated. If the familysize m is not too high then this is a very useful technique because of its simplicity and versatility. (Also see our discussion of the Bonferroni procedure in Sec. 3.) For the statements may be dependent. An example is an ANOVA where each statement is an F-test on the significance of the main effects of factors A, B, etc., of the two-factor interactions, etc. All these F-tests have a common denominator, viz. the mean square for error. [The numerators will be independent in an orthogonal design (see Chap. IV).] Another example is provided by the estimators of the main effects of a particular factor, say $\underline{\hat{\alpha}}_i^A$, which are negatively correlated (see Exercise 8). The individual observations may be nonnormal as we can apply nonparametric procedures per statement, each with an error size α_i. We may even use different procedures for different families! For example, Tukey's studentized range can be applied for main effects and Scheffé's procedure for interactions. Part of the total error rate may be reserved for Scheffé's MCP as his technique makes it possible first to examine the data and next to decide which comparisons will be made ("data snooping"). However, since the remaining α for Scheffé's procedure will be small, very wide confidence intervals will result [see Miller (1966, p. 62) and Scheffé (1964, p. 80)]. Dunn (1961, pp. 61-63) gives an application of the Bonferroni approach in a two-way layout.[21)] We shall also apply this technique in the case-study of Chap. VI.

Multiple F-Tests

It is well-known that in ANOVA of higher-way layouts possible main effects and interactions are tested by F-statistics. As an example consider a two-way layout with factor A at a levels, B at b levels and n observations per combination of levels. To test the hypothesis that factor A has no main effect, i.e.

$$H^A: \alpha_1^A = \cdots = \alpha_a^A = 0 \tag{58}$$

we divide the mean square for A by the mean square for error and compare this ratio with the upper α-point of the F-statistic with degrees of freedom $(a - 1)$ and $(n - 1)ab$. To test the hypothesis that both A and B have no main effect, i.e.

$$H^{A+B}: \alpha_1^A = \cdots = \alpha_a^A = \alpha_1^B = \cdots = \alpha_b^B = 0 \tag{59}$$

we pool the sums of squares of A and B, calculate their mean square, divide by the mean square for error, and compare with the upper α-point of the F-statistic with degrees of freedom $(a + b - 2)$ and $(n - 1)ab$. This procedure can be found in any book on ANOVA; also in Chap. IV, Eq. (33).

We have to realize that the hypothesis H^{A+B} in Eq. (59) is a single (composite) hypothesis. Now consider the multiple hypotheses

$$H^A : \alpha_1^A = \cdots = \alpha_a^A = 0 \tag{60}$$

$$H^B : \alpha_1^B = \cdots = \alpha_b^B = 0 \tag{61}$$

possibly extended with the hypothesis on interactions

$$H^{AB} : \alpha_{11}^{AB} = \alpha_{12}^{AB} = \alpha_{21}^{AB} = \cdots = \alpha_{ab}^{AB} = 0 \tag{62}$$

Each individual hypothesis can be tested applying the F-tests traditional in ANOVA. Obviously the experimentwise error rate, i.e. the probability of falsely rejecting a hypothesis, is now higher than α; α being used for each individual F-test. We may wish to keep the experimentwise error rate below a level α_E. Then we may apply the Bonferroni inequality, i.e. each of the three F-tests uses $\alpha = \alpha_E/3$.[22] A second technique is based on an improved Bonferroni inequality which will be presented next.

Miller (1966, pp. 101-102) gives the following lemma which goes back to a lemma proved by Kimball. Suppose there are k ratios $\underline{U}_i = \underline{V}_i/\underline{W}$ $(i = 1, \ldots, k)$, all $\underline{V}_i$ and $\underline{W}$ being independent and $\underline{W}$ having only positive values. Define the constants $-\infty \leq a_i \leq 0$ and $0 \leq b_i \leq \infty$ $(i = 1, \ldots, k)$. Then from Kimball (1951) it follows that

$$P(a_i < \underline{U}_i < b_i \quad \text{for all} \quad i) \geq \prod_{i=1}^{k} P(a_i < \underline{U}_i < b_i) \tag{63}$$

Observe that this inequality is better than the Bonferroni inequality since

$$\prod_{i=1}^{k} (1 - \alpha_i) > 1 - \sum_{i=1}^{k} \alpha_i \quad \text{for} \quad 0 < \alpha_i < 1,\ k > 1 \qquad (64)$$

The improved inequality can be applied in an orthogonal layout where main effects and interactions among 2, ..., k factors have independent (or completely confounded) mean squares $\underline{MS}_i$ (or $\underline{V}_i$ in the above lemma), and are tested against the independent mean square for error $\underline{MSE}$ (or $\underline{W}$).[23)]

Next we shall study multiple F-tests in a special type of designs, viz. 2^{k-p} designs. In Chap. IV we saw that each effect in a 2^{k-p} design can be tested individually by applying F-tests with one degree of freedom for the numerator and v degrees of freedom for the denominator. (Notice that statements on effects in a 2^{k-p} design can be classified as multiple F-tests or as tests on individual effects, since $\underline{F}_{1,v} = \underline{t}_v^2$, where $\underline{t}_v^2$ is applied in two-sided confidence intervals on one parameter or the difference between two parameters.) Since in our opinion 2^{k-p} designs are very useful in simulation, we shall show that it is simple to derive simultaneous tests for effects in these designs. As we saw in Chap. IV in a 2^{k-p} design ($p \geq 0$) main effects and interactions (but not the grand mean) are estimated by

$$\hat{\underline{\alpha}}_j = \frac{2}{N} \sum_{i=1}^{N} x_{ij} \underline{y}_i \qquad (j = 1, \ldots, J) \qquad (65)$$

Let σ^2 be the variance of the individual normally distributed independent observations $\underline{y}_i$. Since the design is orthogonal all estimated effects are independent so that

$$\hat{\underline{\alpha}}_j : \text{NID}\left(\alpha_j, \frac{4\sigma^2}{N}\right) \qquad (66)$$

Consequently we can use the studentized maximum modulus for obtaining the simultaneous confidence intervals

$$|\hat{\underline{\alpha}}_j - \alpha_j| \leq m_{J,v}^{\alpha} \frac{2}{\sqrt{N}} \hat{\underline{\sigma}} \qquad (j = 1, \ldots, J) \qquad (67)$$

A second procedure uses the maximum of the J F-statistics, each F-statistic having one degree of freedom in its numerator. This maximum F-statistic has been calculated by Nair (1948, p. 26) for α 1% and 5% and is also tabulated in Pearson and Hartley (1966, p. 176). The maximum F-statistic and the (squared) maximum modulus give about the same value. (No statistic is uniformly better; we checked for $J = 2, 8$, $v = 10, \infty$, and $\alpha = 0.05$ using the tables in Miller (1966, p. 239) and Pearson and Hartley (1966, p. 176).) A third alternative is the improved Bonferroni inequality (63) discussed for general designs above. The fourth technique yielding wide confidence intervals, would be Scheffé's S-procedure with $q = J$. This technique is applicable since, as we say in Chap. IV, the 2^{k-p} design corresponds with the linear regression model

$$\underline{y}_i = \gamma_0 + \sum_{j=1}^{J} \gamma_j x_{ij} + \underline{e}_i \qquad (i = 1, \ldots, N) \qquad (68)$$

where $\gamma_j = \alpha_j/2$. Scheffé's technique will be discussed in more detail for the general case of an arbitrary layout to which we proceed now.

Individual Effects

In Eqs. (58) through (64) we saw how we can test whether a factor has a main effect or whether interactions among factors exist. This may be a first step in the analysis. Usually it is more interesting to find which individual effects are important and hence implied rejection of one or more hypotheses tested by the F-statistics, e.g. the hypothesis of no main effect of A may be rejected because a particular level of A produces an outlier. For an experiment with a single factor we have already discussed several techniques for investigating means individually, contrasts among means, etc. For a higher-way layout the Bonferroni inequality is a useful technique [see e.g. Dunn (1961)]. Another technique is Scheffé's S-procedure. Equation (38) showed how his procedure can be applied to a one-way layout. Actually the technique is even more general.

For Scheffé (1964, pp. 68-70) proved that the procedure is applicable to any linear model

$$\vec{\underline{y}} = \vec{X}'\vec{\beta} + \vec{\underline{e}} \tag{69}$$

where $\vec{\underline{e}}$ is the vector of normally distributed error terms with covariance matrix $\sigma^2\vec{I}$. [See Miller (1966, p. 53) for $\vec{\underline{e}}$ with covariance matrix $\sigma^2\vec{\Sigma}$ with $\vec{\Sigma}$ known and σ^2 unknown.] Suppose that we want to investigate a "q-dimensional space L of estimable functions generated by a given set of independent estimable functions $(\psi_1, \ldots, \psi_q)$."[24] For instance, the k means in a one-way layout yield $q = k$ and in a two-way layout $q = ab$ while the contrasts among the ab means give $q = ab - 1$ (minus one since there is one constraint $\Sigma c_i = 0$). Then the probability is $(1 - \alpha)$ that Eq. (70) simultaneously holds for all ψ in the q-dimensional space L.

$$|\hat{\underline{\psi}} - \psi| \le S\hat{\underline{\sigma}}_\psi \tag{70}$$

with

$$S = (qF^{\alpha}_{q,v})^{1/2} \tag{71}$$

v being the degrees of freedom of the estimator of σ^2. For contrasts among the k means in a one-way layout Eqs. (70) and (71) are equivalent to Eqs. (38) and (39). Application of the technique in, say, a two-way layout would run as follows. If we want to investigate the main effects of factor A then there are a main effects $\alpha_i^A = \eta_{i..} - \eta_{...}$ $(i = 1, \ldots, a)$ which satisfy one side-condition $\Sigma \alpha_i^A = 0$ [see Sec. 2 in Chap. IV]. Hence if we want simultaneous confidence intervals for the main effects of A we put $q = a - 1$ in Eq. (71). Instead of investigating main effects $\eta_{i..} - \eta_{...}$ we can look at the a row means $\eta_{i..}$ and compare these row means with each other, i.e. we look at the contrasts $\eta_{i..} - \eta_{i'..}$ $(i < i')$ and this again yields $q = a - 1$. If we also want confidence intervals for the row means $\eta_{i..}$ themselves then $q = a$. If the family of interest consists of the a main effects of A plus the b main effects of B then $q = a + b - 2$ (minus 2 since $\Sigma\, \alpha_i^A = 0$ and $\Sigma\, \alpha_j^B = 0$). Another family of interest may consist of the ab interactions α_{ij}^{AB}. These interactions satisfy $\Sigma_i\, \alpha_{ij}^{AB} = 0$ $(j = 1, \ldots, b)$ and $\Sigma_j\, \alpha_j^{AB} = 0$ $(i = 1, \ldots, a)$, together $a + b$ side-conditions; however, one side-condition is not independent (see Exercise 10). Hence, for the family of interactions $q = ab - (a + b - 1) = (a - 1)(b - 1)$. Other examples of the wide application range of Scheffé's technique can be found in Miller (1966, pp. 54-62) and Scheffé (1964, pp. 104, 110, 119, 273). Note that Puri and Sen (1971, pp. 308-318, 328-331) derived nonparametric asymptotic procedures for contrasts among the effects in a two-way layout.

Scheffé's technique is very versatile; the price to be paid is wide confidence intervals. Therefore, for special situations special techniques may be applied [see, e.g. Eq. (67)]. Such special situations also occur if we are not interested in main effects and interactions but in the cell means themselves and in some particular functions of these cell means, as we shall see now.

Individual Cell Means

If we are interested only in the cell means (rather than row and column means, interactions, etc.) then we can use the studentized maximum modulus since the cell means are independent with common variance σ^2/n [see Eqs. (43) through (45) above]. Functions of the cell means being contrasts, can be studied applying Tukey's studentized range [see Miller (1966, pp. 41-42) and Konijn (1959, pp. 62, 64) who gives a detailed example of the application of Tukey's test and the studentized maximum modulus in a two-way layout]. Comparisons between a control mean (i.e. some particular cell in the layout) and the experimental means can be based on Dunnett (1955). Konijn (1960, pp. 16-17) applied Dunnett's technique in a two-way layout [also see Dunnett (1964, pp. 485-486)].[25] The crucial point is that when proceeding to higher-way layouts the cell means remain independent. The appropriate estimator of the common variance of the cell means and its degrees of freedom follow from the ANOVA formulas, e.g. $v = (n - 1)ab$ in a two-way layout or, if no interactions are assumed and $n = 1$, $v = (a - 1)(b - 1)$, i.e. v is the degrees of freedom of the interaction mean square.

V.B.4. EFFICIENCY AND ROBUSTNESS OF MCP IN SIMULATION

Introduction

In Sec. V.B.3 we discussed a multitude of MCP. (We shall mention even more MCP in the next section.) In this section we investigate which procedures are useful under which circumstances in simulation. The application of MCP in simulation studies has been advised by Conway et al. (1959, p. 107), Fishman and Kiviat (1967, p. 28), and Naylor et al. (1967a, p. 1327). Two MCP, viz. Tukey's and Dunnett's techniques, were applied by Naylor et al. (1967b; 1968) to construct simultaneous confidence intervals for the differences $\bar{\underline{x}}_i - \bar{\underline{x}}_j$ and $\bar{\underline{x}}_i - \bar{\underline{x}}_0$ in their simulated multi-station queuing system and national economic system. We do not know any other simulation (or Monte Carlo) studies using MCP. This may be explained by the fact that in most simulation studies a formal statistical analysis is lacking and the fact that most textbooks on statistics do not discuss MCP so that these procedures are unknown to most simulation users.
are unknown to most simulation users.

For the selection of a particular procedure several factors are relevant. [For a different approach consult Gabriel (1969a) and O'Neill and Wetherill (1971, pp. 231-232).]

(i) The experimental objective and design. As we saw in the preceding section experiments may have different objectives, e.g. comparisons with a control may be wanted, or all pairwise comparisons, etc. A procedure suitable for a particular set of comparisons is also usable for a subset but not for a larger set. Hence if we want to make all pairwise comparisons then Dunnett's procedure is not applicable, Tukey fits exactly and Scheffé is also usable. A different type of objective is the selection of a subset containing the best population. The experimental designs may also differ, e.g. we may have a one-way layout or a higher-way layout and the number of observations per population may be the same or not.

(ii) The efficiency of the MCP. The efficiency can be measured by the lengths of the resulting confidence intervals. Under (i) we mentioned that a procedure developed for a particular set of comparisons may also be applied to a subset. However, MCP developed especially for that subset are more efficient since they can utilize the particularities of the subset. Hence Tukey is more efficient than Scheffé for pairwise comparisons but less efficient than Dunnett for comparisons with a control; Scheffé is worse than the studentized maximum modulus for the means themselves [see Scheffé (1964, pp. 75-79)]. Another type of efficiency comparisons can be made for parametric versus nonparametric procedures. As with nonmultiple comparisons the parametric MCP are more efficient if the assumed distribution indeed holds [also see Conover (1971, p. 281)]. As the deviations from the assumed distribution increase the efficiency of the parametric MCP decreases and the parametric procedures ultimately break down. Unfortunately the critical sizes of these deviations are unknown. Deviations from the assumed distribution lead us to the next factor.

(iii) The robustness of the MCP. By definition the robustness of a procedure increases as the procedure is less sensitive to violations of its underlying assumptions. The assumptions usually made in MCP are: normal distributions (except in nonparametric MCP), independent observations, and equal variances. [The procedures also assume a particular design, e.g. a one-way layout with equal sample sizes, and experimental objective; e.g. comparisons with a control, as we saw under (i).] Robustness is a difficult problem since there are many types and sizes of deviations from the underlying assumptions.

This may explain why the sensitivity of most statistics in MCP has not been investigated. Exceptions are the F- and t-statistics. The sensitivity of these two statistics is rather well known since they are much used in mathematical statistics and have been studied in some detail. These two statistics are robust when investigating means as we saw in Sec. 2 of Chap. IV. No studies, however, are available on the sensitivity of the other statistics so that our recommendations must be intuitive. Miller (1966, p. 108) states: "If one has to hazard a guess, one would guess that [the studentized range, the studentized maximum modulus and Dunnett's multiple t-statistics] were more sensitive to the assumptions [of normality, equal variances and independence] than the F-statistic. Each depends on an extreme statistic, i.e. the largest of a set of variables or the difference between the largest and the smallest of the set. The distribution of an extreme statistic is a good deal more sensitive to the form of the tails of the distribution (nonnormality), the largest variance (heteroscedasticity), and the interdependence between variables (dependence), than the distribution of a sum of squares... . The nonrobustness of these statistics is probably not catastrophic, but it is likely to be worse than for the F statistic." For a discussion of nonnormality and heteroscedasticity we also refer to Tukey (1953). In simulation and Monte Carlo experiments independent observations can be obtained by using different sequences of random numbers. An idea about the degree of nonnormality and heteroscedasticity can be gained by making frequency diagrams per population, n observations being taken per population. The usual tests for normality and homogeneous variances may be made.[26)] Common variances can be realized in the simulation of steady-state systems through appropriate run-lengths; transformation of the observations can also yield a common variance (see Chap. IV, especially Table 4). In case of nonnormality we might still want to apply a parametric procedure because such a procedure is the asymptotic equivalent of a permutation test[27)] under mild restrictions on the moments of the distribution [see Miller (1966, p. 181)].

We shall now examine which MCP are attractive (i.e. efficient and/or robust) for a particular experimental objective and design.

One-Way Layouts

(i) Comparisons With a Control

In Sec. 3 we discussed several MCP especially devised for confidence intervals or tests on $\mu_i - \mu_0$ $(i = 1, \ldots, k)$. Two MCP were shown to be most suitable, viz. Dunnett's parametric and Steel's nonparametric procedure. As in all other multiple comparisons situations the Bonferroni approach is a potential competitor. Dunnett's technique gives smaller confidence intervals than Tukey's or Scheffé procedure as the latter two procedures are meant for larger sets of contrasts. In general we favor Steel's testing procedure since as Miller (1966, pp. 146-147, 155) states "The only assumption imposed on the underlying distributions is that they have densities [i.e. no normality or common variances assumed]... For normal, or nearnormal, distributions the [Dunnett] t-test will perform more sharply, but the [Steel] rank test will not lag far behind... For nonnormal distributions the rank test is apt to be more efficient than the t test, in fact, very much more efficient... Note that for [Steel's procedure] the confidence intervals [for means as opposed to the tests for medians] are applicable only to rigid translations of the distributions. There can be no distortion of the distribution except for location change." See also Steel (1959a, p. 562). For a discussion of the assumption of equal variances in Dunnett's procedure we refer to the Appendix.

In those situations where no tables are available (e.g. because of unequal sample sizes or an experimentwise error rate higher than 5%) or where confidence intervals are desired we recommend a Bonferroni approach. If extreme nonnormality (like exponential distributions) does not occur the t-statistic can be applied per comparison; otherwise a suitable nonparametric statistic should be used per comparison. (Also see our discussion on the Bonferroni approach in Sec. 3.)[28]

(ii) All Pairwise Comparisons

We compare the following MCP in some detail: Tukey's and Scheffé's parametric MCP, Steel's and Peritz's nonparametric techniques, and the Bonferroni approach. Since Tukey's procedure was especially derived for the particular set of pairwise comparisons it yields smaller confidence intervals than Scheffé. Scheffé's procedure has the advantage of being known to be robust since the F-statistic is known to be insensitive to nonnormality and heteroscedasticity [see Scheffé (1964, pp. 75-78, 331-364)]. A serious competitor is a Bonferroni approach using individual t-tests. The t-statistic is known to be insensitive to nonnormality; in case of unequal variances we can apply, e.g. the Cochran and Cox t-statistic discussed in Sec. V.A.3; for dependent observations we can take the differences $\underline{d}_j = \underline{x}_{ij} - \underline{x}_{i'j}$ (j = 1, ..., n). (For seriously nonnormal observations we can use a nonparametric procedure in the Bonferroni approach.) Apart from its versatility the Bonferroni approach is quite efficient. Tables with detailed efficiency comparisons are given by Dunn (1961). She found that if all $k(k-1)/2$ pairwise comparisons are to be made, then Tukey's technique is more efficient than a Bonferroni t-approach [see Dunn (1961, Table 6)]. The efficiency of this Bonferroni approach compared with Scheffé's technique increases as the number (m) of statements becomes small relative to the number (k) of populations; we have reproduced her table 4 in our Table 3. If extreme nonnormality (e.g. exponential distributions) is suspected then Steel's nonparametric procedure is more efficient than Tukey's procedure; see Miller (1966, p. 155) and Sherman (1965). Miller (1966, p. 155) further states that Steel's test is a reasonable procedure to use for testing medians (as opposed to confidence intervals for means) when the distributions may differ in both location and other parameters (like variances). Later on, however, Gabriel and Lachenbruch (1969) found that Steel's procedure is very conservative for small samples. Peritz (1971) does not give tables for his ranking procedure but proves that (only) asymptotically a χ^2-table applies.

TABLE 3

Values of m, the Maximum Number of Statements on k Means, for Which Scheffé's Procedure Gives Longer Confidence Intervals than the Bonferroni t-Approach

	$1 - \alpha = .95$			$1 - \alpha = .99$		
k	$v = 7$	$v = 20$	$v = \infty$	$v = 7$	$v = 20$	$v = \infty$
2	0	0	0	0	0	0
3	2	3	3	3	3	4
4	5	7	9	6	10	13
5	10	17	24	12	23	36
6	16	33	55	20	46	99
7	26	63	129	32	104	241
8	37	110	281	49	19×10	59×10
9	53	189	614	71	38×10	14×10^2
10	71	30×10	126×10	100		316×10
11	95	56×10	27×10^2			696×10
12	123	89×10	56×10^2			149×10^2
13	158	14×10^2	108×10^2			310×10^2
14	190		223×10^2			694×10^2
15	26×10		426×10^2			150×10^3
16	30×10		872×10^2			312×10^3
17	38×10		182×10^3			66×10^4
18	45×10		329×10^3			12×10^5
19	54×10		635×10^3			26×10^5
20			132×10^4			6×10^6

Summarizing we have <u>Tukey</u>: most efficient if all assumptions hold; <u>Scheffé</u>: more robust than Tukey; <u>Bonferroni</u> parametric or non-parametric: robust and often more efficient; <u>Steel and Peritz</u>: no tables for small n.

(iii) Linear Contrasts

For linear contrasts the choice is rather simple as we discussed only one parametric procedure (Scheffé) and one nonparametric procedure (Dunn). Only for extreme nonnormality we recommend Dunn's ranking technique (based on Bonferroni). Otherwise we use the robust Scheffé procedure or a Bonferroni t-approach, depending on the results in Table 3 above.

(iv) Linear Functions

As with linear contrasts the choice is between Scheffé's technique and a Bonferroni approach. If the assumptions of normality, independence, and common variance hold then we can consult Table 3 to find out if a Bonferroni t-approach gives shorter confidence intervals than Scheffé. If extreme deviations from these assumptions are suspected then we can apply the Bonferroni approach with appropriate parametric or nonparametric, nonmultiple techniques per linear function.

(v) Means (or Regression Coefficients)

If the assumptions of normality, independence, and homogeneity of variances hold, then the studentized maximum modulus is most efficient. If the unknown variances are suspected to differ then we can use independent t-statistics;the t-statistics are insensitive to nonnormality. In case of extreme nonnormality we should use a nonparametric test. If the means or regression coefficients are dependent we can apply the Bonferroni inequality.

(vi) Subset Containing Best Population

If we want to select a subset containing the population with the largest (or smallest) median then Rizvi and Sobel's nonparametric procedure is most attractive since it does not require normality (by definition) or equal variances. If we think that the assumptions of normality and common variance hold then it is more efficient to use Gupta and Sobel's parametric procedure for selecting the population with the highest mean (equal to the median for normal populations). The robustness of the latter procedure is unknown. If the performance criterion of the systems is not the mean (but, e.g. the variance or binomial probability) then for simplicity we may rely on the above procedures, possibly after a transformation of the observations, or use a special technique (references for these techniques were given in Sec. 3).

(vii) Subset Containing all Populations Better than a Standard

Gupta and Sobel give a procedure for selecting a subset containing all populations with means not worse than the standard population. Their procedure assumes normality, independence, and a common unknown variance. Its robustness is unknown. If nonnormality is suspected we may try a transformation of the data. Nonnormality existing when variances or binomial probabilities are to be selected, can be remedied by applying procedures especially devised for these situations [see Gupta and Sobel (1958, pp. 239-244)]. Unequal unknown variances may be handled by a transformation of the observations or the following heuristic approach. Gupta and Sobel (1958) do give a procedure for different but known variances. Now we might simply replace σ_i^2 by $\underline{s}_i^2$ in their formulas, where $\underline{s}_i^2$ is an estimate based on prior knowledge, another pilot sample, or the observations used to estimate the means μ_i. However, in case of nonnormality and/or heteroscedasticity an attractive alternative is Rizvi, Sobel, and Woodworth's procedure.

Higher-Way Layouts

In a layout with more than one factor we may use: (i) an over-all test per type of effect (e.g. an F-test for the hypothesis that factor A has no main effect); (ii) after rejection of the hypothesis under (i) we may find out which individual effect (e.g. the pth main effect of A, α_p^A) caused this rejection; (iii) we may also be interested in cell means and comparisons among cell means (rather than, e.g. row means that measure main effects).

(i) Multiple F-Tests

In an orthogonal design (not necessarily a 2^{k-p} design) with observations showing normality, independence and a common variance we can use the improved Bonferroni inequality. If any of these four conditions is not satisfied then the numerator mean squares become dependent.[29] With nonnormal and/or heteroscedastic observations we can still apply an F-test per type of effect since this test is not sensitive to these two types of deviations from the assumptions. The dependence between the individual F-statistics can be handled by the simple Bonferroni inequality. Scheffé's technique is also applicable as long as we have a linear model; the F-statistic on which his technique hinges is insensitive to nonnormality and heteroscedasticity. His technique becomes less efficient as the dimensionality q of the linear space of effects increases; consult Table 3 with $k - 1 = q$ (e.g. for $q = 2$ read the row $k = 3$). In 2^{k-p} designs we can, moreover, apply the studentized maximum modulus or the maximum F-test provided all assumptions hold; otherwise the choice is again between a Bonferroni approach and a Scheffé approach.

(ii) Individual Effects

We can apply the robust Scheffé procedure or a Bonferroni approach.

(iii) Individual Cell Means

Cell means in a higher-way layout do not present any new viewpoints compared with a one-way layout since these means remain independent. Hence the discussion for the one-factor design applies here when making a choice among Bonferroni, Scheffé, Tukey, Dunnett, studentized maximum modulus, etc.

The reader may also want to consult the practical recommendations on the use of various MCP as given by Seeger (1966, pp. 142-148, 156-157). His main conclusion is that a Bonferroni approach (applying either the t-statistic with common or different variances, or a nonparametric statistic) is the best procedure in most circumstances.

V.B.5. MORE TECHNIQUES AND EXPERIMENTAL SITUATIONS

In this section we shall give a brief exposé on some more MCP for the above experimental situations and some MCP for other experimental objectives and designs. [Also consult O'Neill and Wetherill (1971) who published a bibliography of over 200 publications, Conover (1971, pp. 263-292, 342-349), Dudewicz (1968), Gupta and Panchapakesan (1972), and Tukey (1953)].

More Multiple Comparison Procedures

(i) A well-known approach to simultaneous testing is Duncan's procedure. He proceeds in stages: If in a stage a mean (or difference between means) is found to be significant then in the next stage the procedure acts as if the number of means, k, has decreased to (k - 1), etc. Related multiple-stage procedures are the Keuls-Newman test and Fisher's "least significant difference" (LSD) test. Typical for Duncan's approach is the use of "p-mean significance levels," a type of error rate increasing with the family size and in our opinion based on rather arbitrary considerations. The multiple-stage tests are based on normal, independent and homoscedastic observations in a one-way layout (except for Fisher's test that can be applied in higher-way layouts). The tests cannot be used for estimation purposes. A critical discussion can be found in Hartley (1955, pp. 57-61), Miller (1966, pp. 24-31, 81-94, 97-98), and Scheffé (1964, p. 78); also see Seeger (1966, pp. 123-127) and Rhyne and Steel (1965, p. 302). Yet Duncan has his advocates; compare the tables and Monte Carlo experiments for his procedure given by Harter (1957; 1960a; 1961) and Balaam (1963), respectively.[30)]

(ii) Duncan (1961) and Waller and Duncan (1969) devised MCP based on loss functions and prior probabilities. If the experimenter is willing to accept the type of prior information and loss functions used in these MCP then he can apply the simple rules and tables derived by these authors[31)] [also see Anscombe (1965) and Kurtz et al. (1965)]. Deely and Gupta (1968) derived a Bayesian solution for the subset approach [also see Gupta and Panchapakesan (1972)].

(iii) Some miscellaneous MCP are: Tukey's gap-straggler-variance test, short-cut techniques (not relevant in simulation where a computer is available), Nemenyi's test based on Friedman's rank statistics, and Nemenyi's median test. These tests are discussed in Miller (1966, pp. 94-97, 172-178, 182-185). Also see Seeger (1966, pp. 115-116, 132-141), the survey in O'Neill and Wetherill (1971, pp. 224-230) and the discussion of their paper.

(iv) Slivka (1970) gives a nonparametric procedure for testing if the median of survival times is larger for an experimental population i $(i = 1, \ldots, k)$ than for a standard population. His approach is attractive for simulation studies of survival systems since it saves computer time.[32)]

More Experimental Objectives and Designs

(i) As mentioned in Sec. 3 the objective of the experiment may be a comparison among variances instead of means. We have already mentioned several techniques for handling this problem. We add the Ramachandran (1956b) procedure for simultaneous confidence bounds on all variance ratios $\sigma_i^2/\sigma_{i'}^2$, $(i, i' = 1, \ldots, k, i \neq i')$. Alternatively we may apply a logarithmic transformation and make comparisons among the σ_i^2 as if it were means; see Scheffé (1964, pp. 83-87) and also Miller (1966, pp. 221-223).

(ii) Instead of means or variance we may compare entire distributions. For one or two populations we can apply the well-known χ^2 or Kolmogorov-Smirnov tests. A solution also exists for the general case of k distributions and is discussed by Conover (1971, pp. 317-326) and Miller (1966, pp. 185-188).

(iii) Miller (1966, pp. 109-128) also gives techniques for simultaneous inferences in the case of regression analysis, e.g. confidence bounds around the line $\underline{y} = ax + b$ that hold for several values of x. Unfortunately most results apply only to the simple case $y = ax + b$.

(iv) The above author further discusses techniques for experimental designs with multivariate responses. Remember that the Bonferroni inequality holds for dependent observations and consequently it is applicable to multivariate situations. Indeed Miller (1966, pp. 189-210) shows that the Bonferroni approach is a serious competitor in case of multivariate responses.

We shall apply the Bonferroni inequality to multivariate responses in Chap. VI. Multivariate procedures are also discussed by Gabriel (1968; 1969b), Gabriel and Sen (1968), Gnanadesikan and Gupta (1970), Gupta and Panchapakesan (1969a; 1972), and Khatri (1967).

(v) The response variable may have a qualitative character. If there are c categories ($c > 2$) then a multinomial distribution results. (If $c = 2$ we have a binomial distribution.) Simultaneous inferences for multinominal populations are given by Miller (1966, pp. 215-221) and Gabriel (1966). Yet these distributions seem of little use in simulation since the computer gives quantitative responses only. (Quantitative responses can be made qualitative, e.g. "good" or "bad," but this means loss of information.)

(vi) Finally we mention some miscellaneous objectives: Sen (1969) gives a procedure for interactions in PBIB designs. Scheffé (1970) discusses estimation of multiple ratios (instead of differences). Gupta and Panchapakesan (1969b; 1972) give a survey of various procedures, especially subset selection procedures; also consult Sobel (1969).

APPENDIX V.B.1. UNEQUAL VARIANCES IN DUNNETT'S PROCEDURE

It is easy to verify that $\rho_{ii'}$, i.e. the correlation between $(\bar{\underline{x}}_i - \bar{\underline{x}}_0)$ and $(\bar{\underline{x}}_{i'} - \bar{\underline{x}}_0)$ $(i \neq i')$, is a constant equal to 1/2 provided Eq. (1.1) is satisfied.

$$\mathrm{var}(\bar{\underline{x}}_i) = \frac{\sigma_i^2}{n_i} = \mathrm{var}(\bar{\underline{x}}_0) = \frac{\sigma_0^2}{n_0} \qquad (i = 1, \ldots, k) \qquad (1.1)$$

If we assume a common variance $\sigma_0^2 = \sigma_i^2 = \sigma^2$ then Eq. (1.1) is met for equal sample sizes $n_0 = n_i = n$. Dunnett's tables are based on a constant correlation coefficient $\rho_{ii'} = 1/2$.[33] If the population variances σ_0^2 and σ_i^2 may be different we can decide to use Dunnett's tables simply hoping that Eq. (1.1) is approximately satisfied so that the experimentwise error rate is approximately realized. Alternatively we may try to take the sample sizes n_0 and n_i in such a way that (1.1) holds (approximately). Consequently we may (i) use prior knowledge on σ_0^2 and σ_i^2; (ii) take a pilot sample to estimate only the variances, not the means (i.e. $\bar{\underline{x}}_0$ and $\bar{\underline{x}}_i$ are not based on the pilot sample); (iii) take a pilot sample, estimate the variances and select the number of additional observations in such a way that Eq. (1.1) is approximately satisfied; the complete sample is used to estimate the means. This results in stochastic sample sizes conflicting with Dunnett's assumption of fixed sample sizes. Hence more research is needed to check the validity of this heuristic approach.

If the populations have no common variance then we replace $\underline{s}_v \sqrt{2/n}$ in Eqs. (17) through (19) by

$$[\widehat{\underline{\mathrm{var}}}(\bar{\underline{x}}_i - \bar{\underline{x}}_0)]^{1/2} = \left(\frac{\underline{s}_i^2}{n_i} + \frac{\underline{s}_0^2}{n_0} \right)^{1/2} \tag{1.2}$$

while the degrees of freedom v in $d_{k,v}^{\alpha}$ and $d'^{\alpha}_{k,v}$ can be approximated using the Cochran and Cox approach given in Sec. V.A.3, i.e.

$$v = \frac{(\underline{s}_i^2/n_i)(n_i - 1) + (\underline{s}_0^2/n_0)(n_0 - 1)}{(\underline{s}_i^2/n_i) + (\underline{s}_0^2/n_0)} \tag{1.3}$$

or using Welch's approach [see Eq. (41) in Sec. V.A]. We remark that for populations with a common variance, Dunnett (1955, pp. 1106-1107) derived that the optimum sample size allocation would be $n_0/n_i \sim \sqrt{k}$, i.e. more observations are taken from the control population that occurs in all k comparisons $\bar{\underline{x}}_i - \bar{\underline{x}}_0$; also see Scheffé

(1964, p. 88). If the populations have different variances then this optimum allocation rule needs adjustment. [Consult Bechhofer and Nocturne (1972) and the references in Dudewicz and Dalal (1971, p. 7.2).] However, if the variances are unequal then it becomes more important to satisfy the basic assumptions of Dunnett's procedure. Hence we would try to satisfy Eq. (1.1).

EXERCISES

1. Let $\underline{S}_i = 0$ denote that statement i is false (i = 1, 2). Show that with $\underline{S}_1$ and $\underline{S}_2$ positively correlated the Bonferroni inequality is cruder than with $\underline{S}_1$ and $\underline{S}_2$ negatively correlated (i.e. with positive correlation α_E is definitely smaller than $\Sigma_1^2 \alpha_i$).

2. Show that Dunnett's confidence limits in Eq. (17) imply Eq. (18).

3. Construct a two-sided confidence interval for $\mu_i - \mu_0$ (i = 1, ..., k) based on Steel's rank sum test, applying both the graphical and the numerical technique described in Miller (1966, pp. 145, 149). Assume $\underline{x}_{01} = 1$, $\underline{x}_{02} = 3$; $\underline{x}_{i1} = 1$, $\underline{x}_{i2} = 2$, $\underline{x}_{i3} = 3.5$ and $r^{*\alpha} = 11$.

4. Why is Dunn's rank statistic $(\underline{d}_j)$ for contrasts [see Eq. (41)] tested, using the normal table instead of the t-table?

5. Prove that a_i is Eq. (48) is the upper $[1/2 + (1 - \alpha)^{1/k}/2]$ point of the t-distribution.

6. Let Dunnett's MCP be used for the selection of the highest mean. Demonstrate that $P(CS) > 1 - \alpha$ if $\mu_i < \mu_0$ and $P(CS) < 1 - \alpha$ if, e.g. $\mu_0 = \mu_1 = \cdots = \mu_{k-1} = \mu_k - \delta$ (δ a positive constant).

7. What is the relation between variances of normal distributions and scale parameters of gamma distributions?

8. Prove that in a layout with equal numbers of observations per cell, the estimated main effects of a particular factor, say $\underline{\hat{\alpha}}_i^A$ (i = 1, ..., a), have covariance $(-1/a)(\sigma^2/n)$.

9. Let the family of statements consist of statements on the ab cell-means of a two-way layout with factor A at a levels and B at b levels. What is the value of q in Scheffé's procedure, Eq. (71)? What is q if we are interested only in contrasts among the ab means?

10. Prove that in a two-way layout the side-condition $\Sigma_{i=1}^{a} \alpha_{ib}^{AB} = 0$ follows from $\Sigma_j \alpha_{ij}^{AB} = 0$ $(i = 1, \ldots, a)$, and $\Sigma_i \alpha_{ij}^{AB} = 0$ $(j = 1, \ldots, b-1)$.

11. Consider simultaneous testing of the hypotheses of no effects in a 2^{k-p} design. Prove that the analogue of the maximum F-test is the squared studentized maximum modulus.

12. Prove that the correlation between $(\bar{\underline{x}}_i - \bar{\underline{x}}_0)$ and $(\bar{\underline{x}}_{i'} - \bar{\underline{x}}_0)$ is $1/2$ if $\sigma_i^2/n_i = \sigma_0^2/n_0$ for $i = 1, \ldots, k$; see Eq. (1.1) in the Appendix.

13. Suppose we have a one-factor layout with five factors and fifty replications per level. We use Scheffé's S-technique for multiple contrasts with $\alpha = .05$. (a) Show that antithetic variates give smaller expected confidence interval widths provided they create a negative correlation higher in absolute value than 0.03. (b) Prove that this negative correlation must rise to -0.30 if the total number of replications reduces from 250 to 24. (Compare Chap. III, footnote 48.)

14. Naylor et al. (1968, p. 190) give a table showing the estimated variance for each of the fifty replications for five different stabilization policies in their simulated economic model. Apply the following MCP to both the original data and their logarithmic transformation: Dunnett, Steel, and Bonferroni (assuming policy I to be the control) and Tukey, Steel, Bonferroni, and Scheffé for all pairwise comparisons.

15. Apply the procedures of exercise 14 to the steady-state expected waiting time in single-server queuing systems with varying parameters.

NOTES

1. For example in regression analysis we may assume a linear response curve, $\underline{y} = \beta_0 + \beta_1 x + \underline{u}$, and test whether $\beta_1 = 0$.

2. However, there are some MCP (e.g. Duncan's multiple range test) that give significance statements but have no confidence interval equivalent; compare Miller (1966, p. 26) and also Ryan (1959, pp. 40-41).

3. Seeger (1966, pp. 117-119) discusses all above error types except Verhagen's caution level. Harter (1957) gives tables showing how the values of several types of error rates vary. Similar tables are provided by Boardman and Moffit (1971) and Gabriel (1964, pp. 470-476). Various error rates are also clearly explained by Balaam and Federer (1965), Hartley (1955, pp. 47-49), O'Neill and Wetherill (1971, pp. 220-223), Ryan (1959, pp. 26-40) and Tukey (1953). Tukey also gives many examples. Further compare Cox (1965), Dunnett (1964, p. 483), Konijn (1959, p. 61), Rhyne and Steel (1965, p. 295), Scheffé (1964, p. 89), and Seeger (1968).

4. Briefly the criticism on the other rates is as follows. Verhagen (1963) defines his error rate only for independent statements. Eklund's rate as presented by Seeger (1966, pp. 134-141) is made operational by introducing many rough approximations. As Miller (1966, pp. 31, 85) points out Duncan's protection level most times reaches its value at the unrealistic value ∞; moreover the level increases with the number of populations involved; also see Scheffé (1964, p. 78). The rate per experiment does not differ much from the experimentwise error rate as is mentioned in Hartley (1955, p. 48), Miller (1966, p. 10) and Ryan (1959, pp. 38-40).

5. This expectation also holds for dependent statements; see Eq. (9) in Miller (1966, p. 7).

6. See Cox (1965), Kurtz et al. (1965, pp. 146-148), Miller (1966, p. 10), and Ryan (1959, pp. 35, 38).

7. The following notations are equivalent: $a \in b \pm c$, or $b - c \leq a \leq b + c$, or $|a - b| \leq c$.

8. $P(\underline{t}_i \leq a$ for all $i) = P(\max_i \underline{t}_i \leq a)$.

9. Steel (1959a, p. 370) also gives such a table. However, Miller's table uses a more recent approximation based on Gupta; see Miller (1966, p. 152).

10. In private communication Miller pointed out that Steel rejects H_0 for small values of $\min[\underline{r}_i, n(2n+1) - \underline{r}_i]$, while Miller rejects H_0 for large values of $\max[\underline{r}_i, n(2n+1) - \underline{r}_i]$. Both approaches give identical conclusions since $\underline{max} = n(2n+1)-\underline{min}$. Miller tabulated the critical max-values while Steel (1959a, p. 571) gives the critical min-values.

11. This contradicts Seeger (1966, p. 130) who claims that the sign test (in a multiple comparison situation) does not require constant variances. (In private communication Seeger explained that he meant to apply the Bonferroni inequality to the non-multiple sign test for two populations, which does not require a common variance.) Seeger (1966, p. 132) further states that the rank test is sensitive to heterogeneity of variances, and this is questioned by Steel (1959a, p. 562) and Miller (1966, p. 146) as far as tests are concerned; confidence intervals require a common variance (see our discussion on the robustness of various tests in Sec. 4).

12. The procedure does realize a certain type of transitivity of its conclusions [see Gabriel (1964, pp. 460-462].

13. This relation between the ANOVA F-test on $H_0: \mu_1 = \cdots = \mu_k$ and tests on individual contrasts holds only for contrast-tests based on Eq. (38). Consequently, if we apply, e.g. Tukey's studentized range tests to the contrasts $\mu_i - \mu_{i'}$ then an insignificant F-test may be followed by some significant statements on $\mu_i - \mu_{i'}$.

14. Tests for more general contrasts can be based on formula (12) in Dunn (1964).

15. Actually Dunn considers two MCP, one where all populations are ranked and one where only the populations involved in a particular constrast are ranked. The efficiency of both procedures has been compared for several situations, but no procedure is uniformly better; see Dunn (1964, p. 252) and Sherman (1965). We do not discuss the first variant since that approach implies that populations not mentioned in a particular contrast, yet are relevant for the ranking.

16. These values can be found in the tables 1 through 4 of Dunn and Massey in the columns where $\rho = 0.0$ and the rows where $1 - \alpha$ is .80 and .90.

17. In Gupta and Sobel (1957) consult the column corresponding with the particular value of k or p, where p is defined as $k - 1$. In Dunnett (1955) consult the column for p $(= k - 1)$. For example, for $P^* = .95$, $k = 5$, $v = 15$. Dunnett gives 2.36 so that D is $(\sqrt{2})(2.36) \approx 3.33$ while Gupta and Sobel give 3.34.

18. a is equal to $(r - c)$ in Rizvi and Sobel.

19. Equation (6.6) in Rizvi and Sobel is identical to their Eq. (3.3). Hence Eq. (6.6) or (3.3) is satisfied by looking up $a = r - c$ in table 3. So in Eq. (6.3) we have $(r + c) = r + (r - a) = 2r - a = (n + 1) - a$.

20. The critical constant a is then 0 and $\underline{x}_{0(i)}$ is defined as $-\infty$; so-called degeneracy [see Rizvi and Sobel (1967, pp. 1789-1790, 1802)].

21. Dunn (1961, p. 61) contains a printing error. The pooled estimator of the variance has $(n - 1)ab$ degrees of freedom instead of $n(a - 1)(b - 1)$.

22. Hartley (1955, p. 50) presents an alternative procedure. However, it can be checked that his procedure boils down to the Bonferroni technique (since he uses $F_\alpha(k;v^*,v) \approx F_{\alpha/k}(1;v^*,v)$ provided the degrees of freedom are equal (as is the case in

2^{k-p} experiments). His procedure is a somewhat revised Bonferroni technique if the degrees of freedom may differ; see Hartley (1955, pp. 52-54). Ramachandran (1956a) gives a solution for testing only two mean squares. Gupta (1963) gives tables for $\max(\underline{F}_i)$ (derived for a different problem) that look relevant at first sight. For $P(\underline{F}_1 < F^\alpha \wedge \underline{F}_2 < F^\alpha \wedge \ldots \wedge \underline{F}_k < F^\alpha) = 1 - \alpha$ is equivalent to $P(\max_i \underline{F}_i < F^\alpha) = 1 - \alpha$. Unfortunately his tables are only valid of all degrees of freedom are the same. [In the example: $a - 1 = b - 1 = (a-1)(b-1) = (n-1)ab$.] This condition is very restrictive.

23. Miller (1966, p. 102) mentions another application involving k t-statistics with independent numerators and a common denominator. As he mentions, exact results can be derived applying the studentized maximum modulus. The improved Bonferroni inequality (63) is still useful if the α_i are desired to be different or if no table for the studentized maximum modulus is available.

24. ψ is an estimable function if it has an unbiased linear estimator [see Scheffé (1964, p. 13)].

25. Konijn (1960, p. 17) also gives an approximation for confidence intervals for $\eta_{ij} - \eta_{1j}$, where η_{1j} is the control varying with the B-columns. A simple approximation may be based on Bonferroni, i.e. apply Dunnett's technique per column with $\alpha = \alpha_E/b$.

26. For small n the power of these tests is small. This is less serious if we like to use one of the MCP and only in case of strong evidence of nonnormality and heteroscedasticity we reject the null-hypotheses.

27. A permutation test can be roughly described as follows. Under H_0 any arrangement of the observed values has the same probability. Hence determine all possible arrangements; calculate the corresponding values of some selected statistic; find which value is exceeded by a proportion α of the values of this

statistic; compare this critical value with the value of the statistic for the particular arrangement resulting from the experiment. More on permutation tests can be found in, e.g. Conover (1971, pp. 357-364) and Miller (1966, pp. 179-182).

28. Seeger (1966, p. 145) gives a table for the sizes of the confidence intervals with the Dunnett and the Bonferroni approach. (He uses the term Fisher's method instead of Bonferroni.) For an experimentwise error rate of 5% both sizes are about the same so that the Bonferroni approach is more attractive as it requires less assumptions.

29. Consider a 2^{k-p} design. The covariance between, say, effect 1 and 2 is

$$\operatorname{cov}(\underline{\hat{\alpha}}_1, \underline{\hat{\alpha}}_2) = \frac{4}{N^2} \sum_{i=1}^{N} \sum_{i'=1}^{N} x_{i1} x_{i'2} \operatorname{cov}(\underline{y}_i, \underline{y}_{i'})$$

If the observations $\underline{y}_i$ are independent this expression reduces to

$$\frac{4}{N^2} \overset{N}{\underset{i=1}{N}} x_{i1} x_{i2} \sigma_i^2$$

If further the observations have a common variance σ^2 we have

$$\frac{4}{N^2} \sigma^2 \sum_i x_{i1} x_{i2}$$

In an orthogonal design $\Sigma_i\, x_{i1} x_{i2} = 0$. For normally distributed observations a zero covariance implies independence; see Fisz (1967, p. 159). Independence of the estimated effects means that their sums of squares are independent [see, e.g. Eq. (99) in Chap. IV].

30. In private communication Gabriel reported on an unpublished Monte Carlo study by Sauerbrun showing equal experimentwise error rates for Duncan, Tukey, Scheffé-Gabriel, and Ryan (variant of Duncan). The power of these methods did not differ much, Ryan's procedure being best. Keul's technique gave invalid results.

31. Duncan's solution results in a critical constant that does not vary with the number of populations (k). This implies a high experimentwise error rate if k is high, but this characteristic is accepted since the prior probability of all k populations being identical is judged to be small. See Duncan (1961, pp. 1029-1030).

32. In a simulation experiment no treatments are simultaneously given to experimental units. Yet Slivka's technique is applicable. For simulate n_0 observations for the standard system. Start simulating population i; stop the simulation as soon as the survival time of population i is longer than the median of the control population; score 0. (In this way computer time is saved as the simulation run is not continued until the system breaks down.) If the survival time is smaller than the control median then we score 1.

33. Dunnett (1964, p. 490) derived a rather complicated correction factor for his two-sided critical constant d' for the special case of $\bar{\underline{x}}_0$ having a variance smaller than the experimental averages $\bar{\underline{x}}_i$, which have common variances.

REFERENCES

Andrews, D.F., P.J. Bickel, F.R. Hampel, P.J. Huber, W.H. Rogers, and J.W. Tukey (1972). Robust Estimates of Location. Princeton University Press, Princeton, N.J.

Anscombe, F.J. (1965). "Comments on Kurtz-Link-Tukey-Wallace paper," Technometrics, 7, 167-168.

Balaam, L.N. (1963). "Multiple comparisons--a sampling experiment," Australian J. Stat., 5, 62-84.

Balaam, L.N., and W. T. Federer (1965). "Answer to query: Error rate bases," Technometrics, 7, 260-262.

Barlow, R.E. and S.S. Gupta (1969). "Selection procedures for restricted families of probability distributions," Ann. Math. Stat., 40, 905-917.

Bartlett, N.S. and Z. Govindarajulu (1968). "Some distribution-free statistics and their application to the selection problem," Ann. Inst. Stat. Math., 20, 79-97.

Bechhofer, R.E. and D.J. Nocturne (1972). "Optimal allocation of observations when comparing several treatments with a control, II: 2-sided comparisons," Technometrics, 14, 423-436.

Boardman, T.J. and D.R. Moffitt (1971). "Graphical Monte Carlo type I error rates for multiple comparison procedures," Biometrics, 27, 738-744.

Conover, W.J. (1971). Practical Nonparametric Statistics, Wiley, New York.

Conway, R.W., B.M. Johnson, and W.L. Maxwell (1959). "Some problems of digital systems simulation," Management Sci., 6, 92-110.

Cox, D.R. (1965). "A remark on multiple comparison methods," Technometrics, 7, 223-224.

Crouse, C.F. (1969). "A multiple comparison of rank procedure for a one-way analysis of variance," South African Stat. J. 3, No. 1, 35-48.

Deely, J.J. and S.S. Gupta (1968). "On the properties of subset selection procedures," Sankhya, Ser., A 30, Pt. 1, 37-50.

Desu, M.M. (1970). "A selection problem," Ann. Math. Stat., 41, 1596-1603.

Desu, M.M. and M. Sobel (1968). "A fixed subset-size approach to the selection problem," Biometrika, 55, 401-410.

Dudewicz, E.J. (1968). A Categorized Bibliography on Multiple-Decision (Ranking and Selection) Procedures, Department of Statistics, The University of Rochester, New York.

Dudewicz, E.J. (1972). Statistical Inference with Unknown and Unequal Variances, Department of Statistics, The University of Rochester, New York.

Dudewicz, E.J. and S.R. Dalal (1971). Allocation of Observations in Ranking and Selection with Unequal Variances, Department of Statistics, The University of Rochester, New York.

Duncan, D.B. (1961). "Bayes rules for a common multiple comparisons problem and related Student-t problems," Ann. Math. Stat., 32, 1013-1033.

Dunn, O.J. (1961). "Multiple comparisons among means," J. Amer. Stat. Assoc., 56, 52-64.

Dunn, O.J. (1964). "Multiple comparisons using rank sums," Technometrics, 6, 241-252.

Dunn, O.J. and F.J. Massey (1965). "Estimation of multiple contrasts using t-distributions," J. Amer. Stat. Assoc., 60, 573-583.

Dunnett, C.W. (1955). "A multiple comparison procedure for comparing several treatments with a control," J. Amer. Stat. Assoc., 50, 1096-1121.

Dunnett, C.W. (1964). "New tables for multiple comparisons with a control," Biometrics, 20, 482-491.

Fishman, G.S. and P.J. Kiviat (1967). Digital Computer Simulation: Statistical Considerations, RM-5387-PR, The Rand Corporation, Santa Monica, California.

Fisz, M. (1967). Probability Theory and Mathematical Statistics. Wiley, New York, third printing.

Gabriel, K.R. (1964). "A procedure for testing the homogeneity of all sets of means in analysis of variance," Biometrics, 20, 459-477.

Gabriel, K.R. (1966). "Simultaneous test procedures for multiple comparisons on categorical data," J. Amer. Stat. Assoc., 61, 1081-1096.

Gabriel, K.R. (1968). "Simultaneous test procedures in multivariate analysis of variance," Biometrika, 55, 489-504.

Gabriel, K.R. (1969a). "Simultaneous test procedures--some theory of multiple comparisons," Ann. Math. Stat., 40, 224-250.

Gabriel, K.R. (1969b). "Comparison of some methods of simultaneous inference in MANOVA," in Multivariate Analysis, Vol. 2 (P.R. Krishnaiah, ed.), Academic, New York.

Gabriel, K.R. and P.A. Lachenbruch (1969). "Non-parametric ANOVA in small samples: A Monte Carlo study of the adequacy of the asymptotic approximation," Biometrics, 25, Pt. 3, 593-596.

Gabriel, K.R. and P.K. Sen (1968). "Simultaneous test procedures for one-way ANOVA and MANOVA based on rank scores," Sankya, Ser. A, 30, Pt. 3, 303-362.

Gnanadesikan, M. and S.S. Gupta (1970). "A selection procedure for multivariate normal distributions in terms of the generalized variances," Technometrics, 12, 103-117.

Goel, P.K. (1972). A Note on the Non-Existence of Subset Selection Procedure for Poisson Populations, Mimeograph Series No. 303, Department of Statistics, Division of Mathematical Sciences, Purdue University, Lafayette, Indiana.

Gupta, S.S. (1963). "On a selection and ranking procedure for gamma populations," Ann. Inst. Stat. Math., 14, 199-216.

Gupta, S.S. (1965). "On some multiple decision (selection and ranking) rules," Technometrics, 7, 225-245.

Gupta, S.S. and G. McDonald (1970). "On some classes of selection procedures based on ranks," in Nonparametric Techniques in Statistical Inference (M.L. Puri, ed.) University Press, Cambridge, England.

Gupta S.S. and G.C. McDonald (1972). "Some selection procedures with applications to reliability problems," in Operations Research and Reliability (D. Grouchko, ed.), Gordon and Breach, New York.

Gupta, S.S. and K. Nagel (1971). "On some contributions to multiple decision theory," in Statistical Decision Theory and Related Topics (S. S. Gupta and J. Yackel, eds.), Academic, New York.

Gupta, S.S., K. Nagel, and S. Panchapakesan (1971). On the Order Statistics from Equally Correlated Normal Random Variables, Mimeographed Series No. 290, Department of Statistics, Division of Mathematical Sciences, Purdue University, Lafayette, Indiana.

Gupta, S.S., and S. Panchapakesan (1968). On a Class of Selection and Ranking Procedures, Mimeographed Series No. 171, Department of Statistics, Division of Mathematical Sciences, Purdue University, Lafayette, Indiana.

Gupta, S.S. and S. Panchapakesan (1969a). Some selection and ranking procedures for multivariate normal distributions," in Multivariate Analysis, Vol. 2,(P.R. Krishnaiah, ed.), Academic, New York.

Gupta, S.S. and S. Panchapakesan (1969b). "Selection and ranking procedures," in The Design of Computer Simulation Experiments (T.H. Naylor, ed.), Duke University Press, Durham, North Carolina.

Gupta, S.S. and S. Panchapakesan (1972). "On multiple decision procedures," J. Math. Physical Sci., 6, 1-71.

Gupta, S.S. and T.J. Santner (1972). Selection of a Restricted Subset of Normal Populations Containing the One with the Largest Mean, Mimeographed Series No. 299, Department of Statistics, Division of Mathematical Sciences, Purdue University, Lafayette, Indiana.

Gupta, S.S. and M. Sobel (1957). "On a statistic which arises in selection and ranking procedures," Ann. Math. Stat., 28, 957-867.

Gupta, S.S. and M. Sobel (1958). "On selecting a subset which contains all populations better than a standard," Ann. Math. Stat., 29, 235-244.

Gupta, S.S. and M. Sobel (1960). "Selecting a subset containing the best of several binomial populations," in Contributions to Probability and Statistics (I. Alkin, S.G. Ghurye, W. Hoeffding, W.G. Madow, and H.B. Mann, eds.), Stanford University Press, Stanford, California.

Gupta, S.S. and M. Sobel (1962). "On selecting a subset containing the population with the smallest variance," Biometrika, 49, 495-507.

Harter, H.L. (1957). "Error rates and sample sizes for range tests in multiple comparisons," Biometrics, 13, 511-536.

Harter, H.L. (1960a). "Critical values for Duncan's new multiple range test," Biometrics, 16, 671-685.

Harter, H.L. (1960b). "Tables of range and studentized range," Ann. Math. Stat., 31, 1122-1147.

Harter, H.L. (1961). "Corrected error rates for Duncan's new multiple range test," Biometrics, 17, 321-324.

Hartley, H.O. (1955). "Some recent developments in analysis of variance," Commun. Pure Appl. Math., 8, 47-72.

Khatri, C.G. (1967). "On certain inequalities for normal distributions and their applications to simultaneous confidence bounds," Ann. Math. Stat., 38, 1853-1867.

Kimball, A.W. (1951). "On dependent tests of significance in the analysis of variance," Ann. Math. Stat., 22, 600-602.

Konijn, H.S. (1959). "Basing decisions on an analysis of variance," Australian J. Stat., 1, 57-68.

Konijn, H.S. (1960). "Multiple comparison with controls," Australian J. Stat., 2, 16-18.

Krishnaiah, P.R. and J.V. Armitage (1966). "Tables for multivariate t distribution," Sankhya, Ser. B, 28, Pt. 1 and 2, 31-56.

Kurtz, T.E., R.F. Link, J.W. Tukey, and D.L. Wallace (1965). "Short-cut multiple comparisons for balanced single and double classifications: Part 1, results," Technometrics, 7, 95-162.

Lehmann, E.L. (1963). "Nonparametric confidence intervals for a shift parameter," Ann. Math. Stat., 34, 1507-1512.

Lewis, P.A.W. (1972). Large-Scale Computer-Aided Statistical Mathematics, Naval Postgraduate School, Monterey, California.

Martin, W.A. (1971). "Sorting," Computing Surveys, 3, 147-174.

Miller, R.G. (1966). Simultaneous Statistical Inference. McGraw-Hill, New York.

Nair, K.R. (1948). "The studentized form of the extreme mean square tests in the analysis of variance," Biometrika, 35, 16-31.

Naylor, T.H., D.S. Burdick, and W.E. Sasser (1967a). "Computer simulation experiments with economic systems: The problem of experimental design," J. Amer. Stat. Assoc., 62, 1315-1337.

Naylor, T.H., K. Wertz, and T.H. Wonnacott (1967). "Methods for analyzing data from computer simulation experiments," Commun. ACM, 10, 703-710.

Naylor, T.H., K. Wertz, and T.H. Wonnacott (1967b). "Methods for evaluating the effects of economic policies using simulation experiments," Rev. Inter. Stat. Inst., 36, 184-200.

O'Neill, R. and G.B. Wetherill (1971). "The present state of multiple comparison methods," J. Roy. Stat. Soc., Ser. B, 33,218-250 (including discussions).

Patterson, D.W. (1965). "A nonparametric population selection procedure," Ann. Math. Stat., 36, 1614-1615.

Pearson, E.S. and H.O. Hartley (1966). Biometrika Tables for Statisticians, Vol. I, Cambridge University Press, Cambridge, England, third edition.

Peritz, E. (1971). "On a statistic for rank analysis of variance," J. Roy. Stat. Soc., Ser. B, 33, 137-139.

Puri, M.L. and P.K. Sen (1971). Nonparametric Methods in Multivariant Analysis, Wiley, New York.

Ramachandran, K.V. (1956a). "On the simultaneous analysis of variance test," Ann. Math. Stat., 27, 521-528.

Ramachandran, K.V. (1956b). "Contributions to simultaneous confidence interval estimation," Biometrics, 12, 51-56.

Rhyne, A.L. and R.G.D. Steel (1965). "Tables for a treatments versus control multiple comparison sign test," Technometrics, 7, 293-306.

Rizvi, M.H. and M. Sobel (1967). "Nonparametric procedures for selecting a subset containing the population with the largest α-quantile," Ann. Math. Stat., 38, 1788-1803.

Rizvi, M.H., M. Sobel, and G.G. Woodworth (1968). "Nonparametric ranking procedures for comparison with a control," Ann. Math. Stat., 39, 2075-2093.

Ryan, T.A. (1959). "Multiple comparisons in psychological research," Psych. Bull., 56, 26-47.

Scheffé, H. (1964). The Analysis of Variance. Wiley, New York, fourth printing.

Scheffé, H. (1970). "Multiple testing versus multiple estimation; improper confidence sets; estimation of directions and ratios," Ann. Math. Stat., 41, 1-29.

Seeger, P. (1966). Variance Analysis of Complete Designs. Almqvist and Wiksell, Uppsala.

Seeger, P. (1968). "A note on a method for the analysis of significance en masse." Technometrics, 10, 586-593.

Sen, P.K. (1969). "A generalization of the T-method of multiple comparisons for interactions," J. Amer. Stat. Assoc., 64, 290-295.

Sherman, E. (1965). "A note on multiple comparisons using rank sums," Technometrics, 7, 255-256.

Slivka, J. (1970). "A one-sided nonparametric multiple comparison control percentile test: treatments versus control," Biometrika, 57, 431-438.

Sobel, M. (1967). "Nonparametric procedures for selecting the t populations with the largest α-quantiles," Ann. Math. Stat., 38, 1804-1816.

Sobel, M. (1969). "Selecting a subset containing at least one of the best populations," in Multivariate Analysis, Vol. 2 (P.R. Krishnaiah, ed.), Academic, New York.

Steel, R.G.D. (1959a). "A multiple comparison rank sum test: treatments versus control," Biometrics, 15, 560-572.

Steel, R.G.D. (1959b). "A multiple comparison sign test: treatments versus control," J. Amer. Stat. Assoc., 54, 767-775.

Steel, R.G.D. (1960). "A rank sum test for comparing all pairs of treatments," Technometrics, 2, 197-207.

Tobach, E., M. Smith, G. Rose, and D. Richter (1967). "A table for making rank sum multiple paired comparisons," Technometrics, 9, 561-567.

Tukey, J.W. (1953). The Problem of Multiple Comparisons, Department of Statistics, Princeton University, Princeton, New Jersey.

Verhagen, A.M.W. (1963). "The caution-level in multiple tests of significance," Australian J. Stat., 5, 41-48.

Waller, R.A. and D.B. Duncan (1969). "A Bayes rule for the symmetric multiple comparisons problem," J. Amer. Stat. Assoc., 64, 1484-1503.

Wilks, S.S. (1963). Mathematical Statistics. Wiley, New York, second printing.

V.C. MULTIPLE RANKING PROCEDURES[1)]

V.C.1. INTRODUCTION AND SUMMARY

In this part of Chap. V we shall present procedures for determining how many observations should be taken from each of k ($k \geq 2$) populations Π_i ($i = 1, \ldots, k$) in order to select the best population. In the main part of the following sections we assume that the "best" population is the one with the largest mean. When looking for the smallest mean we can use the procedures for selection of the largest mean, applying them to the observations multiplied by minus one. In Sec. V.C.5 we shall briefly consider selection criteria other than the mean, e.g. the variance, and problem formulations other than selection of the best population, e.g. complete ranking of all k populations. Procedures for the above complete or incomplete rankings are called multiple ranking procedures (or MRP); in the literature they can also be found under such headings as multiple selection and decision procedures. In simulation, an observation can be defined as one run of a particular system, i.e one sequence of system outputs that together give one unbiased estimator of the system characteristic, of interest to the experimenter (e.g. a simulated history of one month yields one observation of monthly profit). Different system variants correspond with different populations. In experimental design terminology, k populations can correspond with one factor at k levels, or, e.g. with f factors each at k_j levels such that $k = \Pi_{j=1}^{f} k_j$. We assume that not all f factors ($f \geq 1$) are quantitative. For purely quantitative factors other procedures (e.g. response surface methodology discussed in Chap. II and IV) are more appropriate for the determination of the optimum combination of factor levels. We do not suggest that, in the simulation of important system alternatives, MRP be applied mechanically. The decision maker will use other considerations besides the simulation output in order to reach a decision.

Nevertheless the MRP enable the experimenter to determine a "reasonable" number of observations.

To the best of our knowledge there is no textbook giving a survey of MRP suitable for simulation. Bechhofer et al. (1968) wrote an excellent book but only "accessible to most advanced graduate students in statistics" and presenting only certain sequential MRP, i.e. not all those sequential and nonsequential MRP that may be important in simulation. We hope that our survey is complete enough. It is based on Statistical Theory and Method Abstracts (1959 through 1972) and on the other publications listed in our references, including Bechhofer et al.(1968) with approximately 320 references, and Dudewicz (1968) with 250 references.

In Sec. V.C.2 we shall discuss a basic concept in MRP, viz. the indifference zone approach, and the effect of various configurations of the population means (such as least favorable, generalized least favorable, equal means, and a new concept, "partially indifferent" configurations). In Sec. V.C.3 we shall present those existing procedures that in our opinion may be applicable in simulation. These procedures are classified according to their assumptions: parametric versus distribution-free and "semidistribution-free" MRP, same form of a distribution or not, known or unknown variances. We also discuss procedures for situations where a standard population exists and for multi-factor designs. In Sec. V.C.4 the application of MRP in simulation will be discussed. The efficiency and robustness of MRP will first be studied in general and then the specific procedures of Sec. V.C.3 will be examined for efficiency and robustness. Instead of relying on the robustness of existing procedures, the heuristic MRP presented in Sec. V.C.4 may be applied. Practical recommendations for selection of a procedure in simulation will be given. In Sec. V.C.5 we shall briefly mention procedures for other distributions (e.g. gamma, binomial), decision theoretic solutions, and other problem formulations (e.g. best t means). Appendixes, simple exercises, and references follow Sec. V.C.5. The reader may wish to skip the detailed exposé in Sec. V.C.3 since in Sec. V.C.4 all relevant MRP are briefly characterized and evaluated. Such a multitude of MRP are discussed since all procedures are based on assumptions that may very well be violated in simulation. Hence the experimenter had to decide which assumptions seem to hold and select a corresponding procedure.

V.C.2. THE INDIFFERENCE ZONE APPROACH

In the introduction we stated that we shall mainly restrict our attention to MRP aimed at selection of the population with the largest mean. A natural statistic for this selection is the sample mean or average $\bar{x}$, i.e. select as the best population the one yielding the highest sample mean, say $\bar{x}_{[k]}$.[2] (We let $\bar{x}_{(i)}$ denote the sample mean from the ith best population and $\bar{x}_{[i]}$ the ith largest sample mean in the sample.) In a small sample the probability is high that the best population does not give the best sample mean ($\bar{x}_{(k)} \neq \bar{x}_{[k]}$). Hence, we want to determine how many observations should be taken to arrive at a correct selection with high probability. If, however, the population means do not differ much, large samples are required to identify the best population. (Compare Fig. 4 in Sec. V.A.4 for two populations.) Usually it is uneconomical to take a great many observations to detect only small differences among the means. Therefore, Bechhofer (1954) proposed the so-called "<u>indifference zone</u>" approach which has, since then, been followed by most other authors. Let us now consider this approach in more detail.

We have k populations Π_i (i = 1, ..., k), also labeled $\Pi_{(i)}$, where $\mu_{(1)} \leq \mu_{(2)} \leq \cdots \leq \mu_{(k-1)} \leq \mu_{(k)}$; evidently we do not know which population corresponds with $\mu_{(k)}$. We want to find the population with the highest mean. The probability of a correct selection (or CS) is desired to be at least a predetermined constant P^*. [This P^* should be specified higher than $1/k$ since random selection of a population without any sampling would guarantee P(CS) $\geq 1/k$.] Large samples would be required to select the best population if the population means differ only slightly; the lost involved in a wrong selection would then be small. Therefore Bechhofer (1954) proposed guaranteeing a CS with probability of at least P^*, only if the best population mean is at least a specified number of units, say δ^* ($\delta^* > 0$), better than the next best mean. Or

$$P(CS) \geq P^* \quad \text{if} \quad \delta \equiv \mu_{(k)} - \mu_{(k-1)} \geq \delta^* \tag{1}$$

This approach is discussed in, e.g. Bechhofer (1954, p. 23; 1958, p. 411); a more general discussion is given by Barr and Rizvi (1966).

Next we shall have a closer look at the effect various <u>configurations of the means</u> have on $P(CS)$.

(i) <u>The least favorable configuration</u> (or LFC):

$$\mu_{(1)} = \cdots = \mu_{(k-1)} = \mu_{(k)} - \delta^* \tag{2}$$

Since the best mean is δ^* units better than the next one we do want to guarantee $P(CS) \geq P^*$. In Eq. (2) competition of the other $k-1$ populations is keenest among all configurations that make it worthwhile to guarantee $P(CS) \geq P^*$, since the best population is only δ^* units better and all other $k - 1$ populations are this close to the best one [also see Eq. (2.4) in Appendix 2]. The true configuration of the means is not known to the experimenter; yet the procedure has to realize $P(CS) \geq P^*$ (provided $\mu_{(k)} - \mu_{(k-1)} \geq \delta^*$) even in the LFC. The MCP reach the minimum value of $P(CS)$ given $\mu_k - \mu_{k-1} \geq \delta^*$, in the LFC.[3] As $\mu_{(k)} - \mu_{(k-1)}$ becomes higher than δ^* and/or one or more inferior populations have means smaller than the next best population, $P(CS)$ increases. See e.g. Bechhofer et al. (1968, p. 5).

(ii) <u>The partially indifferent configuration</u> (PIC): Consider the following configuration:

$$0 \leq \mu_{(k)} - \mu_{(k-1)} < \delta^* \tag{3a}$$

$$\mu_{(1)} = \cdots = \mu_{(k-2)} \leq \mu_{(k-1)} - \delta^* \tag{3b}$$

We introduce the term PIC for this configuration. The formulation in the literature, e.g. Bechhofer (1954; 1958), may give the impression that in the configuration Eqs. (3a) and (3b) a correct selection is not desired, since the best mean is not at least δ^* units better. Nevertheless, the best two populations are δ^* better than the other k-2 populations so that selection of Π_k or Π_{k-1} is desirable. Now we claim that a procedure which guarantees the problem requirement, Eq. (1), also guarantees P(CS) if Eqs. (3a) and (3b) hold, provided we define CS as selection of either the best population Π_k or Π_{k-1}, the population less than δ^* worse than Π_k. For consider a situation with one of the best two populations, say $\Pi_{(k)}$, eliminated. Then there remain k-1 populations with the best one at least δ^* better. Hence, in that situation the procedure indeed guarantees selection of the best mean with probability of at least P^*. Next include the population that was eliminated. This inclusion does not decrease the probability that the previously best population $\Pi_{(k-1)}$ gives a higher sample mean than the inferior populations j (j = 1, ..., k-2). (All observations are independent, an increase of the number of populations from k-1 to k does not decrease the number of observations to be sampled under the rule of the procedure.) The population Π_k now included, may or may not be selected instead of the previously superior population $\Pi_{(k-1)}$ but both selections are defined to be correct. A more formal derivation is given in Appendix 1.

(iii) <u>The generalized least favorable configuration</u> (GLF):

$$\mu_{(1)} = \cdots = \mu_{(k-1)} = \mu_{(k)} - \delta \qquad (\delta \geq 0) \tag{4}$$

The GLF comprises the LFC; put $\delta = \delta^*$ in Eq. (4). Further, condition (3a) does not conflict with Eq. (4) but Eq. (3b) does (since $\delta^* > 0$). Now suppose that Eq. (4) holds with δ being "small"; suppose even that $\delta = 0$ so that we have the <u>equal means configuration</u> (EMC):

$$\mu_{(1)} = \cdots = \mu_{(k-1)} = \mu_{(k)} \tag{5}$$

The EMC, the GLF with δ small, or the PIC has unpleasant consequences for open end sequential MRP. Open end sequential procedures do not limit the number of observations that may be sampled. Hence, such MRP can result in extremely large sample sizes, as the best population $\Pi_{(k)}$ cannot easily be identified since two or more populations keep competing. Therefore it is practical to specify an upper bound for the sample sizes. Such a bound may result from the available computer time. We may also calculate the sample sizes that single-sample or two-stage MRP would require and use these sample sizes (possibly adjusted with a correction factor) as upper limit [compare Bechhofer et al. (1968, p. 227)]. Clearly such an adaptation of the original open end procedure may effect P(CS) but it is a practical solution.

An advocate of decision theory would like to have P(CS) increase as $\mu_{(k)} - \mu_{(k-1)}$ increases, instead of the "all or nothing" formulation in Eq. (1). As we have just made plausible, a consequence of the indifference zone approach is that P(CS) increases as $\mu_{(k)} - \mu_{(k-1)}$ grows larger than δ^*, but the increase of P(CS) is not controlled. We shall briefly return to decision-theoretic solutions in Sec. V.C.5 [also see Bechhofer et al. (1968, p. 253)].
p. 253)].

V.C.3. EXISTING PROCEDURES

In this section we shall give a survey of those existing MRP that in our opinion may be applicable in simulation experiments. Since our major concern is whether a procedure can indeed be applied in simulation, we shall classify these MRP according to their assumptions. All procedures assume independent observations (within and between populations). Among the parametric procedures we shall discuss only MRP for normal distributions. We suppose that the latter type of procedures can be more widely applied in simulation than other parametric procedures. The normal distribution is an adequate approximation to many other distributions, and deviations between the actual unknown distribution and the normal distribution are hoped not to be disastrous. We shall return to the problem of robustness of MCP in the next section. Procedures based on other distributions will be briefly discussed in Sec. V.C.5. In the present section we shall also examine distribution-free and "semi-distribution-free" MRP. Besides the distinction parametric versus nonparametric another relevant distinction is that between distributions differing only in location and distributions differing in location and form. For normal populations the latter difference is equivalent to common variance versus unequal variances. A further subclassification is known versus unknown variances. All MRP in this section use the indifference zone approach [specified in Eq. (1) or a modification as, e.g. Sobel (1967) used]. After sampling has been terminated the parametric procedures select as best population the one yielding the highest sample mean, and the (semi-)distribution-free procedures the one with the highest rank (or more general the highest score) or highest modus. We shall also present some procedures for situations where a standard population exists and shall discuss higher-way layouts.

Normal Populations

Known Variances

Various MRP exist for known variances. The variances may be different or not, so that these procedures also apply to common known variances. If the experimenter would assume a common but unknown variance σ^2 and, moreover, he would specify the indifference quantity δ^* as a percentage of σ, say $\delta^* = \lambda^*\sigma$ then he could still apply MRP for a common but known variance. [For Eq. (6) yields $n = (d/\lambda^*)^2$ and Eq. (12) contains $\delta^*/c = \lambda^*$.] Such an approach, however, seems very unattractive in simulation. Next we shall present one single-sample and one sequential procedure.

The Bechhofer (1954) single-sample procedure. From the above it follows that Bechhofer assumed normally distributed independent observations with known variances, say σ_i^2 $(i = 1, \ldots, k)$. From population i, n_i observations are to be taken where n_i is so chosen that the probability requirement in Eq. (1) is satisfied. In Appendix 2 we have given a simple derivation for the determination of n_i; our derivation does not follow Bechhofer but we arrive at the same result:[4)] Take the sample sizes n_i as in Eq. (6)

$$n_i = (\sigma_i d/\delta^*)^2 \tag{6}$$

where δ^* is the specified constant in Eq. (1) and d is a critical constant increasing with k (more competitors) and P^*. Bechhofer (1954, pp. 30-34) tabulated d for $k = 2(1)10$ and P^* between (roughly) $1/k$ and 0.9995; d is the solution of Eq. (2.13) in Appendix 2. Other tables exist, e.g. Milton (1963), but the most elaborate table has been calculated by Gupta (1963a, pp. 800, 810), with $k = n + 1$ and $h = d/\sqrt{2}$ for $k = 1(1)49$ and P^* is 0.75, 0.90, 0.95, 0.975 and 0.99; also see Gupta et al. (1971, p. 18) who listed h with one more decimal place. If Eq. (6) does not result in an integer value, we take the next largest integer. In the derivation of Eq. (6) the sample sizes are supposed to be taken in the proportion

$$\frac{n_i}{n_{i'}} = \frac{\sigma_i^2}{\sigma_{i'}^2} \qquad (i, i' = 1, \ldots, k) \tag{7}$$

so that the sample means have a common variance. The sample allocation in Eq. (7) is not meant to be most efficient but simplifies the derivation of the sample sizes needed to satisfy the probability requirement in Eq. (1); compare Bechhofer (1954, p. 24). Dudewicz and Dalal (1971) compared the usual allocation in Eq. (7) with the optimal allocation $n_i/n_{i'} = \sigma_i/\sigma_{i'}$ for the case $k = 2$ (also consult Sec. V.A.4). For $k > 2$ no optimal allocation rule is available. They further found that (asymptotically) the usual rule Eq. (7) requires a smaller total sample size than taking an equal number of observations from each population [and still realizing the probability requirement in Eq. (1)]. (Also consult the references in Dudewicz and Dalal (1971, p. 7.2).) We observe that several approximative formulas for the single-sample size (not requiring the tabulated constant d) are discussed in Dudewicz and Zaino (1971). For an application of Eq. (6) see Exercise 1.

The Bechhofer, Kiefer, and Sobel (1968) sequential procedure. Bechhofer et al. (1968, pp. 258-259, 264-265) give a sequential procedure where in each stage s $(s = 1, 2, \ldots)$ r_i observations are taken from population i $(i = 1, \ldots, k)$, the r_i being the smallest integers for which Eq. (8) holds

$$\frac{\sigma_1^2}{r_1} = \cdots = \frac{\sigma_i^2}{r_i} = \cdots = \frac{\sigma_k^2}{r_k} = c^2 \qquad (c > 0) \tag{8}$$

[Observe the correspondence between Eq. (8) and the single-sample analogue Eq. (7).] The constant c^2 is the common variance of $\bar{\underline{x}}_{is}$, the average of the observations sampled in stage s from Π_i, i.e.

$$c^2 = \operatorname{var}(\bar{\underline{x}}_{is}) = \operatorname{var}\left(\sum_{g=1}^{r_i} \frac{\underline{x}_{isg}}{r_i} \right) \tag{9}$$

where $\underline{x}_{isg}$ is the gth individual observation in stage s from Π_i. After stage m $(m = 1, 2, \ldots)$ calculate the "cumulative" statistic

$$\underline{y}_{im} = (cr_i)^{-1} \sum_{s=1}^{m} \sum_{g=1}^{r_i} \underline{x}_{isg} \tag{10}$$

Let $\underline{y}_{[i]m}$ denote the ranked $\underline{y}_{im}$, i.e. $\underline{y}_{[1]m} \leq \cdots \leq \underline{y}_{[k]m}$. Then after each stage m calculate the $k - 1$ differences

$$\underline{d}_{jm} = \underline{y}_{[k]m} - \underline{y}_{[j]m} \qquad (j = 1, \ldots, k-1) \tag{11}$$

Terminate the experiment at the first value of $\underline{m}$ for which Eq. (12) holds.[5)]

$$\sum_{j=1}^{k-1} \exp[-(\delta^*/c)\, \underline{d}_{jm}] \leq (1 - P^*)/P^* \tag{12}$$

As long as Eq. (12) does not hold, sampling is continued. Once the experiment is stopped the population with the largest $\underline{y}_{im}$ (or the largest sample mean) is selected as best. Observe that smaller samples are needed as δ^* increases or P^* decreases.

If all populations have the same known variance, say $\sigma_i^2 = \sigma^2$, then the procedure simplifies since in each stage a single observation is taken $(r_i = 1)$. The authors also give procedures for many other distributions as we shall see in Sec. V.C.5. In contrast with Bechhofer's single-sample procedure the sample size is unknown until experimentation is terminated. It is desirable to have some ideas about the expected sample size in order to plan the computer time needed for the experiment. Bechhofer et al. (1968, pp. 301-313, 364) derived several asymptotic results for the <u>expected sample size</u> (or average sample number, ASN).

(i) If δ^*/σ[6] is close to zero, use the tables in Bechhofer et al. (1968, pp. 353-362) showing Monte Carlo estimates of ASN in the LFC and the EMC for various P^* and k [with $\delta^*/\sigma = 0.2$ so that the estimates should be multiplied by $(0.2\ \sigma/\delta^*)^2$; see Bechhofer et al. (1968, p. 303)]. They conjecture that $\delta^*/\sigma \leq 0.2$ is close enough to zero to make this approximation work.

(ii) For P^* close to one Bechhofer et al. (1968, pp. 306-307) give a table for ASN but only for the LFC. [Multiply the tabulated values by $(\sigma/\delta^*)^2$.] An example is given by Bechhofer et al. (1968, p. 312).

(iii) For the LFC they also fitted regression equations to the data in the tables mentioned in (i). The coefficients of these equations are shown in their table 14.2.3.

Known Variance Ratios

The Bechhofer, Dunnett, and Sobel (1954) two-stage procedure. Bechhofer et al. (1954) derived a procedure for unknown variances σ_i^2 with known ratios a_i, say $\sigma_i^2 = a_i\sigma^2$ with unknown σ^2 and known integers a_i. The procedure runs as follows.

(i) Take a first sample of $a_i n_0$ observations from population i; n_0 may be any integer; no practical rule for the optimal choice of n_0 is known.

(ii) Calculate $\underline{s}_0^2$, the unbiased estimator of σ^2 based on the first stage:

$$\underline{s}_0^2 = \frac{1}{v} \sum_{i=1}^{k} \frac{1}{a_i} \sum_{g=1}^{a_i n_0} \left(\underline{x}_{ig} - \frac{\sum_{g=1}^{a_i n_0} \underline{x}_{ig}}{a_i n_0} \right)^2 \qquad (13)$$

where

$$v = \sum_{i=1}^{k} (a_i n_0 - 1) = n_0 \sum_{i=1}^{k} a_i - k \qquad (14)$$

is the degrees of freedom.

(iii) Take a second sample of $a_i(\underline{n} - n_0)$ observations from Π_i, where

$$\underline{n} = \max\{n_0, \,]2\underline{s}_0^2(h/\delta^*)^2[\} \qquad (15)$$

the square brackets][denoting the smallest integer equal to or greater than the rational number between the brackets, and h being a critical constant increasing with P^* and k and decreasing v. We shall return to this constant after step (v).

(iv) Calculate the over-all sample means

$$\bar{\underline{x}}_i = (a_i \underline{n})^{-1} \sum_{g=1}^{a_i \underline{n}} \underline{x}_{ig} \qquad (16)$$

(v) Select the population with the largest sample mean.

The critical constant h is the solution of a multivariate t-distribution (with correlation 1/2) and is tabulated in tables 1a and 1b in Dunnett (1955, with $p = k - 1$) and Krishnaiah and Armitage (1966, pp. 41, 51, with $n = v$, $p = k - 1$) while $h\sqrt{2}$ is given by Gupta and Sobel (1957).[7] An application is mentioned in Exercises 3 and 8.

Equal Unknown Variances

The Bechhofer, Dunnett, and Sobel (1954) two-stage procedure. A common unknown variance is a special case of known variance ratios, viz. $a_i = 1$ in $\sigma_i^2 = a_i \sigma^2$. Hence, the procedure in Eqs. (13) through (16) can be applied.

The Robbins, Sobel, and Starr (1968) sequential procedure. If the variances $\sigma_i^2 = \sigma^2$ were known, we could apply Bechhofer's single-sample procedure, i.e. from Eq. (6) it would follow that

$$n = (\sigma\ d/\delta^*)^2 \tag{17}$$

where d is tabulated in, e.g. Bechhofer (1954). Now Robbins et al. (1968) simply estimate the common variance σ^2 and update their estimate in each stage, using the well-known estimator based on all observations available after stage m $(m \geq 2)$:

$$\underline{s}_m^2 = \frac{1}{k} \frac{1}{(m-1)} \sum_{i=1}^{k} \sum_{s=1}^{m} \left(\underline{x}_{is} - \frac{\Sigma_{s=1}^{m} \underline{x}_{is}}{m} \right)^2 \tag{18}$$

From Eq. (17) it follows that sampling is terminated for the first value m of $\underline{m}$ for which Eq. (19) holds.[8)]

$$\underline{m} \geq (\underline{s}_m\ d/\delta^*)^2 \tag{19}$$

The authors proved that their procedure guarantees the probability requirement in Eq. (1) asymptotically (i.e. for $\delta^* \to 0$ and consequently $n \to \infty$). For small samples their analysis also shows satisfactory results for the cases studied [see the table in Robbins et al. (1968, p. 91)]. As we remarked in Sec. V.A, it may be convenient for computer applications to base the calculation of s_m^2 on Helmert's transformation of the sum of squares:

$$\sum_{s=1}^{m} (\underline{x}_{is} - \bar{\underline{x}}_i)^2 = \sum_{s=1}^{m-1} \underline{u}_s \tag{20}$$

where

$$\underline{u}_s = \frac{1}{s(s+1)} \left(s\underline{x}_{s+1} - \sum_{t=1}^{s} \underline{x}_t \right)^2 \qquad (s = 1, \ldots, m-1) \qquad (21)$$

so that a new observation x yields a new u without changing the old u's [see, e.g. Kendall and Stuart (1963, p. 250) or Tocher (1963, p. 114)].

The Bechhofer and Blumenthal (1962) sequential procedure. An alternative sequential procedure was proposed by Bechhofer (1958); later on an equivalent but algebraically simpler rule was formulated in Bechhofer and Blumenthal (1962, pp. 54-57). It runs as follows.

(i) Let $\underline{y}_{im}$ denote the sample sum of population i after stage m:

$$\underline{y}_{im} = \sum_{s=1}^{m} \underline{x}_{is} \qquad (22)$$

[Compare Eq. (10) above when $\sigma_i^2 = c^2 = 1$ and consequently $r_i = 1$.] Denote the ranked $\underline{y}_{im}$ by $\underline{y}_{[i]m}$. Let $\underline{s}_{v_m}^2$ be the best unbiased estimator of the common variance σ^2 based on v_m degrees of freedom after stage m; if no other information is available in a one-way layout then $\underline{s}_{v_m}^2$ is identical to $\underline{s}_m^2$ in Eq. (18) with $v_m = k(m-1)$. The numerator of $\underline{s}_{v_m}^2$, i.e. a sum of squares, is denoted by $(\underline{SS}_{v_m})$.

(ii) Compute the value of

$$\underline{G}_m = \frac{\Sigma_{i=m}^{k} \underline{y}_{im}^2}{m} - \frac{(\Sigma_i \underline{y}_{im})^2}{km} + \frac{2\delta^* \Sigma_i y_{im}}{k} + \frac{m(k-1)(\delta^*)^2}{k} \qquad (23)$$

(iii) Compute the $(k-1)$ values of

$$\underline{L}_{jm} = \left[\frac{(\underline{SS}_{v_m}) + \underline{G}_m - 2\delta^* \underline{y}_{[k]m}}{(\underline{SS}_{v_m}) + \underline{G}_m - 2\delta^* \underline{y}_{[j]m}} \right]^{(v_m+k-1)/2} \qquad (24)$$

$$(j = 1, \ldots, k-1)$$

(iv) Compute the value of the stopping statistic at stage m:

$$\underline{z}_m = \sum_{j=1}^{k-1} \underline{L}_{jm} \tag{25}$$

(v) After stage m $(m \geq 2)$ terminate sampling if

$$\underline{z}_m \leq (1 - P^*)/P^* \tag{26}$$

and select as the best population the one with the highest sample sum, $y_{[k]m}$; otherwise proceed to stage (m+1).

In a one-way layout with no additional information [i.e. $v_m = k(m-1)$] we can substitute

$$\underline{H}_m = (\underline{SS}_{v_m}) + \underline{G}_m$$
$$= \frac{1}{km}\left[km \sum_{i=1}^{k} \sum_{s=1}^{m} \underline{x}_{is}^2 - (\sum_i \underline{y}_{im})^2 + 2m\delta^* \sum_i \underline{y}_{im} + (k-1)(m\delta^*)^2\right] \tag{27}$$

in Eq. (24). For additional comments on the computation formulas and a computation example we refer to Bechhofer and Blumenthal (1962). Bechhofer (1958, pp. 413-414) shows how the formula for the stopping statistic $\underline{z}_m$ simplifies as the degrees of freedom v_m increase; if $v_m = \infty$ then $\underline{z}_m$ reduces to the stopping statistic in the Bechhofer, Kiefer, and Sobel procedure for common but known variances. [Compare Eq. (12) above with $\approx{z}_m(d_m)$ in Bechhofer (1958, p. 414).]

Recently Bechhofer (1970) pointed out that it has not been proved analytically that the Bechhofer-Blumenthal procedure guarantees the probability requirement in Eq. (1) though "he conjectures that it does asymptotically as $\delta^* \to 0$ for fixed P^*." We add that in quite extensive Monte Carlo studies it was found that indeed $P(CS) \geq P^*$ in the LFC [see Bechhofer and Blumenthal (1962, pp. 58-64) and our own results in Chap. VI]. Bechhofer (1970) remarks that unlike the suggestion in the original publications the procedure cannot capitalize on a more favorable configuration of the means; actually the sample size then increases. We shall return to this remark in Sec. V.C.4, where the efficiency of MRP is compared.

The Paulson (1964a) closed sequential procedure with elimination. Paulson (1964a) derived a sequential procedure where populations are eliminated during experimentation if they yield sample means that are at least a certain amount worse than the best average. He gives a procedure for common unknown and for common known variances but the latter assumption does not hold in simulation so that we present only the procedure for common unknown variances.[9)]

(i) Take n_0 observations from each population and estimate σ^2 by $\underline{s}_v^2$, v being the degrees of freedom. There is no simple practical rule for an efficient choice of n_0; for a tentative rule see Paulson (1964a, p. 179). The estimator $\underline{s}_v^2$ is based only on these n_0 observations; later observations are not used. The sample sums, however, are based on all observations.

(ii) Calculate the values of the following auxiliary variables.

$$g = \{[(k-1)/(1-P^*)]^{2/v} - 1\}v/2 \quad (28)$$

$$\underline{a}_\lambda = \frac{\underline{s}_v^2 g}{\delta^* - \lambda} \quad (29)$$

$$\underline{W}_\lambda =]\underline{a}_\lambda/\lambda - 1[\quad (30)$$

where λ is a constant satisfying $0 < \lambda < \delta^*$. Originally Paulson (1964a, p. 177) suggested that $\lambda = \delta^*/4$, but later on Paulson (1964b, p. 1050) proposed $\lambda = 3\delta^*/8$ as a more efficient value for this type of sequential procedures; Ramberg (1966, p. 65) studied Paulson's procedure for known variance σ^2 in Monte Carlo experiments and found that $\lambda = 0$ is optimal in the LFC (but not necessarily in other configurations). $\underline{W}_\lambda$ is the largest integer less than $\underline{a}_\lambda/\lambda$; $\underline{W}_\lambda + 1$ is the maximum number of stages, i.e. the sequential procedure is a closed procedure.

(iii) If $n_0 > W_\lambda$ we stop the experiment and select as best population the one with the largest sample mean, $\bar{x}_{[k]}$.

(iv) If $n_0 \leq W_\lambda$ then we eliminate any population j for which Eq. (31) holds.

$$\sum_{s=1}^{n_0} x_{js} < \max_{1 \leq i \leq k} \left(\sum_{s=1}^{n_0} x_{is} \right) - a_\lambda + n_0 \lambda \tag{31}$$

If a single population remains after application of the elimination rule, Eq. (31), we terminate the experiment and select that population; otherwise we proceed to the next stage and take one observation from each population not eliminated before that stage.

(v) In general, after stage $n_0 + t$ $(t = 1, \ldots, W - n_0)$ we eliminate any population j satisfying

$$\sum_{s=1}^{n_0+t} x_{js} < \max_{i'} \sum_{s=1}^{n_0+t} x_{i's} - a_\lambda + (n_0 + t)\lambda \tag{32}$$

where the maximum is taken over all populations i' not being eliminated after stage $n_0 + t - 1$, etc. We have summarized the procedure in the flow chart of Fig. 1.

Hoel and Mazumdar (1968) extended Paulson's procedure to so-called Koopman-Darmois populations [a particular class of populations having a single unknown parameter, e.g. exponential, binomial, Poisson, normal with known variances; see Bechhofer et al. (1968, pp. 62, 251)]. pp. 62, 251)].

Unequal and Unknown Variances

The Dudewicz and Dalal (1971) two-stage procedure. Dudewicz and Dalal (1971, Section 4) derived the following procedure where, however, steps (iii) and (iv) are taken from a more recent publication by Dudewicz (1972, p. 3).

(i) Take a first sample of n_0 observations $(n_0 \geq 2)$ from each population and calculate the first-stage sample mean

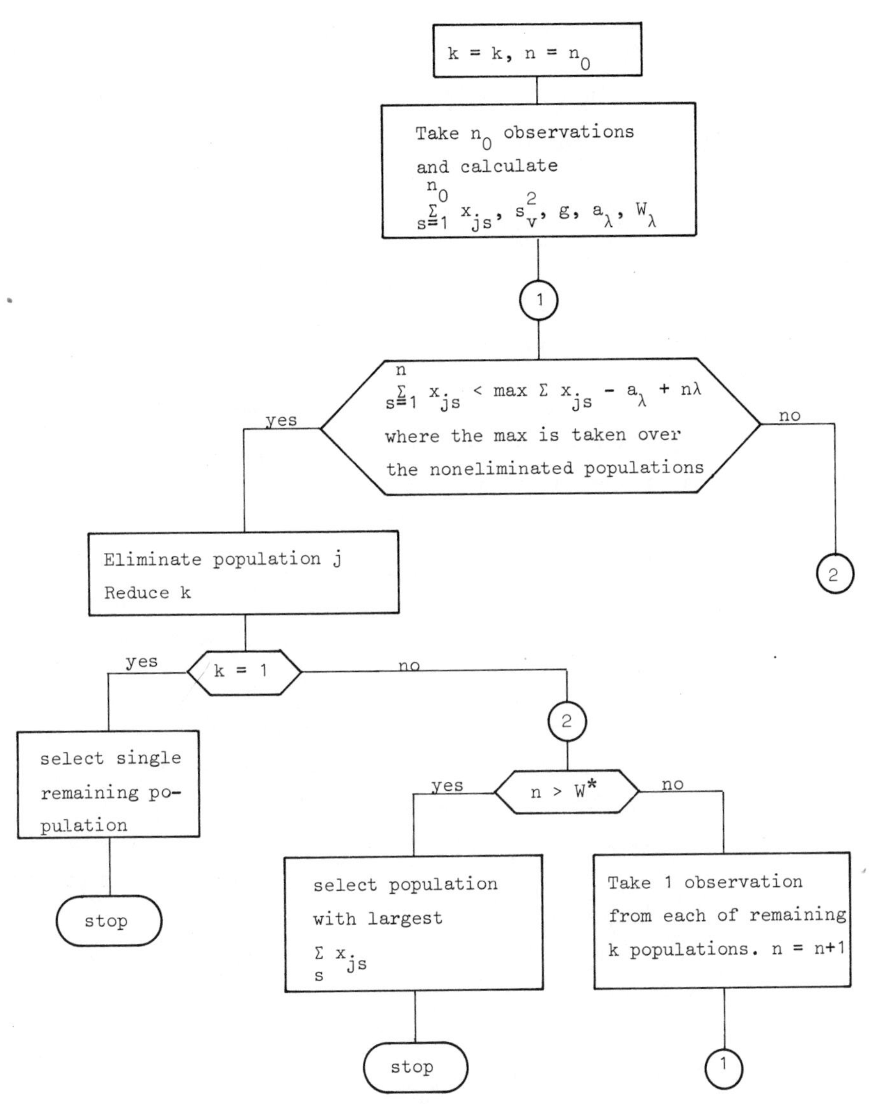

Fig. 1. Paulson's closed sequential procedure with elimination.

$$\bar{\underline{x}}_{i0} = \sum_{g=1}^{n_0} \underline{x}_{ig}/n_0 \qquad (i = 1, \ldots, k) \tag{33}$$

and the estimators of the variances

$$\underline{s}_i^2 = \sum_{g=1}^{n_0} \frac{(\underline{x}_{ig} - \bar{\underline{x}}_{i0})^2}{n_0 - 1} \tag{34}$$

(ii) Calculate the total sample sizes $\underline{n}_i$ required for population i:

$$\underline{n}_i = \max\left[n_0 + 1, \left]\left(\frac{\underline{s}_i h_D}{\delta^*}\right)^2\right[\,\right] \tag{35}$$

where the constant h_D is the solution of the Student analogue of Bechhofer's equation for the constant d [cf. Eq. (2.13) in Appendix 2], i.e. h_D is the solution of

$$\int_{-\infty}^{\infty} [F_{n_0-1}(t + h)]^{k-1} f_{n_0-1}(t)\, dt = P^* \tag{36}$$

F_{n_0-1} and f_{n_0-1} being the distribution function and density function of Student's t-variable with $n_0 - 1$ degrees of freedom. The constant h_D is tabulated in Dudewicz and Ramberg (1972) for k = 2(1)5, P^* = 0.75 (0.05) 0.95, 0.975, 0.99, and n_0 = 2(1)10(5)30. Further Dudewicz and Dalal (1971) calculated the resulting P^* for fixed h_D = 0.0(0.1)5.1, k = 2(1)25, and n_0 = 2(1)15(5)30. For $k \geq 26$ Dudewicz and Dalal (1971, Section 5) derived a formula giving lower bounds on P(CS). Table III in Dudewicz and Dalal (1971) shows that for increasing n_0 the constant h_D rapidly approaches its limit, viz. Bechhofer's d.

(iii) Take $\underline{n}_i - n_0$ additional observations from population 1 and calculate the second-stage mean $\bar{\underline{x}}_{i1}$:

$$\bar{\underline{x}}_{i1} = \sum_{j=n_0+1}^{\underline{n}_i} \frac{\underline{x}_{ij}}{\underline{n}_i - n_0} \tag{37}$$

(iv) Calculate $\tilde{\bar{x}}_i$, the weighted average of the two sample means:

$$\tilde{\bar{x}}_i = b_1 \bar{x}_{i0} + (1 - b_1)\bar{x}_{i1} \tag{38}$$

where the weight b_1 is

$$b_1 = \frac{n_0}{n_i}\left[1 + \left(1 - \frac{n_i}{n_0}\left[1 - \frac{n_i - n_0}{(h_D s_i/\delta^*)^2}\right]\right)^{1/2}\right] \tag{39}$$

(v) Select as best population the one yielding the highest $\tilde{\bar{x}}_i$.

A variant of this procedure is given in Dudewicz and Dalal (1971). Instead of steps (iii) and (iv) the following linear combination of the n_i observations is taken:

$$\tilde{\bar{x}}_i = \sum_{g=1}^{n_i} a_{ig} x_{ig} \tag{40}$$

where

$$a_{i1} = \cdots = a_{i(n_i-1)} = c_i$$

$$a_{in_i} = 1 - (n_i - 1)c_i$$

$$c_i = \frac{(n_i-1) \pm [(n_i-1)^2 - (n_i-1)n_i\,(1 - (\delta^*/h_D)^2/s_i^2)]^{1/2}}{(n_i - 1)n_i} \tag{41}$$

Actually any a_{ig} satisfying the following conditions are valid.

$$\sum_{g=1}^{n_i} a_{ig} = 1, \quad a_{i1} = \cdots = a_{in_0}, \quad s_i^2 \sum_{g=1}^{n_i} a_{ig}^2 = (\delta^*/h_D)^2 \tag{42}$$

Finally, the authors also discuss using the over-all sample means but they recommend the exact procedure based on $\tilde{\bar{x}}_i$. Later on Dudewicz (1972, p. 7) states that use of the sample means "seems safe."

(Semi-) Distribution-Free Procedures

Unknown Distributions with Common Unknown Variance

The Srivastava (1966) sequential procedures A and B. Srivastava (1966) derived two classes of sequential procedures which guarantee the probability requirement in Eq. (1) asymptotically, i.e. for $\delta^* \to 0$. For small samples Srivastava and Ogilvie (1968) studied the cases $P^* = 0.95$ and $k = 2, 4, 6$ and derived analytically that Eq. (1) is again satisfied. Let us consider a particular procedure belonging to one class and one belonging to the other class of procedures.[10)]

Procedure A based on the t-distribution. From Srivastava (1966, p. 372) and Srivastava and Ogilvie (1968) we found that this procedure can be described as follows.

(i) In each stage s take one observation from each population and calculate $\underline{s}_v^2$, the estimated variance after m stages:

$$\underline{s}_v^2 = [k(m-1)]^{-1}\left[\sum_{i=1}^{k}\sum_{s=1}^{m}(\underline{x}_{is} - \bar{\underline{x}}_i)^2 + 1\right] \tag{43}$$

The factor 1 in the square brackets may be deleted for continuous distributions so that Eq. (43) reduces to Eq. (18) in that case. In his Eq. (2.5) Srivastava (1966) gives the estimator

$$\underline{s}_v^2 = (km)^{-1}\left[\sum_{i=1}^{k}\sum_{s=1}^{m}(\underline{x}_{is} - \bar{\underline{x}}_i)^2 + 1\right] \tag{44}$$

but replacing m by $m-1$ in the denominator as Srivastava and Ogilvie (1968, p. 1041) did, does not disturb the asymptotic behavior.

(ii) Stop sampling if

$$\underline{m} \geq \left(\frac{\underline{s}_v a_m \sqrt{2}}{\delta^*}\right)^2 \tag{45}$$

and select the population with the largest sample mean; a_m is a critical constant determined from the t-distribution with $v = k(m-1)$ degrees of freedom, say $f_v(t)$, in such a way that[11)]

$$\int_{-\infty}^{-a_m} f_v(t)\, dt = \frac{1 - P^*}{k - 1} \tag{46}$$

Procedure B based on the multivariate t-distribution. This procedure is identical to A except for its critical constants. In Appendix 3 we have shown that these constants, say a_m^*, can now be determined from the tables in Dunnett (1955) or Krishnaiah and Armitage (1966), and from Gupta and Sobel (1957). Look up d' in Dunnett and put $a_m^* = d'$, or look up the constant q in Gupta and Sobel and put $a_m^* = q/\sqrt{2}$. [The values of d' and q vary with $v = k(m-1)$, P^* and k.][12)]

Unknown Distributions, Identical Except for Location

MRP exist for unknown distribution functions that are identical except for a possible location shift, i.e. the $\underline{x}_i$ have distribution functions F_i satisfying

$$F_i(x) = F(x - \mu_i) \qquad (i = 1, \ldots, k) \tag{47}$$

Unfortunately, some of these procedures have certain drawbacks explained below, that make them less attractive. Therefore we shall describe these MRP only briefly; for more details we refer to the literature.

The Bechhofer and Sobel (1958) nonparametric procedure. Bechhofer and Sobel's procedure is based on selection of the population with the highest probability of yielding the largest observation. Consider

$$p_i = P(\underline{x}_i = \max_{i'} \underline{x}_{i'}) \qquad (i, i' = 1, \ldots, k) \tag{48}$$

so that $\Sigma_1^k p_i = 1$. Then find the population corresponding with $p_{(k)}$, where $p_{(1)} \leq \cdots \leq p_{(k-1)} \leq p_{(k)}$. Because of Eq. (47) the population corresponding with $p_{(k)}$ is also the population with the largest location parameter $\mu_{(k)}$. (Also see Exercise 4.) Define the multinomial variable $\underline{y}$:

$$\underline{y}_{is} = 1 \qquad \text{if} \quad \underline{x}_{is} = \max_{i'} \underline{x}_{i's} \qquad (s = 1, 2, \ldots)$$
$$= 0 \qquad \text{otherwise} \tag{49}$$

From Eq. (49) it follows that the multinomial variable with the highest probability of occurrence is the one corresponding with the population with the highest p_i (which in turn is associated with the population with the highest μ_i). Procedures for selection of the multinomial event with the highest probability are, e.g. the single-sample procedure of Bechhofer et al. (1959) and the sequential procedure of Bechhofer and Sobel (1956); see these publications for details. Additional references are given by Dudewicz (1968) and Gupta and Panchapakesan (1969a, p. 475). Unfortunately the above approach realizes

$$P(CS) \geq P^* \quad \text{if} \quad p_{(k)}/p_{(k-1)} \geq \theta^* \tag{50}$$

where $\theta^* > 1$. Correct selection occurs if we select the population with the highest p_i which is equivalent to selection of the highest μ_i but the indifference zone is now defined in terms of the p_i, not the μ_i. The μ_i are the quantities that really matter. The μ_i and p_i are related with each other. Analogous to Eq. (2) the configuration

$$p_{(1)} = \cdots = p_{(k-1)} = p_{(k)}/\theta^* \tag{51}$$

is the LFC for Eq. (50), i.e. it minimizes $P(CS)$ subject to the condition $p_{(k)}/p_{(k-1)} \geq \theta^*$. From Eq. (47) it follows that Eq. (51) implies $\mu_{(1)} = \cdots = \mu_{(k-1)}$ and, since $\theta^* > 1$, $\mu_{(k)} > \mu_{(k-1)}$. Denote $\mu_{(k)} - \mu_{(k-1)}$ associated with θ^* by δ_{θ^*}. The correspondence between θ^* and δ_{θ^*} depends on the number of populations, the form of the distributions and the nuisance parameters; see Dudewicz (1971, tables 2 and 7) who lists δ_{θ^*} as a function of θ^* and k for normal populations assuming a common known variance σ^2 [or, equivalently, replacing the indifference zone in Eq. (1) by

$\mu_{(k)} - \mu_{(k-1)} \geq \lambda^* \sigma]$ and for uniform distributions assuming a range 1. Summarizing, if we want to adhere to the original problem formulation in Eq. (1) with the indifference zone approach $\mu_{(k)} - \mu_{(k-1)} \geq \delta^*$, knowledge of the distribution form (and nuisance parameters) is necessary. The purpose of Bechhofer and Sobel's procedure, however, is to circumvent the requirement of knowledge of F_i.

The Lehmann (1963) single-sample ranking and related procedures. Lehmann (1963) proposed a single-sample procedure based on "expected scores," e.g. expected normal scores or ranks. [For a general discussion of expected and random scores see, e.g. Bradley (1968, pp. 146-163) or Conover (1971, pp. 281-282, 290-292).] The procedures select the population that yields the highest expected score (instead of highest sample mean). As in truly distribution-free procedures this selection criterion does not require knowledge of the distribution function F. But, when the sample size n is to be so determined that (asymptotically) the probability requirement in Eq. (1) is satisfied, then the formula for n contains the density function f(x). Therefore, we have to guess which density function the variables $\underline{x}_i$ have, e.g. approximately normal with a particular value of the variance σ^2. Lehmann's procedures are less sensitive to deviations from the assumed distribution than procedures based on the means. Unfortunately his procedures require specification of nuisance parameters, such as the variance σ^2 in the normal approximation. Other procedures permit estimation of these parameters. We remark that Bartlett and Govindarajulu (1968) studied the use of either expected or random scores [also see Puri and Puri (1969) and Randles (1970)].

The Sen and Srivastava (1972b) sequential rank-sum procedure. Sen and Srivastava (1972a and b) derived several sequential, asymptotically correct and efficient procedures based on Lehmann (1963) [also see Srivastava and Sen (1972)]. The procedures assume continuous distributions with location shift only (and some technical, regularity conditions). They also obtained Monte Carlo estimates. Since no rule was found to require fewer observations in all cases [and all rules satisfied the probability requirement in Eq. (1)], we shall present only the rule that requires the smallest amount of computer time for application; moreover it does not assume the distributions to be symmetric, whereas the other rules do. This procedure runs as follows.

(i) Take a pilot sample of n_0 observations from each of the k populations, n_0 being such that c in Eq. (52) is positive for $n = n_0$.

$$c = m_{k-1,\alpha}(n-1)\,(n/2)^{1/2} - n \tag{52}$$

where $m_{k-1,\alpha}$ is found in, e.g. Gupta (1963, p. 810). [Enter Gupta's table with $n = k - 1$, and $\alpha = 1 - P^*$ and find $m_{n,\alpha} = h\sqrt{2}$; see our comment on Eq. (6) above.]

(ii) Stop taking additional observations from each population as soon as the sample size $\underline{n}$ satisfies Eq. (53).

$$c \leq k^{-1} \sum_{i=1}^{k} \sum_{j=1}^{n} \sum_{j'=j+1}^{n} I(-2\delta^* \leq \underline{x}_{ij} - \underline{x}_{ij'} \leq 2\delta^*) \tag{53}$$

where I is the so-called indicator function, i.e.in general

$$\begin{aligned} I(A) &= 1 \qquad \text{if condition A holds} \\ &= 0 \qquad \text{if A is false} \end{aligned} \tag{54}$$

Hence, in Eq. (53) $\underline{I}$ is 1 if $-2\delta^* \leq \underline{x}_{ij} - \underline{x}_{ij'} \leq 2\delta^*$, etc.

(iii) Once sampling has stopped select as the best population the one with the highest rank sum $\underline{r}_i$, all populations being ranked together, i.e.

$$\underline{r}_i = \sum_{j=1}^{n} \underline{r}_{ij} = \sum_{j=1}^{n} \left[\sum_{g=1}^{n} \sum_{i'=1}^{k} I(\underline{x}_{ij} \geq \underline{x}_{i'g}) \right] \tag{55}$$

where the indicator form is an alternative way for expressing the rank of observation $\underline{x}_{ij}$.

Note that sequential ranking procedures imply that all individual observations are stored in computer memory (they are reranked as new observations become available). Since simulation may require large samples (as we saw in Part A of this chapter), much computer memory may be needed. This problem can be solved by not storing the data in core but on a tape or disk. In simulation the time for ranking and reranking observations may be of minor importance compared with the time required to generate those observations (one observation requires one simulation run). [Also see Lewis (1972, pp. 9-10), Martin (1971) and the discussion in Chap. V, Part A, under Eq. (59).]

Stochastically Ordered Distributions

As we know population i is called stochastically larger than population i' $(i \neq i')$ if

$$F_i(x) \leq F_{i'}(x) \qquad \text{for } \underline{\text{all}}\ x \tag{56}$$

Obviously distributions with location-shift-only form a subclass of Eq. (56).

The Hoel (1971) sequential ranking procedure with elimination. Recently Hoel derived a procedure in which all k (continuous) populations are compared with each other pairwise, and, as in the Paulson (1964a) procedure, populations can be eliminated during the sampling process. Hoel's rule is based on the sequential probability ratio test (SPRT) discussed in Sec. V.A.4. The indifference zone formulation in Eq. (1) is redefined in relative form, i.e.

$$P(CS) \geq P^* \qquad \text{if} \quad \mu_{(k)}/\mu_{(k-1)} \geq \theta^* \tag{57}$$

where obviously $\theta^* \geq 1$. The right-hand side of Eq. (57) implies

$$P(\underline{x}_{(k)} \geq \underline{x}_{(k-1)}) \geq \theta^*/(1 + \theta^*) \tag{58}$$

[for the specification of θ^* also consult Hoel (1971, p. 637)].

From Hoel (1971, pp. 632, 637) we deduced that the following procedure is applied in each stage to all pairs of populations. [When sampling starts we have $k(k-1)$ pairs, not $k(k-1)/2$.]

(i) Determine the rank $\underline{r}_{ii'}(j)$ of observation j from population i (i.e. $\underline{x}_{ij}$) when ranked with population i', ($i \neq i'$ and $1 \leq j \leq n$). As sampling continues this rank may change so that actually we have to determine $\underline{r}_{ii'}(j,n)$, i.e. the rank of observation j from population i with respect to population i' in a sample of n observations from the two populations.

(ii) Calculate the logarithm of the likelihood ratio when n observations are available from the populations i and i' ($n = 1, 2, \ldots$):

$$\underline{\ell}_{ii'}(n) = n \ln \frac{\tau_1}{\tau_0} + \sum_{j=1}^{n} \left(\ln \frac{\Gamma[\underline{r}_{ii'}(j,n) + j(\tau_1 - 1)]}{\Gamma[\underline{r}_{ii'}(j+1,n) + j(\tau_1 - 1)]} - \ln \frac{\Gamma[\underline{r}_{ii'}(j,n) + j(\tau_0 - 1)]}{\Gamma[\underline{r}_{ii'}(j+1,n) + j(\tau_0 - 1)]} \right) \tag{59}$$

where $\tau_0 = 1/\theta^*$ and it is recommended to take $\tau_1 = (1 + \theta^*)/2$; by definition $\underline{r}_{ii'}(n+1,n) = 2n+1$. For the computation of Eq. (59) the reader may wish to consult Bradley (1967, p. 596).

(iii) Eliminate population i' if for some i

$$\underline{\ell}_{ii'}(n) \geq \ln[(k-1)/(1 - P^*)] \tag{60}$$

(iv) Continue sampling from the remaining populations until only one population remains. If, however, in a stage, all remaining populations are eliminated then select from these populations the population i for which $\min_{i \neq i'} \underline{\ell}_{ii'}(n)$ is a maximum.

The procedure reaches a decision in a finite number of stages so that we know that "extremely" large samples do not occur [compare our comment below Eq. (5) in Sec. V.C.2]. The same type of procedure can be applied to other problems; Hoel (1971, p. 634) has specified his procedure for selection of the normal population with the smallest variance. We point out that Fu (1970, pp. 54-63, 96-116) also discusses the application of the SPRT to problems with k (≥ 2) alternatives, possibly in a nonparametric setting.

Unknown Distributions

The Sobel (1967) single-sample nonparametric procedure. Sobel assumes only that the distributions functions F_i are continuous. Unfortunately, he has to redefine the original problem formulation in Eq. (1) to derive a solution. He defines correct selection as selection of the population with the highest median. (For a symmetric distribution, mean and median coincide.) His procedure selects the population yielding the largest sample median and guarantees $P(CS) \geq P^*$ provided a particular indifference relation is met. Unfortunately the indifference quantity is not the "natural" one, the difference between the best two medians, as we shall see next.

Denote the α-quantile of $F(x)$ by $x_\alpha(F)$, i.e. the value[13] of $\underline{x}$ satisfying

$$F(x) = \alpha \tag{61}$$

The median corresponds with $\alpha = .5$ so that the median is denoted by $x_{.5}(F)$. The natural indifference relation would be

$$x_{.5}(F_{(k)}) - x_{.5}(F_{(k-1)}) \geq \delta^* \tag{62}$$

where $F_{(i)}$ denotes the distribution function with the ith smallest median. Sobel uses the following two indifference relations, both differing from Eq. (62).

Approach (1)

$$P(CS) \geq P^* \qquad \text{if} \quad d \geq d^* \tag{63}$$

where

$$d = \min_{x \in I} [F_{(1)}(x) - F_{(k)}(x)] \tag{64}$$

I being the interval

$$I = [x_{.5-\epsilon^*}(F_{(k)}),\ x_{.5+\epsilon^*}(F_{(k)})] \tag{65}$$

[Compare Fig. 2.] The constants P^*, d^*, and ϵ^* are specified by the experimenter. (Obviously $1 > P^* > 1/k$, $d^* > 0$, $0 < \epsilon^* < .5$). When the distribution functions lie close together then d decreases. To keep $d \geq d^*$ [so that $P(CS) \geq P^*$] we may reduce d^*; Sobel's tables show that the sample size increases as d^* decreases. Another way to keep d larger than d^* is to reduce ϵ^* so that I in Eq. (68) decreases and x can be less varied to find the minimum in Eq. (64); a smaller ϵ^* requires a larger sample size. The exact relation between δ^* in Eq. (62) and the new constants d^* and ϵ^* is unknown so that we cannot use Eq. (63) to solve the original problem formulation (modified for medians).

Approach (2)

$$P(CS) \geq P^* \qquad \text{if} \quad d' \geq 0 \tag{66}$$

where

$$d' = x_{.5-\epsilon^*}(F_{(k)}) - x_{.5+\epsilon^*}(F_{(1)}) \tag{67}$$

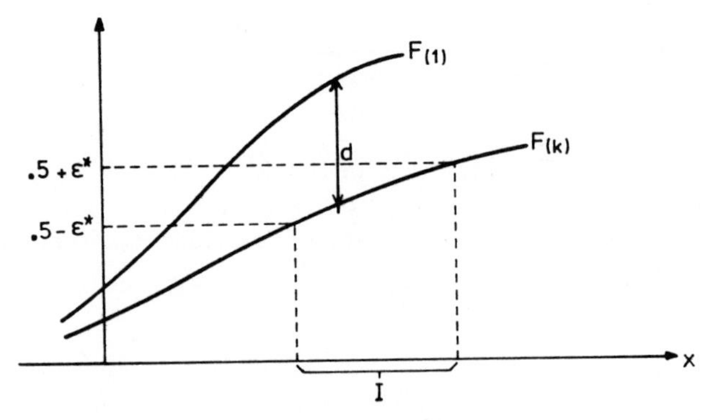

Fig. 2. Sobel's problem formulation--Approach (1).

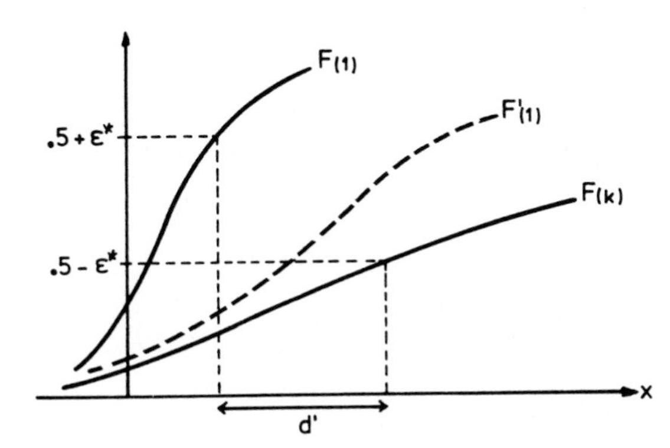

Fig. 3. Sobel's problem formulation--Approach (2).

[See Fig. 3.] From Fig. 3 it follows that when the distributions lie close to each other then d' decreases and may become negative (see dotted distribution $F'_{(1)}$) so that $P(CS) \geq P^*$ is not guaranteed. To prevent a negative d' we can choose a smaller ϵ^* but we do not know how small an ϵ^* we should take.

Procedures with a Standard Population

We shall present some procedures for situations with a standard or control population (e.g. the existing system in simulation experiments).

The Paulson (1962) procedure for selection of best mean. Assuming normal populations with a common unknown variance σ^2 Paulson derived a sequential procedure with elimination of inferior populations during experimentation. The purpose is to determine if any of the experimental populations are better than the standard and, if so, to select the best one. Similar to the above MRP he guarantees $P(CS) \geq P^*$ if the best experimental population $\Pi_{(k)}$ is at least δ^* better than either the other experimental populations $\Pi_{(j)}$ $(j = 1, \ldots, k-1)$ or the standard population Π_0, i.e.[14)]

$$P(CS) \geq P^* \quad \text{if} \quad \mu_{(k)} - \max(\mu_{(k-1)}, \mu_0) \geq \delta^* \tag{68}$$

The special feature of his procedure is the extra protection for the standard population. The standard is selected with probability at least P_0^* even if it is only as good as the best experimental population (instead of δ^* better), i.e.[15]

$$P(CS) \geq P_0^* \quad \text{if} \quad \mu_0 - \mu_{(k)} \geq 0 \tag{69}$$

The procedure runs as follows.

(i) Take a preliminary sample to estimate σ^2. Denote this estimator by $\underline{s}_v^2$, where v is the degrees of freedom. This sample is not used for the estimation of the means, in contrast with the Paulson (1964a) procedure above.

(ii) Calculate the boundaries a and b:

$$a = [(\lambda\alpha/k)^{-2/v} - 1]v/2 \tag{70}$$

$$b = -\{[\beta - (k-1)\lambda\alpha/k]^{-2/v} - 1\}v/2 \tag{71}$$

with

$$\lambda = \min[1, \beta/(\alpha k - \alpha)] \tag{72}$$

and $\alpha = 1 - P_0^*$, $\beta = 1 - P^*$.

(iii) Let $\underline{x}_{i's}$ denote the sth observation (after the above preliminary sample) from a population i' that is not yet eliminated in stage s ($s = 1, 2, \ldots; i' = 1, \ldots, k$). In stage s calculate for each population i':

$$\underline{z}_{i's} = \delta^*(\underline{x}_{i's} - \underline{x}_{0s} - \delta^*/2)/(2\underline{s}_v^2) \tag{73}$$

(iv) After m stages ($m = 1, 2, \ldots$) act as follows:

(a) If

$$\sum_{s=1}^{m} \underline{z}_{i's} \leq b \tag{74}$$

eliminate population i'; if no experimental population remains select the control as best population.

(b) If

$$\max_{i'} \left(\sum_{s=1}^{m} \underline{z}_{i's} \right) \geq a \tag{75}$$

then select the population yielding that maximum.

(c) If

$$b < \max_{i'} \left(\sum_{s=1}^{m} \underline{z}_{i's} \right) < a \tag{76}$$

then proceed to stage m+1 taking one observation from each remaining population; repeat steps (iii) and (iv).

Paulson (1962, pp. 439-440) also studied known μ_0 and known (arbitrary) density functions.

<u>The Sen and Srivastava (1972b) nonparametric procedure for selection of best mean</u>. Recently Sen and Srivastava (1972b) developed four nonparametric procedures guaranteeing Eqs. (68) and (69) [also see Srivastava and Sen (1972)]. They assumed location shift only (and some technical regularity conditions). Their rules are asymptotically correct and efficient and are sequential but do not eliminate inferior populations. The Monte Carlo results for their small-sample behavior showed that they do guarantee Eqs. (68) and (69). No procedure requires fewer observations in all cases studied. Therefore we shall present only the rule that runs analogous to Eqs. (52) through (55) (the situation without a standard population).

(i) Take a pilot sample of n_0 observations from each of the k+1 populations, n_0 being chosen such that c in Eq. (77) is positive for $n = n_0$.

$$c = m_{k-1,b}(n-1)(n/3)^{1/2} - n \tag{77}$$

the constant $m_{k-1,b}$ being determined by the following set of six equations with six variables $(m_{k,\alpha}, b, m_{k-1,b}a, x_a, d)$:

$$P(\max_{1 \leq i \leq k-1} \underline{x}_i > m_{k,\alpha}) = \alpha \quad (\equiv 1 - P^*) \tag{78a}$$

$$P(\max_{1 \leq i \leq k-1} \underline{x}_i > m_{k-1,b}) = b \tag{78b}$$

$$P(\max_{i=1} \underline{x}_i > x_a) = a \tag{78c}$$

$$m_{k,\alpha} = x_a \; d(d-1)^{-1} \tag{78d}$$

$$m_{k-1,b} = x_a(d-1)^{-1} \tag{78e}$$

$$a + b = \beta \quad (\equiv 1 - P_0^*) \tag{78f}$$

where the $\underline{x}_i$ are standard normal variables with correlation 1/2 for which Gupta (1963a, p. 810) tabulated

$$P(\max_{i \leq i \leq n} \underline{x}_i > m_{n,\alpha}) = \alpha \tag{79}$$

So Eqs. (78a), (78b) and (78c) can be looked up in Gupta's table with $n = k, k - 1, 1$, and $\alpha = 1 - P^*, b, a$.

(ii) Stop taking one additional observation from each population as soon as $\underline{n}$ satisfies Eq. (53) above where in Eq. (53) c is defined by Eq. (77), k becomes $k + 1$ and i runs from 0 to k.

(iii) Once sampling has stopped, select the standard population if

$$\max_{1 \leq i \leq k} (\underline{r}_i - \underline{r}_0) \leq (12)^{-1/2} \, m_{k,\alpha}(k+1) \, n^{3/2} \tag{80}$$

where $\underline{r}$ is the rank sum in the case of $k+1$ populations defined analogous to Eq. (55). If Eq. (80) does not hold then the experimental population with the highest rank sum is selected.

The Paulson (1964b) classification procedure. The goal of an experimenter may be not to select the best population but only to classify the experimental populations as superior $(\mu_i > \mu_0)$ or inferior $(\mu_i \leq \mu_0)$. We met procedures for this goal in the part on multiple comparison procedures. Paulson (1964b, pp. 1052-1054) derived a procedure whereby the number of observations is not fixed but can be chosen in a sequential way. The (joint) probability of classifying all experimental populations with means $\mu_i \leq \mu_0$ or $\mu_i > \mu_0 + \delta^*$ correctly, is at least P^* (or $1 - \alpha$). (Notice the indifference quantity δ^*.) He assumed that all populations have normal distributions. For later reference we first present his procedure for known possibly different variances σ_i^2 $(i = 1, \ldots, k)$ and σ_0^2; next his procedure for unknown common variance σ^2.

(a) Known variances.[16] Let $\underline{z}_{is}$ denote the differences between the experimental and the control observations, i.e.

$$\underline{z}_{is} = \underline{x}_{is} - \underline{x}_{0s} \qquad (i = 1, \ldots, k;\ s = 1, 2, \ldots) \tag{81}$$

so that the average difference after m stages is

$$\bar{\underline{z}}_{im} = \sum_{s=1}^{m} \frac{\underline{z}_{is}}{m} \tag{82}$$

and the variance of an individual difference is

$$\sigma_{zi}^2 = \sigma_i^2 + \sigma_0^2 \tag{83}$$

Calculate the "allowances"

$$d_i = \lambda + [\sigma_{zi}^2 \ln(k/\alpha)]/(2\lambda m) \tag{84}$$

where $\lambda = 3\delta^*/8$ is recommended by Paulson, and calculate the average difference corrected for the allowances:

$$\underline{u}_i = \bar{\underline{z}}_{im} + d_i \tag{85a}$$

$$\underline{v}_i = \bar{\underline{z}}_{im} - d_i \tag{85b}$$

After $\underline{m}$ stages an experimental population i is classified as inferior if $\underline{u}_i < \delta^*$ and as superior if $\underline{v}_i > 0$. In the next stage one observation is taken from each population not yet classified and from the control. The experiment is terminated when all populations are classified.

(b) Common unknown variances. Take a first sample of n_0 observations from each population and calculate the usual pooled estimator with $v = (k+1)(n_0-1)$ degrees of freedom. Proceed as under (a) but replace σ_{zi}^2 by $2\underline{s}_v^2$ and $\ln(k/\alpha)$ by $[(k/\alpha)^{2/v}-1]v/2$. Observe that σ_{zi}^2 is estimated from the first sample only but the means are estimated from all available observations. The choice of n_0 is briefly discussed in Paulson (1964b, pp. 1053-1054).

(c) Confidence intervals. An additional requirement may be that once the populations are classified we desire confidence intervals for the differences between the superior means, say μ_p, and the control mean μ_0 with joint confidence level $1 - \gamma$ and with lengths not exceeding L $(L > \delta^*)$. Then, after the populations are classified using (a) or (b), sampling is continued for the populations classified as superior until

$$\underline{u}_{pm} - \underline{v}_{pm} \leq L \qquad (p = 1, \ldots, P \leq k) \tag{86}$$

where for known variances $\underline{u}_{pm}$ is the minimum value of $\underline{u}_p$ in any of the stages s $(1 \leq s \leq \underline{m})$ provided α in Eq. (84) is replaced by $\gamma/2$ (two-sided intervals with confidence level γ are constructed and $\underline{v}_{pm}$ is the maximum of $\underline{v}_p$ over $\underline{m}$ stages; for unknown variances this minimum and maximum are taken only over the stages after the n_0 pilot observations. The required confidence intervals are $\underline{u}_{pm} - \underline{v}_{pm}$.

Tong (1969) also gives a classification procedure for normal populations with a common variance σ^2. He derived a single-sample procedure for known σ^2, and a two-stage and asymptotic sequential procedure for unknown σ^2. Contrary to Paulson he does not eliminate populations during the sampling process [also see Sobel and Tong (1970)].

The Puri and Puri (1969) subset procedure. Puri and Puri give a single-sample procedure based on ranks, or more generally scores. They assume identical populations except for location and give an asymptotic solution for the determination of the sample size. Their procedure selects a subset including all "superior" populations ($\mu_i \geq \mu_0 + \delta^*$) with prescribed probability P^*. A population is included in this subset if its score is higher than the score of the control population. (Compare the related subset procedures in the part on multiple comparison procedures.) Their solution, however, requires knowledge of the density function; see Lehmann's procedure for selecting the best mean.

The Srivastava (1966) procedure for correct selection of the standard. Srivastava derived a sequential procedure for the sample size determination that guarantees Eq. (69) asymptotically for unknown distributions with common unknown variance. Contrary to Paulson's procedure he does not guarantee Eq. (68).

The Dudewicz and Ramberg (1972) procedure for simultaneous confidence intervals. Dudewicz and Ramberg (1972) and Dudewicz (1972, p. 9) give a two-stage procedure for normal populations with unequal and unknown variances. The procedure is completely analogous to the Dudewicz and Dalal (1971) procedure given in Eqs. (33) through (42) above. However, δ^* (or their symbol a) now denotes "the least sample difference one would want to be sure was significant at level α" ($\alpha = 1 - P^*$). In private communication Dudewicz stated that δ^* may also be interpreted as the least population difference, i.e., if $\mu_0 \geq \mu_i - \delta^*$ then one is P^* sure of finding $\tilde{\bar{x}}_0 \geq \tilde{\bar{x}}_i$ (i = 1, ..., k-1). (Notice that there are k populations in total.) The (k-1) one-sided confidence intervals with joint probability P^* are:

$$\mu_i - \mu_0 \geq (\tilde{\bar{\underline{x}}}_i - \tilde{\bar{\underline{x}}}_0) - \delta^* \qquad (i = 1, \ldots, k-1) \tag{87}$$

so that we shall conclude that $\mu_i > \mu_0$ only if $\tilde{\bar{\underline{x}}}_i - \tilde{\bar{x}}_0 \geq \delta^*$. In this way we may select a subset of populations better than the standard, but the procedure is not specifically devised for that purpose.

Multi-Factor Designs

Until now we have restricted our discussion to experimental designs with one factor, possibly with a different number of observations per level; see, e.g. Bechhofer, Kiefer and Sobel's procedure. In multi-factor designs we may want to find the best level of factor A, the best level of B, etc. If we look only at factor A with a levels, we have a one-way layout with $k = a$ populations. In the way discussed above we can determine the number of observations, say n_A, guaranteeing selection of the best level of A, i.e. $\alpha^A_{(a)}$, provided it is at least δ^*_A better than the next best level. Or

$$P(CS_A) \geq P^*_A \qquad \text{if} \quad \alpha^A_{(a)} - \alpha^A_{(a-1)} \geq \delta^*_A \tag{88}$$

The number of observations per level of A must be distributed among the b levels of factor B (see Table 1). If per cell we take an equal number of observations, n_A/b, then we have (an_A/b) observations per level of B. We can then check the resulting value of P^*_B for (an_A/b) observations in

$$P(CS_B) \geq P^*_B \qquad \text{if} \quad \alpha^B_{(b)} - \alpha^B_{(b-1)} \geq \delta^*_B \tag{89}$$

As Bechhofer (1954, p. 27) showed, if the cell means have equal variances then the probabilities in Eqs. (88) and (89) are independent so that the joint probability of correct selection of the best level of A and B is the product of the individual probabilities (also see Exercise 6). We may not be sure of this independence because

TABLE 1

A Two-Factor Design

		Factor a			
		1	2	...	a
	1				
	2				
Factor B	⋮				
	b				
Total number of observations		n_A	n_A	...	n_A

the cell means have no equal variances or the distributions are not normal (zero correlations imply independence for normal distributions). Then we can use the Bonferroni inequality to derive bounds for the joint probability of correct selection:

$$P(CS) \geq 1 - \sum_{f=1}^{F} (1 - P_f^*)$$
$$\text{if} \quad \alpha_{(k_f)}^f - \alpha_{(k_f-1)}^f \geq \delta_f^* \qquad (f = 1, \ldots, F) \tag{90}$$

where P_f^* is the desired minimum probability of correct selection of the best level of factor f $(f = 1, \ldots, F)$ and $\alpha_{(k_f)}^f$ is the best level of factor f, etc. For details on the Bonferroni approach see Sec. V.B,3. If we want to fix the minimum value of the joint probability of CS then we can split this minimum value P^* into parts P_A^*, P_B^*, etc. and determine the required number of observations per cell in an iterative way; for an example see Bechhofer (1954, pp. 37-38).

If the factors interact with each other it makes no sense to look for the best level of A and that of B. Instead we can look at the design as a one-way layout with $k = ab$ levels and find the best combination of levels. In an experiment without interaction the observations in a particular cell can be used to make inferences about both A and B, i.e. each observation works twice [also see Bawa (1972)]. Note that in multi-factor designs with or without interaction the common variance of an individual observation can be estimated from the traditional ANOVA formulas. For MRP in higher-way layouts also see Bechhofer (1954, pp. 25-29, 37-38) and (1958, pp. 414-417, 425-426).

V.C.4. EFFICIENCY, ROBUSTNESS, AND APPROXIMATIVE PROCEDURES

MRP in Simulation

We shall next investigate the appropriateness of the above MRP in simulation experiments. We strongly recommend the reader also to consult the introduction of Sec. V.B.4, named "Efficiency and robustness of MCP in simulation." The desirability of applying MRP in simulation has been briefly mentioned by a few authors, viz. Conway et al. (1959, p. 107), Fishman and Kiviat (1967, p. 28), and Naylor et al. (1967a, p. 1327). Actual applications have been provided by Naylor et al. (1967b; 1968) who used the Bechhofer et al. (1954) two-stage procedure in their simulated multi-station queuing systems and national economic systems respectively, assuming a common unknown variance; the first application is also presented in Kleijnen et al. (1972, pp. 251-253).[17)] Recently Sasser et al. (1970) applied the Bechhofer and Blumenthal (1962) sequential procedure for finding the largest mean and the Paulson (1964b) classification procedure to their simulated multi-item inventory system (without known analytical solution), assuming a common unknown variance. They also applied a heuristic variant of Bechhofer and Blumenthal's procedure to which we shall return later on. They found considerable computer time savings compared with nonsequential sample-size procedures. As with multiple comparison procedures the number of application of MRP to simulations is still very limited.

General Considerations of Efficiency and Robustness

For the selection of a particular procedure several factors are important.

(1) The experimental objective and design. We have limited our survey primarily to procedures for selection of the best mean. A brief survey was also given of procedures for situations with a control population. More objectives will be mentioned in Sec. V.C.5. Both one-factor and multi-factor designs were discussed in Sec. V.C.3.

(2) The efficiency of the MRP. The efficiency can be measured by the number of observations required to guarantee the probability requirement in Eq. (1). The following three dichotomies are relevant.

(a) Known versus unknown variances. If the variance(s) must be estimated the sample size increases. (Compare the analogy: the t-statistic gives wider confidence intervals than the normal variable.)

(b) Sequential versus nonsequential procedures. Sequential MRP tend to need fewer observations since they update the estimates of the unknown parameters. These savings are analogous to the savings in sequential hypothesis testing based on the SPRT [sequential probability ratio test (see Sec. V.A.4)]. Some sequential MRP can also take advantage of a more favorable configuration of the means. Nonsequential procedures cannot capitalize on a MFC since the sample size is determined before observations are taken that can reveal the MFC. On the other hand an open end sequential MRP may require excessive sample sizes in the EMC. Also see Bechhofer et al. (1968, p. 5).

(c) Parametric versus nonparametric procedures. The parametric procedures are at least as efficient as the distribution-free procedures if the assumed distribution does indeed occur in the particulare experiment. As the actual distribution deviates from the assumed one the efficiency of the parametric procedure decreases and finally the procedure breaks down completely. This leads us to the following factor.

(3) The robustness of MRP. Consider the following three assumptions.

(a) Independent observations. Independence is an assumption on which all MRP are based. Violation of this assumption is likely to have serious effects; see Scheffé (1964, pp. 331-369) for a general discussion of the assumption of dependence (and normality and constant variance). Independence can be realized in simulation by generating each run with a new sequence of random numbers, the response of one run being one observation. Unfortunately this practice excludes the use of common random numbers for increased reliability of comparisons among systems (see Chap. III).

(b) Normal distributions. Obviously distribution-free procedures do not use this assumption. Some procedures assume only that the sample means are normally distributed and this assumption holds for "small" sample sizes if the deviations from normality of the individual observations is not "too large" (central limit theorem). There are procedures that are based on some other specific distribution, e.g. gamma distributions for variances, binomial distributions for probabilities, etc. The parametric MRP that are hoped to have general applicability in simulation, however, are based on normally distributed individual observations. If the simulation response is an average then this average may tend to normality asymptotically, even if the individual observations are serially correlated (see the discussion of the "stationary r-dependent central limit theorem" in Sec. V.A.2). Transformation of the observations can also be tried to obtain normal observations (see Secs. IV.2 and V.B.4). We shall return to this assumption later on.

(c) The variances. The MRP based on normality assume either known possibly different variances or unknown variances with known ratios (possibly equal to one, i.e. unknown common variances), the one exception being Dudewicz and Dalal's procedure. Most distribution-free (or semi-distribution-free) MRP not only assume equal variances but identical distributions except for a location shift. Common variances may be realized by a suitable transformation of the observations or by appropriate lengths of the runs in steady-state simulations. Observe that the assumption of distributions identical except for location can be met by normal distributions ($\mu_i \neq \mu_{i'}$, $\sigma_i^2 = \sigma_{i'}^2$) but not by gamma distributions (including exponential distributions) (see Exercise 7).

As we saw in Sec. V.B.4 the F-statistic and t-statistic are known to be quite insensitive to nonnormality and heterogeneous variances; Dunnett's multiple t-statistic, being an extreme statistic, was conjectured to be more sensitive. Bechhofer's d is also the critical point of an extreme statistic, for it hinges on Eq. (91) [also see Eq. (79)]:

$$P(\bar{\underline{x}}_{(j)} < \bar{\underline{x}}_{(k)} \text{ for all } j;\ j = 1, \ldots, k-1)$$

$$= P(\max_{1 \leq j \leq k-1} \bar{\underline{x}}_j - \bar{\underline{x}}_k < 0) \tag{91}$$

Transformations of the observations can be used to attain approximately normal observations and/or common variances. Another effect of transformations, particularly important in MRP, is the change of the indifference relation. Consider, e.g. the logarithmic transformation

$$\underline{z}_i = \ln \underline{x}_i \qquad (i = 1, \ldots, k) \tag{92}$$

so that

$$\eta_i \equiv E(\underline{z}_i) \approx \ln E(\underline{x}_i) = \ln \mu_i \tag{93}$$

Consequently, when applying the ranking procedure to the transformed variables $\underline{z}$, the $P(CS)$ is guaranteed (approximately) if the following indifference relation holds.

$$\eta_{(k)} - \eta_{(k-1)} \geq \delta^* \tag{94}$$

or from Eq. (93)

$$\mu_{(k)}/\mu_{(k-1)} \geq \exp(\delta^*) = \delta_0^* \tag{95}$$

Therefore a logarithmic transformation is useful if we like to specify our indifference relation in relative differences $\mu_{(k)}/\mu_{(k-1)}$. We observe that Bechhofer et al. (1968, p. 273) give the effects of various transformations when ranking variances of normal populations, parameters of exponential distributions, or Poisson distributions. Naylor et al. (1968) applied the log transformation with the Bechhofer, Dunnett and Sobel procedure for their simulated national economy.[18)]

In general we simulate complex systems with a stochastic output having a distribution that is unknown. (If the response is an average it may tend to normality.) The variance of the distribution is unknown (as is the mean) and the variances of the various simulated systems are conjectured to differ (as their means do so that we want to select the best one). We conjecture that the distributions are not identical-except-for-location. Therefore, we would like to have distribution-free MRP accepting (unknown and) possibly different variances. Since simulation takes much computer time the procedures should also be efficient. This efficiency increases with sequentialized designs, capitalization on MFC, and elimination of inferior populations. [Moreover incorporation of prior knowledge and various types of loss functions would be desirable but no practical solution is available (also see Sec. V.C.5).] Let us examine whether the existing MRP meet the requirements of robustness and efficiency and, if not, how they can possibly be adjusted to meet these requirements.

ROBUSTNESS AND EFFICIENCY OF EXISTING PROCEDURES; APPROXIMATIVE PROCEDURES

1. The Bechhofer (1954) Single-Sample Procedure for Normal Distributions with Known Variances

Since in simulation the variances σ_i^2 are not known it is necessary to estimate the σ_i^2 by estimators $\underline{s}_i^2$ of the form

$$\underline{s}_i^2 = \sum_{s=1}^{m} \frac{(\underline{x}_{is} - \bar{\underline{x}}_i)^2}{m - 1} \tag{96}$$

Actually the procedure derived by Robbins et al. (1968) applies Bechhofer's rule replacing the unknown (and as Robbins et al. assumed) common variance $\sigma_i^2 = \sigma^2$ by an estimate; this estimate is sequentially revised [see Eqs. (17) and (19)]. We propose to extend the Robbins et al. procedure to unequal variances: Apply Bechhofer's rule in Eq. (6) with the σ_i^2 replaced by their estimators; to improve the efficiency update the estimators sequentially. This approximative procedure is also related to Chambers and Jarratt (1964) who replaced the unknown unequal variances[19)] by estimates based on the first stage of their double-sampling procedure.

Besides the assumption of known variances the assumption of normal distributions may be violated in simulation. As shown in Appendix 2 the procedure does not require the individual observations to be normal but it suffices that the sample means are normal. Because of the central limit theorem this is a rather weak assumption. [Consequently this procedure can also be used to rank other parameters than the means of normal populations, provided the average of the estimators of these parameters are, approximately, normally distributed; also see Bechhofer (1954, p. 29).] Observe that the estimators of the variances remain unbiased for nonnormal distributions.

Concerning the efficiency, the MRP of Bechhofer, Robbins et al., Chambers and Jarratt and our approximative variant cannot take advantage of a MFC. The tables in Robbins et al. (1968, p. 91) show that their procedure requires only slightly more observations than Bechhofer's procedure even while the variance is unknown.

2. The Bechhofer, Kiefer, and Sobel (1968) Sequential Procedure for Normal Distributions with Known Variances

As in the Bechhofer (1954) procedure we have to replace the unknown variances σ_i^2 by their estimators $\underline{s}_i^2$ if we want to apply this procedure. Again the estimate of σ_i^2 can be based on a preliminary sample or sequentially updated. (We conjecture that the sequential approach requires fewer observations but it is more complicated since in each stage $\hat{\underline{r}}_i$ observations are taken from population i but these $\hat{\underline{r}}_i$ depend on $\underline{s}_i^2$ so that revised values of the $\underline{s}_i^2$ imply revision of the $\hat{\underline{r}}_i$; the factor $\hat{\underline{c}}$ must also be recalculated after each stage.) We are not aware of any procedure closely related to such an adapted Bechhofer, Kiefer, Sobel procedure. We can only say that in other MRP (and in many other types of procedures but not in all) substitution of an estimator for a nuisance parameter gives satisfactory results. More research is definitely necessary to check if our intuitive procedure guarantees the probability requirement in Eq. (1).

Bechhofer et al. briefly studied the effect of deviations from the assumed normal distributions. They performed a Monte Carlo experiment with uniform distributions having common known variances [see Bechhofer et al. (1968, pp. 266-267, 348, 363)]. For $k = 3$, $\delta^* = 0.2$, and $0.40 \leq P^* \leq 0.99$ they found that in the LFC the probability requirement in Eq. (1) was easily satisfied. They concluded that, in general, their procedure is quite insensitive to nonnormality, since it depends only on the differences between sums of independent variables, so that the central limit theorem applies. In our heuristic variant, variance estimators are used but these remain unbiased for nonnormal distributions.

Bechhofer et al. (1968, pp. 274, 289) point out that their procedure gives overprotection, i.e. it yields a $P(CS)$ exceeding P^* even in the LFC. Overprotection means that more observations are sampled than are strictly necessary to realize Eq. (1).[20) Yet, compared with Bechhofer's single-sample procedure, their procedure is much more efficient for P^* larger than, say, 70% in the LFC, but not in the EMC [see Bechhofer et al. (1968, pp. 279,281, 364)]. Inefficiency in the EMC occurs since the sequential procedure continues its search for the best population while there is no superior population [compare our comments on Eq. (5)]. The procedure is further conjectured to capitalize on a MFC, because if the means are far apart the probability of high d values in Eq. (11) increases so that Eq. (12) is satisfied earlier [see Bechhofer et al. (1968, p. 348)].

3. The Bechhofer, Dunnett, and Sobel (1954) Two-Stage Procedure for Normal Populations with Known Variance Ratios

When the variance ratios a_i are unknown we might follow one of the following two approaches.

(i) Estimate the a_i using $\underline{s}_i^2$, the estimators of σ_i^2. The $\underline{\hat{a}}_i$ may be based on a preliminary sample [used only to estimate a_i or also used to calculate $\underline{s}_0^2$ and $\underline{\bar{x}}_i$ in Eqs. (13) and (16) above] or can be sequentialized [for details see Kleijnen (1968, pp. 12-13) or Kleijnen and Naylor (1969, pp. 611-612)]. This variant needs further research to see if Eq. (1) is still satisfied.

(ii) Put $a_i = 1$ $(i = 1, \ldots, k)$, i.e. assume that the populations have a common (unknown) variance. The original procedure can then be applied without any modification. The question remains as to how sensitive the procedure is to heterogeneous variances. Its critical constant h is the constant tabulated by, e.g. Dunnett (1955) and Gupta and Sobel (1957). Consequently it depends on an extreme statistic [see footnote 7 in this part, footnote 8 in Part B, and Gupta and Sobel (1957, p. 957)]. Miller (1966, p. 108) guessed that such an extreme statistic is more sensitive to heterogeneity of variances (and nonnormality) than, e.g. the F-statistic.

The choice between the approaches (i) and (ii) may be based on a pilot sample. If extreme inequality of variances is found then apply (i); otherwise apply (ii). (Such advice may be given for other MRP based on common variance. Take more observations from the populations with the higher variances only if extreme inequality of variances is suspected.) Since both approaches use the constant h both are conjectured to be more sensitive to nonnormality but fortunately only the sample means need be normally distributed; the central limit theorem applies to these averages.

We expect that the efficiency of the adjusted procedure is less than that of the adjusted Bechhofer, Kiefer, and Sobel procedure, since the latter is fully sequential. Moreover, the Bechhofer, Dunnett, and Sobel procedure and its heuristic variant cannot take advantage of a MFC.

4. The Robbins, Sobel, and Starr (1968) Sequential Procedure for Normal Distributions with Common Unknown Variance

As with procedure 3 there are two approaches.

(i) Assume that the original procedure is insensitive to heterogeneous variances. Since the procedure uses Bechhofer's d it depends on an extreme statistic and is conjectured to be sensitive to unequality of variances.

(ii) As Robbins et al. did, replace the σ_i^2 in Bechhofer's rule by their estimators $\underline{s}_i^2$. This yields procedure 1 above.

The sensitivity to nonnormality and the efficiency of both variants were discussed under procedure 1.

5. The Bechhofer and Blumenthal (1962) Sequential Procedure for Normal Distributions with Common Unknown Variance

Again we have two approaches.

(i) Assume that the original procedure is insensitive to heterogeneous variances (and nonnormality). Bechhofer (1958, p. 426) claims that the P(CS) of this procedure reacts in the same way as the power of the traditional ANOVA F-test; he assumes that it is relatively insensitive to nonnormality and more sensitive to inequality of variances. For our discussion of the reactions of the F-statistic to nonnormality and heterogeneity of variances we refer to Sec. IV.2; for unequal variances we recommended equal sample sizes except for the case of small number of observations and strong heterogeneity, where more observations from the populations with higher variances were advised [compare Eqs. (7) and (8) above]. We have also made a Monte Carlo study of the robustness of the Bechhofer-Blumenthal procedure and shall report on the result extensively in the next chapter. Here we state only roughly that the procedure is quite robust; extreme skewness of the distributions (and "high" k, say $k \geq 7$) lead to $P(CS) < P^*$ in the LFC. As far as we know, our experiment is the only extensive study on the robustness of MRP and since our results indicate that the Bechhofer-Blumenthal procedure is quite robust, we tend to recommend its use in simulation.

Unfortunately Bechhofer (1970) pointed out recently that heuristic arguments and Monte Carlo experiments show that the procedure has the undesirable characteristic of increasing sample sizes with improving configurations of the means and he recommends not using it. (One might, however, argue that for large sample sizes the procedure approaches the Bechhofer, Kiefer and Sobel procedure for common known variances and consequently it may capitalize on a MFC asymptotically.)

(ii) The original procedure assumes that in each stage one observation is taken from each of the k populations having a common variance. To meet this common variance assumption we propose to define an observation from Π_i as the <u>average</u> of r_i observations sampled per stage from Π_i, the r_i satisfying

$$\frac{\sigma_i^2}{r_i} = \sigma^2 \qquad (i = 1, \ldots, k) \tag{97}$$

where σ^2 is a constant [compare Eq. (8)]. Hence the averages $\bar{\underline{x}}_{is}$ (s = 1, 2, ...) have a common variance σ^2 (and the original means μ_i) so that we can apply the Bechhofer-Blumenthal procedure to the $\bar{\underline{x}}_{is}$. The next step is to replace the σ_i^2 in Eq. (97) by estimators $\underline{s}_i^2$, which can be based on a pilot sample or can be sequentially revised. Whether this variant guarantees Eq. (1) is unknown. (We know that for large sample sizes the original procedure approaches the Bechhofer, Kiefer, and Sobel procedure for common known variances; applying Eq. (97) with σ_i^2 replaced by $\underline{s}_i^2$ makes this variant approach the Bechhofer-Kiefer-Sobel variant asymptotically, but we do not know their performances.) For the sensitivity to <u>nonnormality</u> we refer to our discussion of approach (i).

About the <u>efficiency</u> we mention the Monte Carlo experiment in Bechhofer and Blumenthal (1962, pp. 58-64) showing that in the LFC (!) the original procedure requires only slightly more observations than the Bechhofer, Kiefer, and Sobel procedure, even though the Bechhofer-Blumenthal procedure has to estimate the variance σ^2; its sample size is much smaller than that of the Bechhofer (1964) procedure.

6. The Paulson (1964a) Sequential Procedure with Elimination for Normal Populations with Common Unknown Variance

Again one approach would try to generate observations with a common variance by using the sample analogue of Eq. (97). The estimators $\underline{s}_i^2$ may be based on a first sample of n_0 observations from each population; these n_0 observations are used later on to estimate the μ_i but the σ_i^2 are not reestimated after the first stage. Such a use of the preliminary sample would follow the original procedure. [Observe, however, that in our variant the observations $\bar{\underline{x}}_{is}$ may still have unequal variances since in Eq. (97) estimators are substituted.] A further adaptation would reestimate the σ_i^2 and consequently the r_i after each stage. We are not aware of any studies on this variant or related procedures, or of any studies on the sensitivity of the original procedure to nonnormality or unequal variances. In the elimination rule Eq. (32) the extreme statistic $\max_i (\Sigma\, \underline{x}_{is})$ occurs so that we conjecture,,analogous to Miller (1966), that the procedure may be sensitive to nonnormality and heterogeneous variances.

Efficiency results are given by Paulson (1964a, p. 177) and Ramberg (1966). Paulson's Monte Carlo results for a common but known variance show that his procedure is much more efficient than Bechhofer's single-sample procedure, in the LFC and also in the EMC. There are no results available for a MFC, the configuration especially important for Paulson's elimination rule. His procedure is conjectured to capitalize on a MFC. (Compare: if the means are further apart the elimination rule tends to work more often.) Ramberg performed Monte Carlo experiments comparing the efficiency of Paulson's procedure, the Bechhofer-Kiefer-Sobel procedure and Bechhofer's single-sample procedure, again for common known variances. From Ramberg (1966, pp. 35-37) we would conclude that Paulson's procedure requires fewer observations than the Bechhofer-Kiefer-Sobel procedure in the LFC and the EMC for $P^* \geq 0.99$; no procedure is uniformly best.[21] An important factor is k, since the more populations there are, the more populations can be eliminated so that Paulson's procedure may become more efficient. For more details on efficiency comparisons we refer to Ramberg and also to the Monte Carlo experiments with the Bechhofer-Kiefer-Sobel procedure in Bechhofer et al. (1968, pp. 344-378). We add out Table 2 which brings together Monte Carlo results from Bechhofer et al. (1968, p. 349), and Paulson (1964a, p. 177). Finally, since both procedures yielded overprotection, for common known variances Ramberg adjusted the procedures using a lower P^* in the stopping rule; the adjusted P^* values were based on his Monte Carlo estimates. We do not know if overprotection exists in the Paulson and Bechhofer-Kiefer-Sobel procedures in the case of unknown and different variances. Therefore we do not discuss these adjustments of P^* in detail but refer to Ramberg (1966). For efficiency comparisons among various MRP remember that the Bechhofer-Blumenthal procedure requires only slightly more observations than the Bechhofer-Kiefer-Sobel procedure in the LFC.

7. <u>The Dudewicz and Dalal (1971) Two-Stage Procedure for Normal Distributions with Unequal and Unknown Variances</u>

Their procedure uses assumptions considerably less restrictive than the above procedures since the variances may be both unknown and unequal. So <u>heterogeneity of variances</u> is <u>no</u> problem.

TABLE 2

Total Number of Observations for Known Variance

$\sigma^2 = 1$ and $\delta^* = 0.2$

			Bechhofer, Kiefer, Sobel, (1968)		Paulson (1964a)	
k	P^*	Bechhofer (1954)	EMC	LFC	EM	LFC
4	.95	850	1101	501	755	443
4	.99	1442	2122	712	1178	644
10	.99	4506	8412	2342	3226	1982

We conjecture that its critical constant h_D has the same sensitive to nonnormality as Bechhofer's constant d has, since it is the studentized analogue of d. Though it depends on an extreme statistic, the central limit theorem suggests that the procedure is rather insensitive to nonnormality.

Regarding the efficiency, the procedure cannot take advantable of a MFC and does not eliminate inferior populations. Dudewicz (1972, p. 4) further refers to Wetherill (1966, pp. 181-182) who conjectures that two-stage procedures can realize most of the efficiency gain that fully sequential plans can when compared with single-stage plans. However, in the literature it can be found that it is very difficult to decide on a reasonable size of the first-stage sample. Further, Starr (1966, p. 38), e.g. studied a related problem, viz. the estimation of a single mean, and found that his sequential rule "is always more efficient than two-stage sampling; the difference in efficiences being sizable whenever, in ignorance of the variance, the first stage sample size is chosen poorly."

8. The Srivastava (1966) Sequential Procedure for Unknown Distributions with a Common Unknown Variance

We may try to generate observations with a common variance applying the sample analogue of Eq. (97). No results are known for such a heuristic procedure. Alternatively we may hope that the original procedures A and B are robust, A being based on the t-statistic and B on the multivariate t-statistic. Srivastava and Ogilvie (1968) showed that B is (roughly) 10% more efficient than A in the case of normal distributions. Yet, for simulation experiments we would prefer A over B since A uses the well-known t-statistic [more specifically, it is a one-sided application of the t-statistic, see, e.g. Eq.(2.17) in Srivastava (1966)]. The t-statistic is known to be insensitive to nonnormality and heterogeneous variances in large samples; in small samples a two-sided t-test remains insensitive, but a one-sided t-test is more sensitive [see Scheffé (1964, pp. 335, 346, 353)].

The efficiency of Srivastava's procedures is asymptotically equal to that of Bechhofer's procedure for known common variances, and it is asymptotically more efficient than the Bechhofer-Dunnett-Sobel two-stage procedures [see Srivastava (1966, pp. 372, 374) and Srivastava and Ogilvie (1968, p. 1041)]. The procedures cannot take advantage of a MFC.

9. The Bechhofer and Sobel (1958) Multinomial Procedure for Unknown Distributions Identical Except for Location

This procedure is unattractive since the original indifference relation is replaced by $p_{(k)}/p_{(k-1)} \geq \theta^*$. The correspondence between $\mu_{(k)} - \mu_{(k-1)}$ and $p_{(k)}/p_{(k-1)}$ depends on nuisance parameters (like σ^2 for normal distributions). These nuisance parameters could be estimated as in the heuristic variants of the other MRP. However, we must also specify the type of distribution function, e.g. normal distributions, and this contradicts the purpose of a nonparametric procedure. It is unknown how the procedure performs if the distributions do not differ only in location but also in form. Dudewicz (1971a) showed that the (original) procedure is very inefficient if the distribution assumed by a particular parametric procedure, holds.

10. The Lehmann (1963) Single-Sample Ranking and Related Procedures for Unknown Distributions Identical Except for Location

As the above Bechhofer and Sobel (1958) procedure the scoring procedures of Lehmann and other authors are not truly distribution-free since sample size determination requires specification of the underlying distribution. It is known that the procedures are less sensitive to deviations from this specified distribution than procedures based on means. A drawback is that they require specification of the values of the nuisance parameters (e.g. σ^2 if normal distributions are specified), whereas many MRP allow estimates of these parameters. It is unknown how the procedures perform if the distributions also differ in form. It would be interesting to investigate the performance of procedures with (i) estimates of the form of the distribution, based on a pilot sample or sequential estimation; (ii) estimates of the nuisance parameters; (iii) small samples; (iv) distributions with varying forms.

In general, the (original) scoring procedures are efficient compared with the Bechhofer (1954) parametric procedure for a known common variance, i.e. for many nonnormal distributions they require fewer observations (especially the normal scores procedures) [see Lehmann (1963, p. 271), Bartlett and Govindarajulu (1968, p. 92) and Puri and Puri (1969, p. 625) for details].

11. The Sen and Srivastava (1972b) Sequential Ranking Procedure for Unknown Distributions Identical Except for Location

Let us first consider the efficiency of this procedure. It cannot capitalize on a MFC and it cannot eliminate inferior populations. It is asymptotically efficient, i.e. for δ^* approaching zero the ASN of this sequential rule is equal to the sample size in Lehmann's single-stage ranking procedure for known density functions. [see Srivastava and Sen (1972)]. We saw above that Lehmann's procedure is efficient compared with Bechhofer's single-sample rule. We also noticed that Lehmann's procedure is robust. Unfortunately we do not know whether Sen and Srivastava's sequential variant is also insensitive to deviations from its assumption of distributions with identical forms. Their procedure uses Gupta's constant h (or Bechhofer's d) which is an extreme statistic and therefore it is conjectured to be sensitive to heterogeneous variances, i.e. difference in form. Remember that the Monte Carlo experiment showed good small-sample behavior (the assumption of distribution shift only, was not violated in this experiment).

12. The Hoel (1971) Sequential Ranking Procedure with Elimination for Stochastically Ordered Distributions

Stochastically ordered distributions comprise distributions with location-shift-only but not normal distributions with unequal variances. We do not know whether the procedure is sensitive to deviations from the assumption of a stochastic ordering.

Regarding its efficiency, the procedure is both sequential and capable of eliminating inferior populations. A drawback is that it requires quite complicated calculations [compare the expression in Eq. (59)].

13. The Sobel (1967) Single-Sample Nonparametric Procedure

This is the only truly distribution-free procedure we know of. The procedure is valid for (unknown) distributions that are not different only in location or stochastically ordered. Unfortunately its problem formulation does not agree with the formulation we might expect in simulation applications.

For procedures assuming the existence of a control population, the same comments can be made as for the MRP above. We add some more remarks. The Paulson (1962) and Sen and Srivastava (1972b) procedures give extra protection for CS of the standard population. It may be that this extra protection is not desired, viz. if the experimenter does not care whether he makes a wrong selection (i) through incorrect selection of the control or an experimental system instead of the experimental system being the best one, or (ii) through incorrect selection of an experimental system instead of the control that is now supposed to be the best system [also see Kleijnen (1968, p. 7) or Kleijnen and Naylor (1969, p. 609)]. Yet imponderables may make it desirable to abandon the existing system only if there is strong evidence that an experimental system is better. For comments on the efficiency of the Paulson procedure we refer to Paulson (1962) and Roberts (1963). The Paulson (1964b) classification procedure can be adapted using either his variant for known, possibly different variances estimating these variances, or his variant for unknown but common variances taking $\hat{r}_i$ observations per stage.

Summarizing this section we see that we are really left with a problem. No existing procedure guarantees the probability requirement in Eq. (1) for unknown, possibly differing distributions. Therefore we can either hope that an existing procedure is insensitive to violations of its underlying assumptions or hope that one of the heuristic variants works well. The relative efficiency of a procedure becomes relevant if we think that more than one procedure will guarantee Eq. (1). To improve the efficiency of a procedure we may add the following heuristic rule, aiming at elimination of populations that are likely to be inferior (compare Paulson's procedures).

Eliminate population i if the difference between the best sample mean $\bar{x}_{[k]}$ and its mean $\bar{x}_i$, minus c times the estimated standard deviation of this difference is still positive, i.e., if

$$(\bar{x}_{[k]} - \bar{x}_i) - c[\widehat{var}(\bar{x}_{[k]} - \bar{x}_i)]^{1/2} > 0 \tag{98}$$

We might choose, e.g. c = 3.[22)] For a discussion of elimination we also refer to Bechhofer et al. (1968, pp. 230-246).

To conclude this section we give the following recommendations (based on intuition so that we are certain that everyone will find some point to disagree).

(i) The Dudewicz and Dalal (1971) procedure seems to be the best choice when probability requirement Eq. (1) is to be guaranteed, unless the central limit theorem is expected not to apply. The procedure, however, is relatively inefficient as it is not sequential, does not benefit from a MFC and cannot eliminate inferior populations.

(ii) The Bechhofer (1954) procedure adjusted through insertion of sequentialized estimators of the variances, is conjectured to guarantee Eq. (1). This variant is nearly identical to the Robbins-Sobel-Starr procedure for common unknown variances and the latter procedure is known to work. The variant is further conjectured to be insensitive to nonnormality since the central limit theorem applies to the sample means and the variance estimators remain unbiased. It is relatively inefficient as it cannot capitalize on a MFC or eliminate populations.

(iii) If the experimenter does not like to rely on conjectures he may apply the original Bechhofer-Blumenthal (1962) procedure since this is the only procedure extensively tested for its sensitivity to nonnormality and unequal variances. The Monte Carlo experiment reported on in the next chapter, shows that the procedure is quite robust. Unfortunately, it cannot capitalize on a MFC and may even require large samples in a MFC. It does not eliminate populations.

(iv) The experimenter who worries most of all about the efficiency of the MRP (since his simulation experiment takes much computer time per run) might select the Paulson (1964a) procedure or one of our heuristic variants. We conjecture that it is the most efficient procedure owing to its elimination of inferior populations during experimentation. Its sensitivity to heterogeneity of variances (or taking $\hat{\underline{r}}_i$ observations per stage) and to nonnormality is unknown but may very well be nonnegligible because of the extreme statistic in the elimination rule. We would not recommend the original Bechhofer, Dunnett, and Sobel (1954) procedure for common unknown variances since its critical constant h is guessed to be sensitive to heterogeneous variances while the procedure is supposed not to be more efficient than the above recommended procedures. The adjusted Srivastava (1966) procedure may very well yield the same results as the adjusted Bechhofer (1954) procedure. The robustness of the sequential nonparametric procedures (Sen and Srivastava, and Hoel) is unknown at present. Anyhow, application of one of the original or revised MRP seems better than the current practice of determining the number of runs purely intuitively.

V.C.5. OTHER DISTRIBUTIONS, SOLUTIONS, AND PROBLEM FORMULATIONS

(i) Other Distributions

We discussed procedures for selection of the normal population with the best mean. As we remarked before these MRP can also be applied to other situations, possibly after a transformation of the data. There are, however, special procedures for other distributions. The best single reference is Bechhofer et al. (1968, pp. 255-274) who give procedures for selection of the best normal population (the means or the variances being the selection criterion), exponential, gamma, Weibull, binomial, negative binomial, and Poisson populations; certain indifference relations must be satisfied. (These relations are not always of the form $\theta_{(k)} - \theta_{(k-1)} \geq \theta^*$ but, e.g. $(\theta_{(k)} - \theta_{(k-1)})/(\theta_{(k)}\ \theta_{(k-1)}) > \theta^*$.) Some additional references not meant to be exhaustive are: Taylor and David (1962) for binomial populations; Bechhofer et al. (1959) and Bechhofer and Sobel (1956) for multinomial distributions, Chambers and Jarratt (1964) if $\sigma_i^2 = \sigma(\mu_i)$ (e.g. binomial or Poisson distributions). Many more references can be found in Bechhofer (1958), Bechhofer et al. (1968), Dudewicz (1968) and O'Neill and Wetherill (1971).

(ii) Other Solutions

Guttman (1963) derived a procedure that, as in Eq. (1), guarantees $P(CS) \geq P^*$ but has no indifference quantity δ^*. His rule is based on subset selection procedures discussed in Part B of this chapter. It continues sampling until the subset has reduced to size one. (The subset, however, may remain larger than one if the available money resources are limited.) The procedure has the disadvantage of being only "pseudo"-multistage, i.e. the observations of previous stages are not used in a subsequent stage. Moreover the necessary tables may not be available [cf. his Eq. (2.5) yielding odd values for P^*]. An advantage is that subset selection procedures are available that have no restrictive assumptions; see, e.g. Rizvi and Sobel's nonparametric procedure in Part B.

Bayesian and decision theoretic solutions of the selection problem can be found in, e.g. Somerville (1954) with extensions in Shirafuji (1956) and Fairweather (1968); Dunnett (1960) and Hayes (1969). More references are found in Bechhofer et al. (1968) and Dudewicz (1968, categories E and P).

(iii) Other Problem Formulations

Instead of the single best population we may want to select the t best populations ($1 \leq t \leq k$), with or without ordering these t populations. See Bechhofer (1955), Bechhofer et al. (1954), Bechhofer et al. (1968), Freeman et al. (1967), and Puri and Puri (1967). Instead of selecting the population associated with the best mean we may want to estimate the value of the best population mean (or more generally, the values of the parameters $\theta_{(i)}$) [see Dudewicz (1968, category Q; 1971b), Dudewicz and Dalal (1971, section 6), Dudewicz and Tong (1971) and Rizvi and Saxena (1972) who give more references]. Some problems related to ranking are: classification of variances relative to the variance of the control population [see Paulson (1964b)]; simultaneous confidence limits of predetermined length, see Healy (1956) with applications to factorial experiments; subset selection procedures, see Barron and Gupta (1970) and Gupta and Panchapakesan (1969b) for sequential rules, and Dudewicz and Dalal (1971, section 6) or Dudewicz (1972, pp. 7-8) for two-stage rules. Ranking of multivariate populations is discussed by Srivastava and Taneja (1972). Also relevant are Somerville (1970) and Zinger (1961). Many more references are given in Dudewicz (1968).

APPENDIX V.C.1. SELECTION OF ONE OF THE BEST POPULATIONS

For simplicity of presentation it is assumed in the appendixes that the populations are so labeled that $\mu_1 \leq \cdots \leq \mu_k$, i.e. $\mu_i = \mu_{(i)}$. The configuration of the means in this appendix is as follows:

$$0 \leq \mu_k - \mu_{k-1} < \delta^* \qquad (1.1a)$$

and

$$\mu_1 = \cdots = \mu_{k-2} \leq \mu_{k-1} - \delta^* \qquad (1.1b)$$

Correct selection is selection of either Π_k or Π_{k-1}. Hence, since the procedure selects the population with the highest sample mean $\bar{x}$ (or some other statistic like the rank sum) we have

$$
\begin{aligned}
P(CS) &= P(\bar{\underline{x}}_k \text{ is maximum}) + P(\bar{\underline{x}}_{k-1} \text{ is maximum}) \\
&= P[\bar{\underline{x}}_k > \bar{\underline{x}}_j \ (j = 1, \ldots, k\text{-}2) \vee \bar{\underline{x}}_k > \bar{\underline{x}}_{k-1}] \\
&\quad + P[\bar{\underline{x}}_{k-1} > \bar{\underline{x}}_j \ (j = 1, \ldots, k\text{-}2) \vee \bar{\underline{x}}_{k-1} > \bar{\underline{x}}_k] \\
&= P(\bar{\underline{x}}_k > \bar{\underline{x}}_{k-1})\, P(\bar{\underline{x}}_k > \bar{\underline{x}}_j \text{ all } j \mid \bar{\underline{x}}_k > \bar{\underline{x}}_{k-1}) \\
&\quad + P(\bar{\underline{x}}_{k-1} > \bar{\underline{x}}_k)\, P(\bar{\underline{x}}_{k-1} > \bar{\underline{x}}_j \text{ all } j \mid \bar{\underline{x}}_{k-1} > \bar{\underline{x}}_k)
\end{aligned}
\tag{1.2}
$$

Use the symbol

$$
P(\bar{\underline{x}}_k > \bar{\underline{x}}_{k-1}) = p \tag{1.3}
$$

so that (assuming continuous variables for simplicity)

$$
P(\bar{\underline{x}}_{k-1} > \bar{\underline{x}}_k) = 1 - p \tag{1.4}
$$

Further

$$
\begin{aligned}
&P[\bar{\underline{x}}_k > \bar{\underline{x}}_j \ (j = 1, \ldots, k\text{-}2) \mid \bar{\underline{x}}_k > \bar{\underline{x}}_{k-1}] \\
&\quad \geq P[\bar{\underline{x}}_k > \bar{\underline{x}}_j \ (j = 1, \ldots, k\text{-}2)]
\end{aligned}
\tag{1.5}
$$

since the condition $\bar{\underline{x}}_k > \bar{\underline{x}}_{k-1}$ implies high values of $\bar{\underline{x}}_k$. Consequently

$$
P(CS) \geq pP[\bar{\underline{x}}_k > \bar{\underline{x}}_j \text{ all } j] + (1 - p)\, P[\bar{\underline{x}}_{k-1} > \bar{\underline{x}}_j \text{ all } j] \tag{1.6}
$$

Consider a situation with only the populations 1, ..., k-2, k, i.e. exclude population k-1. The procedure guarantees

$$
P(\bar{\underline{x}}_k > \bar{\underline{x}}_j \text{ all } j) \geq P^* \quad \text{if} \quad \mu_k - \mu_{k-2} \geq \delta^* \tag{1.7}
$$

Including population k-1 does not decrease the probability in Eq. (1.7) since all observations are independent and the number of observations in the sample means does not decrease. Using Eq. (1.8) and its analogue for $\underline{\bar{x}}_{k-1}$ gives

$$P(CS) \geq pP^* + (1 - p)P^* = P^* \qquad (1.8)$$

APPENDIX V.C.2. SINGLE-SAMPLE PROCEDURE FOR KNOWN VARIANCES

All observations are independent and the sample means $\underline{\bar{x}}_i$ $(i = 1, \ldots, k)$ are assumed to be normally distributed. (If the individual observations are normal this assumption holds for any sample size n_i; otherwise it holds asymptotically for finite variances σ_i^2 and means μ_i.) The probability that the largest sample mean comes from the population with the largest population mean, μ_k, is equal to the probability that the best population gives a sample mean $\underline{\bar{x}}_k$ and all other populations give sample means smaller than $\underline{\bar{x}}_k$, where $\underline{\bar{x}}_k$ can range from $-\infty$ to $+\infty$. Or

$$P(CS) = \int_{-\infty}^{\infty} \varphi\left(x; \mu_k, \frac{\sigma_k^2}{n_k}\right) \cdot \Phi\left(x; \mu_1, \frac{\sigma_1^2}{n_1}\right) \cdots \Phi\left(x; \mu_{k-1}, \frac{\sigma_{k-1}^2}{n_{k-1}}\right) dx \qquad (2.1)$$

where $\varphi(x; \mu_k, \sigma_k^2/n_k)$ is the density function of a normal variable with mean μ_k and variance σ_k^2/n_k; $\Phi(x; \mu_j, \sigma_j^2/n_j)$ is the (cumulative) distribution function of a normal variable with mean μ_j and variance σ_j^2/n_j $(j = 1, \ldots, k-1)$. Now $P(CS) \geq P^*$ should be guaranteed only for configurations satisfying

$$\mu_j \leq \mu_k - \delta^* \qquad (j = 1, \ldots, k-1) \qquad (2.2)$$

It is easily checked that, for $\epsilon > 0$,

$$\Phi\left(x;\ \mu_k - \delta^*,\ \frac{\sigma_j^2}{n_j}\right) < \Phi\left(x;\ \mu_k - \delta^* - \epsilon,\ \frac{\sigma_j^2}{n_j}\right) \tag{2.3}$$

Hence P(CS) in Eq. (2.1) reaches its minimum, subject to Eq. (2.2), if the means are in the configuration

$$\mu_j = \mu_k - \delta^* \qquad (j = 1, \ldots, k-1) \tag{2.4}$$

whatever the values of σ_j^2 and σ_k^2 are. This minimum value of P(CS) is

$$P_{min} = \int_{-\infty}^{\infty} \varphi\left(x;\ \mu_k,\ \frac{\sigma_k^2}{n_k}\right) \cdot \Phi\left(x;\ \mu_k - \delta^*,\ \frac{\sigma_1^2}{n_1}\right) \cdots \Phi\left(x;\ \mu_k - \delta^*,\ \frac{\sigma_{k-1}^2}{n_{k-1}}\right) dx \tag{2.5}$$

Substitution of

$$z = \frac{x - \mu_k}{\sigma_k / \sqrt{n_k}} \tag{2.6}$$

into Eq. (2.5) gives

$$P_{min} = \int_{-\infty}^{\infty} \varphi(z) \cdot \Phi\left(\frac{\sigma_k}{\sqrt{n_k}} z + \mu_k;\ \mu_k - \delta^*,\ \frac{\sigma_1^2}{n_1}\right) \cdots \Phi\left(\frac{\sigma_k}{\sqrt{n_k}} z + \mu_k;\ \mu_k - \delta^*,\ \frac{\sigma_{k-1}^2}{n_{k-1}}\right) dz \tag{2.7}$$

where $\varphi(z)$ denotes the density of the standard normal variable, i.e.

$$\varphi(z) = \varphi(x;0,1) \tag{2.8}$$

In Eq. (2.7) we have for $j = 1, \ldots, k-1$:

$$\Phi\left(\frac{\sigma_k}{\sqrt{n_k}} z + \mu_k;\ \mu_k - \delta^*, \frac{\sigma_j^2}{n_j}\right)$$

$$= P\left(\underline{\bar{x}}_j < \frac{\sigma_k}{\sqrt{n_k}} z + \mu_k \,\middle|\, E(\underline{\bar{x}}_j) = \mu_k - \delta^*,\ \mathrm{var}(\underline{\bar{x}}_j) = \frac{\sigma_j^2}{n_j}\right)$$

$$= P\left\{\underline{z} < \left[\left(\frac{\sigma_k}{\sqrt{n_k}} z + \mu_k\right) - (\mu_k - \delta^*)\right] \Big/ (\sigma_j/\sqrt{n_j})\right\}$$

$$= P\left(\underline{z} < \frac{\sigma_k}{\sigma_j}\frac{\sqrt{n_j}}{\sqrt{n_k}} z + \frac{\sqrt{n_j}}{\sigma_j}\delta^*\right) = \Phi\left(\frac{\sigma_k}{\sigma_j}\frac{\sqrt{n_j}}{\sqrt{n_k}} z + \frac{\sqrt{n_j}}{\sigma_j}\delta^*\right) \tag{2.9}$$

where $\Phi(z)$ denotes the probability that the standard normal variable $\underline{z}$ is smaller than z. Let us take the sample sizes in the proportion

$$\frac{n_j}{n_k} = \frac{\sigma_j^2}{\sigma_k^2} \qquad (j = 1, \ldots, k-1) \tag{2.10}$$

Then Eq. (2.9) becomes

$$\Phi\left(\frac{\sigma_k}{\sigma_j}\frac{\sqrt{n_j}}{\sqrt{n_k}} z + \frac{\sqrt{n_j}}{\sigma_j}\delta^*\right) = \Phi\left(z + \frac{\sqrt{n_k}}{\sigma_k}\delta^*\right) \tag{2.11}$$

From Eqs. (2.9) and (2.11) it follows that Eq. (2.7) reduces to

$$P_{min} = \int_{-\infty}^{\infty} \varphi(z) \left[\Phi\left(z + \frac{\sqrt{n_k}}{\sigma_k}\delta^*\right)\right]^{k-1} dz \tag{2.12}$$

Since we require $P_{min} \geq P^*$ we have to solve Eq. (2.13) for d:

$$P^* = \int_{-\infty}^{\infty} \varphi(z)\,[\Phi(z + d)]^{k-1}\,dz \tag{2.13}$$

with

$$d = \frac{\sqrt{n_k}}{\sigma_k}\delta^* \tag{2.14}$$

or

$$n_k = (\sigma_k \, d/\delta^*)^2 \tag{2.15}$$

Our Eq. (2.13) is equivalent to Eq. (20) in Bechhofer (1954) (if in his equation we replace $d = \sqrt{N}\,\delta/\sigma$ by d defined in Eq. (2.14), and put $t = 1$ since we are looking only for the one best population). Bechhofer tabulated the values of d satisfying Eq. (2.13). Using these values, Eqs. (2.15) and (2.10) we find the required values of n_j $(j = 1, \ldots, k-1)$ and n_k.

APPENDIX C.3. THE CRITICAL CONSTANTS OF SRIVASTAVA'S PROCEDURE B

Srivastava (1966, p. 374) defined the sequence $(a_m) = (a_1, a_2, \ldots)$ as any positive constants satisfying

$$\lim_{m \to \infty} a_m = a \tag{3.1}$$

where a follows from

$$P(\underline{R} > a\sqrt{2}) = 1 - P^* \tag{3.2}$$

or equivalently

$$P(\underline{R} \leq a\sqrt{2}) = P^* \tag{3.3}$$

and $\underline{R}$ is defined as

$$\underline{R} = \max_{1 \leq j \leq k-1} (\underline{z}_j - \underline{z}_k) \tag{3.4}$$

the $\underline{z}_j$ and $\underline{z}_k$ being independent standard normal variables. Srivastava mentions that $\underline{R}$ is tabulated in Bechhofer (1954). In Srivastava and Ogilvie (1968, p. 1043, Eq. 2.6) a formula is given that can be checked to be identical to Bechhofer's formula (20) with $t = 1$ and $d = x\sqrt{2}$. Hence for Srivastava's a we have $a = d/\sqrt{2}$, d being taken from Bechhofer (1954). Next consider a_m.

Gupta and Sobel (1957) tabulated q where

$$P(\underline{y} < q) = P^{*} \tag{3.5}$$

and

$$\underline{y} = \left(\max_{1 \leq j \leq p} \underline{x}_j - \underline{x} \right) / \underline{s}_v \tag{3.6}$$

the $\underline{x}_j$ and $\underline{x}$ being independent normal variables with common mean μ and common variance σ^2, while $\underline{s}_v^2$ is an independent estimator of σ^2. Now we set $\underline{z}_j = (\underline{x}_j - \mu)/\sigma$ $(j = 1, \ldots, p)$, $\underline{z}_k = (\underline{x} - \mu)/\sigma$ and $p = k-1$. Then Eq. (3.6) becomes

$$\underline{y} = \sigma\left[\max_{1 \leq j \leq k-1} (\underline{z}_j) - \underline{z}_k \right] / \underline{s}_v \tag{3.7}$$

the $\underline{z}_j$ and $\underline{z}_k$ being independent standard normal variables. As $v \rightarrow \infty$, $\underline{s}_v^2$ approaches σ^2 and consequently $\underline{y}$ approaches $\underline{R}$. (Indeed the entries in Gupta and Sobel for $v = \infty$ agree with Bechhofer's values.) Therefore we take $a_m = q/\sqrt{2}$, where $q(P^{*}, k, v)$ with $v = k(m-1)$ is tabulated in Gupta and Sobel and $q/\sqrt{2}$ in Dunnett (1955) and Krishnaiah and Armitage (1966).

EXERCISES

1. Apply Bechhofer's single-sample procedure for the determination of the sample sizes n_i if the known variances are $\sigma_1^2 = 90$, $\sigma_2^2 = 130$, $\sigma_3^2 = 191$ and $P^{*} = .75$, $\delta^{*} = 4$. [Source: Bechhofer (1954, p. 37).]

2. Prove that $\underline{s}_0^2$ in Eq. (13) is an unbiased estimator of σ^2 in $\sigma_i^2 = a_i \sigma^2$.

3. Naylor et al. (1967b) simulated 50 runs for each of 5 variants of a queuing system. Assuming a common variance they calculated a sum of squares equal to 12,715,825. The system variant with the best mean is to be selected with probability at least 90% if the best variant is at least 100 units better than the other

variants. Apply the Bechhofer, Dunnett, and Sobel (1954) procedure to determine whether enough runs were generated.

4. Define $p_1 = P(\underline{x}_1 > \underline{x}_2)$ and $p_2 = P(\underline{x}_2 > \underline{x}_1)$. Let $\underline{x}_1$ and $\underline{x}_2$ be continuous variables with distribution functions $F_i(x) = F(x - \mu_i)$, and $\mu_1 > \mu_2$. Prove $p_2 > p_1$.

5. Prove that "$F_1(x) \geq F_2(x)$ for all x" holds if the distribution functions are, e.g. normal with $\mu_1 < \mu_2$ and $\sigma_1^2 = \sigma_2^2$, or exponential with $E(\underline{x}_1) < E(\underline{x}_2)$.

6. Prove that in a two-way layout P(CS) for factor A and B, respectively, are independent if the variances of the cell means, σ_{ij}^2/n_{ij}, are constant.

7. Let the two parameters of the gamma distribution be denoted by p and b [see, e.g. Fisz (1967, p. 152)]. Show that it is impossible to have gamma distributions identical except for location.

8. Naylor et al. (1968) wanted to select the economic policy that minimizes the variance of national income, with $P(CS) \geq 90\%$ if the best variance is no more than 90% of the next best variance. Let $\underline{s}_i^2$ denote the sample variance of policy i $(i = 1, \ldots, 5)$ in run g $(g = 1, \ldots, 50)$. How can the Bechhofer, Dunnett, and Sobel (1954) procedure be applied?

9. Prove that $\underline{s}^2 = \Sigma_{s=1}^{m} (\underline{x}_s - \bar{\underline{x}})^2/(m-1)$ is unbiased for any distribution.

10. Apply the MRP to the selection of the queuing system with the smallest expected steady-state waiting time for various queuing disciplines (e.g. first-in-first-out, first-in-last-out, random).

NOTE: Many more potential applications of MRP can be found in the literature, e.g. Schmidt and Taylor (1970, pp. 516-575).

NOTES

1. Chapter V.C is a revision of Kleijnen (1968), Kleijnen and Naylor (1969), and Kleijnen et al. (1972).

2. The use of the sample mean for selection of the population with the best mean is indeed efficient in many problems but not in all. For example, the midrange is a better statistic for uniform distributions [see Dudewicz (1971a)]. As we mentioned in Sec. V.A an extensive study of about 70 different estimators of the location of a distribution was done by Andrews et al. (1972). They found that the sample mean is a very nonrobust estimator. Rejecting outliers or using the median is a simple solution; for more sophisticated estimators see Andrews et al.

3. Some MCP, e.g. Sobel's (1968) nonparametric procedure, reach a minimum P(CS) for a "least favorable configuration" defined analogous to Eq. (2).

4. Bechhofer (1954, p. 24) uses the following more complicated formulation. Put $\sigma_i^2 = a_i\sigma^2$ where the a_i are known variance ratios and σ^2 is a common known constant. Take n_i in such a way that σ_i^2/n_i is constant. Look up d in the table and calculate $n = (\sigma d/\delta^*)^2$ when σ^2 is the constant in $\sigma_i^2 = a_i\sigma^2$. Put $n_i = a_i n$. Also see Eqs. (22) and (17) in Bechhofer (1954).

5. Bechhofer et al. (1968, Eq. 12.3.7) use δ^* in the exponent, where we use δ^*/c. In their equation (12.6.3), however, they state $P(CS) \geq P^*$ if $\mu_{(k)} - \mu_{(k-1)} \geq \delta^* c$. Hence their $\delta^* c$ is equivalent to our δ^*. In our presentation of their procedure we use δ^* everywhere as defined in Eq. (1).

6. σ^2 is the common variance. If the variances are unequal define an observation from Π_i as the average of r_i individual observations where the r_i satisfy $\sigma_i^2/r_i = \sigma^2$; see Eq. (8). Hence ASN should be multiplied by r_i to obtain the number of individual observations from Π_i.

7. Some confusion about the tables may arise. Bechhofer et al. (1954, p. 173) refer to Dunnett and Sobel (1954) for $k = 3$ and state that there are no tables yet for $k > 3$. Later on Bechhofer (1958, p. 410) refers to Dunnett and Sobel (1954; 1955) and mentions Dunnett (1955) and Gupta and Sobel (1957). From Bechhofer et al. (1954, p. 172) it follows that $P(CS|\delta \geq \delta^*) \geq P[\underline{z}_j/\underline{s}_0 < h \quad (j = 1, \ldots, k-1)]$, where the $\underline{z}_j$ are multivariate normal with means zero, variances σ^2 and correlation $\rho_{jj'} = 1/2 \quad (j \neq j')$. Dunnett (1955, p. 1102) used the same expression with $h = d'$ and $k-1 = p$. Gupta and Sobel (1957) investigate $P[\underline{z}_j/\underline{s}_0 < q/\sqrt{2} \quad (j = 1, \ldots, k-1)]$. The latter two publications make Dunnett and Sobel (1954; 1955) obsolete. For example for $k = 5$, $P^* = 0.99$, $v = 15$ Dunnett (1955) gives 3.20 and Gupta and Sobel (1957) give $q = 4.54$ so that $h = 4.54/\sqrt{2} \approx 3.2$. Also see the discussion and bibliography in Gupta (1963a, pp. 806-809; 1963b, pp. 836-837). A general bibliography of tables relevant for MRP (and MCP) is given by Dudewicz (1968, category R).

8. The requirement in Robbins et al. (1968, p. 89) that m be an odd integer ≥ 5 is imposed only for additional derivations.

9. We present Paulson's procedure for a general λ rather than the special procedure for $\lambda = \delta^*/4$ given by Paulson (1964a, pp. 178-179).

10. We present both procedures A and B though Srivastava and Ogilvie derived that B requires less observations. Procedure A, however, may be more robust. [See Sec. V.C.4.]

11. The constants a_m in this class of procedures can be any sequence of positive constants satisfying $\lim a_m = a$ with $\Phi(-a) = (1 - P^*)/(k-1)$, Φ denoting the standard normal distribution function. Following Srivastava and Ogilvie [1968, Eq. (2.2)] we determined a_m from the t-distribution. It is well-known that the t-distribution approaches the standard normal distribution as v increases.

12. (i) Srivastava and Ogilvie (1968, p. 1046) do not refer to Krishnaiah and Armitage (1966), or Gupta and Sobel, and mention Dunnett only for their first approximation to a^*. They used an approach of their own to calculate a^* but do not list the resulting values of a^*. (ii) We observe the following similarities among Bechhofer, Dunnett, and Sobel (1954), Robbins et al. (1968), and Srivastava (1966): In all three procedures the sample size is determined by a rule $\underline{m} = (\hat{\underline{\sigma}}\, c/\delta^*)^2$, where c is a constant tabulated in, e.g. Dunnett (1955), Bechhofer (1954), and the table for the t-statistic (or the standard normal for degrees of freedom > 30). Gupta and Sobel (1957) derived how these three constants are related: Their Eq. (4.6) shows the relation between Dunnett's d' $(= h = q/\sqrt{2})$ and Bechhofer's d; their Eq. (4.9) shows the relation between Dunnett's h and the standard normal variable. [Also see our Appendix 3.] (iii) Finally notice that Lehmann (1963, p. 269) proved that for unknown distributions with common known variance an asymptotic procedure takes $n = (\sigma\, d/\delta^*)^2$ exactly as Bechhofer (1954) did. Compare also our Appendix 2, where only asymptotic normality of the sample means is required to find the same solution.

13. We assume that the quantiles are unique, i.e. only one value satisfies Eq. (61). For more general definitions see Sobel (1967), or Sec. V.A, Eq. (55).

14. Equation (68) follows from Paulson (1962, p. 437): $P(D_j | R_j) \geq 1 - \beta$, where R_j denotes the configuration $[\mu_{j'} = \mu_0$ $(j' = 1, \ldots, k$ but $j' \neq j)$ and $\mu_j = \mu_0 + \delta^*]$ and D_j denotes the decision that Π_j is the best experimental population and is better than the control.

15. Compare Paulson (1962): $P(D_0 | R_0) \geq 1 - \alpha$ where R_0 denotes $\mu_0 = \mu_i$ $(i = 1, \ldots, k)$ and D_0 denotes selection of Π_0.

16. The following relations hold between Paulson's symbols and our symbols: $\Delta = \delta^*$, $d = \lambda$, $d + [\sigma_{zj}^2 \log(k/\alpha)]/2\, dn = d_i$.

17. Naylor et al. used the critical constant $h = 1.58$, taken from Dunnett and Sobel (1954, Table 3). This value, however, is correct only for $k = 3$. For the correct approach see footnote 7 and Exercise 3.

18. Naylor et al. wanted to select the economic policy that minimizes fluctuations in national income. These fluctuations were measured by $\underline{s}_i^2 = \Sigma_{t=1}^{200} (\underline{x}_{it} - \bar{\underline{x}}_i)^2/199$ $(i = 1, \ldots, 5)$. We point out that the estimator $\underline{s}_i^2$ would be an unbiased estimator of the population variances σ_i^2 if the $\underline{x}_{it}$ were independent observations from one particular population Π_i. Naylor et al. neglected the bias caused by serial correlation and transient state and used $E(\underline{s}_i^2)$ as selection criterion. The ranking procedure was applied to $\underline{z}_{ig} = \ln \underline{s}_{ig}^2$ $(g = 1, \ldots, 50)$ guaranteeing $P(CS) \geq 90\%$ if the best "variance" is no more than 90% of the next best one (see Exercise 8).

19. Chambers and Jarratt assumed unknown variances σ_i^2 that are a known and common function of the means μ_i, say $\sigma_i^2 = \sigma(\mu_i)$. In Chambers and Jarratt (1964, p. 51) $\sqrt{n}\, \delta^*/\sigma(\mu_1)$ is the analogue of $\sqrt{n_1}\, \delta^*/\sigma_1$ in our Appendix 2 (setting γ^* in Chambers and Jarratt equal to 1).

20. We might try to eliminate overprotection by using a P^*-value lower than we really want. How much lower P^* should be taken, can be based on Monte Carlo estimates of the overprotection for certain k and P^* values. However, we do not know if overprotection also exists in our intuitive variant.

21. Ramberg himself suggests that most times the Bechhofer-Kiefer-Sobel procedure performs better. Nevertheless, we think that the total number of observations is more relevant than the number of stages in a simulation experiment and that P^* and k values as low as 0.75 and 3, respectively, are not as relevant as P^* equal to 0.99 (or .999) and k equal to 10. We further restricted our conclusions to the EMC and LFC.

22. Sasser et al. (1970, p. 292) applied the heuristic elimination rule $(\bar{\underline{x}}_{[k]} - \bar{\underline{x}}_i) - 2[\hat{var}(\bar{\underline{x}}_{[k]})]^{1/2} - 2[\hat{var}(\bar{\underline{x}}_i)]^{1/2} > 0$ where we interpret their "standard deviation associated with plan i" as the standard deviation of the average $\bar{\underline{x}}_i$, not the individual observation $\underline{x}_i$. They found considerable computer time savings when they applied this rule to a simulated inventory system [see Sasser et al. (1970, p. 295)].

REFERENCES

Andrews, D.F., P.J. Bickel, F.R. Hampel, P.J. Huber, W.H. Rogers, and J.W. Tukey (1972). Robust Estimates of Location, Princeton University Press, Princeton, New Jersey.

Barr, D.R. and M.H. Rizvi (1966). "An introduction to ranking and selection procedures," J. Amer. Stat. Assoc. 61, 640-646.

Barron, A. and S.S. Gupta (1970). A Class of Non-Eliminating Sequential Multiple Decision Procedures, Mimeograph Series No. 247, Department of Statistics, Purdue University, Indiana.

Bartlett, N.S. and Z. Govindarajulu (1968). "Some distribution-free statistics and their application to the selection problem," Ann. Inst. Stat. Math. 20, 79-97.

Bawa, V.S. (1972). "Asymptotic efficiency of one R-factor experiment relative to R one-factor experiments for selecting the best normal population," J. Amer. Stat. Assoc. 67, 660-661.

Bechhofer, R.E. (1954). "A single-sample multiple decision procedure for ranking means of normal populations with known variances," Ann. Math. Stat. 25, 16-39.

Bechhofer, R.E. (1958). "A sequential multiple-decision procedure for selecting the best one of several normal populations with a common unknown variance, and its use with various experimental designs," Biometrics 14, 408-429.

Bechhofer, R.E. (1970). "Correction note: an undesirable feature of a sequential multiple-decision procedure for selecting the best one of several normal populations with a common unknown variance," Biometrics 26, 347-349.

Bechhofer, R.E. and S. Blumenthal (1962). "A sequential multiple-decision procedure for selecting the best one of several normal populations with a common unknown variance, II: Monte Carlo sampling results and new computing formulae," Biometrics 18, 52-67.

Bechhofer, R.E., C.W. Dunnett, and M. Sobel (1954). "A two-sample multiple decision procedure for ranking means of normal populations with a common unknown variance," Biometrika 41, 170-176.

Bechhofer, R.E., S. Elmaghraby, and N. Morse (1959). "A single-sample multiple-decision procedure for selecting the multinomial event which has the highest probability," Ann. Math. Stat. 30, 102-119.

Bechhofer, R.E., J. Kiefer, and M. Sobel (1968). Sequential Identification and Ranking Procedures, the University of Chicago Press, Chicago.

Bechhofer, R.E. and M. Sobel (1956). "A sequential multiple-decision procedure for selecting the multinomial event with the largest probability (preliminary report)," Ann. Math. Stat. 27, 861.

Bechhofer, R.E. and M. Sobel (1958). "Non-parametric multiple-decision procedures for selecting the one of k populations which has the highest probability of yielding the largest observation (Preliminary report)," Ann. Math. Stat. 29, 325.

Bradley, J.V. (1968). Distribution-Free Statistical Tests, Prentice-Hall, Englewood Cliffs, New Jersey.

Bradley, R.A. (1967). "Topics in rank-order statistics," in Proceedings of the Fifth Berkeley Symposium on Mathematical Statistics and Probability, Vol. I (L.M. LeCam and J. Neyman,eds.), University of California Press, Berkeley, pp. 593-607.

Chambers, M.L. and P. Jarratt (1964). "Use of double sampling for selecting best population," Biometrika 51, 49-64.

Conover, W.J. (1971). Practical Nonparametric Statistics, Wiley, New York.

Conway, R.W., B.M. Johnson, and W.L. Maxwell (1959). "Some problems of digital systems simulation," Management Sci. 6, 92-110.

Dudewicz, E.J. (1968). "A Categorized Bibliography on Multiple-Decision (Ranking and Selection) Procedures, Department of Statistics, The University of Rochester, Rochester, New York.

Dudewicz, E.J. (1969). Estimation of Ordered Parameters, Technical Report No. 60, Department of Operations Research, College of Engineering, Cornell University, Ithaca, New York.

Dudewicz, E.J. (1971a). "A nonparametric selection procedure's efficiency: largest location parameter case," J. Amer. Stat. Assoc. 66, 152-161.

Dudewicz, E.J. (1971b). "Maximum likelihood estimators for ranked means," Zeitschrift Wahrscheinlichkeitstheorie und Verwandte Gebiete 19, 29-42.

Dudewicz, E.J. (1972). Statistical Inference with Unknown and Unequal Variances, Department of Statistics, The University of Rochester, Rochester, New York.

Dudewicz, E.J. and S.R. Dalal (1971). Allocation of Observations in Ranking and Selection with Unequal Variances, Department of Statistics, The University of Rochester, Rochester, New York.

Dudewicz, E.J. and J.S. Ramberg (1972). Multiple Comparisons with a Control: Unknown Variances, Department of Statistics, The University of Rochester, Rochester, New York.

Dudewicz, E.J. and Y.L. Tong (1971). "Optimal confidence intervals for the largest location parameter," in Statistical Decision Theory and Related Topics (S.S. Gupta and J. Yackel, eds.), Academic, New York.

Dudewicz, E.J. and N.A. Zaino (1971). "Sample size for selection," in Statistical Decision Theory and Related Topics, (S.S. Gupta and J. Yackel, eds.), Academic, New York.

Dunnett, C.W. (1955). "A multiple comparison procedure for comparing several treatments with a control," J. Amer. Stat. Assoc. 50, 1096-1121.

Dunnett, C.W. (1960). "On selecting the largest of k normal population means," J. Roy. Stat. Soc., Ser. B, 22, 1-40.

Dunnett, C.W. and M. Sobel (1954). "A bivariate generalization of Student's t-distribution with tables for certain special cases," Biometrika 41, 153-169.

Dunnett, C.W. and M. Sobel (1955). "Approximations to the probability integral and certain percentage points of a multivariate analogue of Student's t-distribution," Biometrika 42, 258-260.

Fairweather, W.R. (1968). "Some extensions of Somerville's procedure for ranking means of normal populations," Biometrika 55, 411-418.

Fishman, G.S. and P.J. Kiviat (1967). Digital Computer Simulation: Statistical Considerations, RM-5387-PR, The RAND Corporation, Santa Monica, California. (Published as: "The statistics of discrete-event simulation," Sci. Simulation 10, 1968, 185-195.)

Fisz, M. (1967). Probability Theory and Mathematical Statistics, Wiley, New York, third printing.

Freeman, H., A. Kuzmack, and R. Maurice (1967). "Multivariate t and the ranking problem," Biometrika 54, 305-308.

Fu, K.S. (1970). Sequential Methods in Pattern Recognition and Machine Learning, Academic, New York, second printing.

Gupta, S.S. (1963a). "Probability integrals of multivariate normal and multivariate t," Ann. Math. Stat. 34, 792-828.

Gupta, S.S. (1963b). "Bibliography on the multivariate normal integrals and related topics," Ann. Math. Stat. 34, 829-838.

Gupta, S.S., K. Nagel, and S. Panchapakesan (1971). On the Order Statistics from Equally Correlated Normal Random Variables, Mimeograph Series No. 290, Department of Statistics, Division of Mathematical Sciences, Purdue University, Lafayette, Indiana.

Gupta, S.S. and S. Panchapakesan (1969a). "Some selection and ranking procedures for multivariate normal populations," in Multivariate Analysis, Vol. 2, (P.R. Krishnaiah, ed.), Academic, New York.

Gupta, S.S. and S. Panchapakesan (1969b). "Selection and ranking procedures," in The Design of Computer Simulation Experiments, (T.H. Naylor, ed.), Duke University Press, Durham, North Carolina.

Gupta, S.S. and M. Sobel (1957). "On a statistic which arises in selection and ranking procedures," Ann. Math. Stat. 28, 957-967.

Guttman, I. (1963). "A sequential procedure for the best population," Sankhya, Ser. A, 25, Pt. 1, 25-28.

Hayes, R.H. (1969). "The value of sample information," in The Design of Computer Simulation Experiments (T.H. Naylor, ed.), Duke University Press, Durham, North Carolina.

Healy, W.C. (1956). "Two-sample procedures in simultaneous estimation," Ann. Math. Stat. 27, 687-702.

Hoel, D.G. (1971). "A method for the construction of sequential selection procedures," Ann. Math. Stat. 42, 630-642.

Hoel, D.G. and M. Mazumdar (1968). "An extension of Paulson's selection procedure," Ann. Math. Stat. 39, 2067-2074.

Kendall, M.G. and A. Stuart (1963). The Advanced Theory of Statistics, Vol. 1. Charles Griffin, London, second edition.

Kleijnen, J.P.C. (1968). The Use of Multiple Ranking and Multiple Comparison Procedures in the Simulation of Business and Economic Systems, Preliminary Report, Los Angeles. (Obtainable at Katholieke Hogeschool, Tilburg, the Netherlands.)

Kleijnen, J.P.C. and T.H. Naylor (1969). "The use of multiple ranking procedures to analyze simulations of business and economic systems," Proceedings of the Business and Economic Statistics Section, Annual Meeting of the American Statistical Association, New York, pp. 605-615.

Kleijnen, J.P.C., T.H. Naylor, and T.G. Seaks (1972). "The use of multiple ranking procedures to analyze simulations of management systems: a tutorial," Management Sci., Appl. Ser. 18, 245-257.

Krishnaiah, P.R. and J.V. Armitage (1966). "Tables for multivariate t distribution," Sankhya, Ser. B. 28, Pts. 1 and 2, 31-56.

Lehmann, E.L. (1963). "A class of selection procedures based on rank ," Mathematische Ann. 150, 268-275.

Lewis, P.A.W. (1972). Large-Scale Computer-Aided Statistical Mathematics, Naval Postgraduate School, Monterey, California.

Martin, W.A. (1971). "Sorting," Computing Surveys, 3, 147-174.

Miller, R.G. (1966). Simultaneous Statistical Inference, McGraw-Hill, New York.

Milton, R.C. (1963). Tables of the Equally Correlated Multivariate Normal Probability Integral, Technical Report No. 27, Department of Statistics, University of Minnesota, Minneapolis.

Naylor, T.H., D.S. Burdick, and W.E. Sasser (1967a). "Computer simulation experiments with economic systems: the problem of experimental design," J. Amer. Stat. Assoc. 62, 1315-1337.

Naylor, T.H., K. Wertz, and T.H. Wonnacott (1967). "Methods for analyzing data from computer simulation experiments," Comm. ACM 10, 703-710.

Naylor, T.H., K. Wertz, and T. Wonnacott (1968). "Some methods for evaluating the effects of economic policies using simulation experiments," Rev. Inter. Stat. Inst. 36, 184-200.

O'Neill, R. and G.B. Wetherill (1971). "The present state of multiple comparison methods," J. Roy. Stat. Soc., Ser. B, 33, 218-241.

Paulson, E. (1962). "A sequential procedure for comparing several experimental categories with a standard or control," Ann. Math. Stat. 33, 438-443.

Paulson, E. (1964a). "A sequential procedure for selecting the population with the largest mean from k normal populations," Ann. Math. Stat. 35, 174-180.

Paulson, E. (1964b). "Sequential estimation and closed sequential decision procedures," Ann. Math. Stat. 35, 1048-1058.

Puri, M.L. and P.S. Puri (1969). "Multiple decision procedures based on ranks for certain problems in analysis of variance," Ann. Math. Stat. 40, 619-632.

Ramberg, J.S. (1966). A Comparison of the Performance Characteristics of Two Sequential Procedures for Ranking Means of Normal Populations, Technical Report No. 4, Department of Industrial Engineering and Operations Research, College of Engineering, Cornell University, Ithaca, New York.

Randles, R.H. (1970). "Some robust selection procedures," Ann. Math. Stat. 41, 1640-1645.

Rizvi, M.H. and K.M.L. Saxena (1972). "Distribution-free interval estimation of the largest α-quantile," J. Amer. Stat. Assoc. 67, 196-198.

Robbins, H., M. Sobel, and N. Starr (1968). "A sequential procedure for selecting the largest of k means," Ann. Math. Stat. 39, 88-92.

Roberts, C.D. (1963). "An asymptotically optimal sequential design for comparing several experimental categories with a control," Ann. Math. Stat. 34, 1486-1493.

Sasser, W.E., D.S. Burdick, D.A. Graham, and T.H. Naylor (1970). "The application of sequential sampling to simulation: an example inventory model," Comm. ACM 13, 287-296.

Scheffé, H. (1964). The Analysis of Variance. Wiley, New York, fourth printing.

Schmidt, J.W. and R.E. Taylor (1970). Simulation and Analysis of Industrial Systems. Richard D. Irwin, Homewood, Illinois.

Sen, A.K. and M.S. Srivastava (1972a). "On a Sequential Selection Procedure Based on Wilcoxon Statistic, University of Connecticut/ University of Illinois at Chicago Circle.

Sen, A. and M.S. Srivastava (1972b). A Monte Carlo Study of Some Sequential Solutions to Ranking and Slippage Problems, Center for Urban Studies, University of Illinois (Chicago Circle), and University of Toronto, Toronto.

Shirafuji, M. (1956). "Note on the determination of the replication numbers for the slippage problem in r-way layout," Bull. Math. Stat. 7, 46-51.

Sobel, M. (1967). "Nonparametric procedures for selecting the t populations with the largest α-quantiles," Ann. Math. Stat. 38, 1804-1816.

Sobel, M. and Y.L. Tong (1970). Optimal Allocation for Partitioning a Set of Normal Populations in Comparison with a Control, Technical Report No. 134, Department of Statistics, University of Minnesota, Minneapolis.

Somerville, P.N. (1954). "Some problems of optimum sampling," Biometrika 41, 420-429.

Somerville, P.N. (1970). "Optimum sample size for a problem in choosing the population with the largest mean," J. Amer. Stat. Assoc. 65, 763-775.

Srivastava, M.S. (1966). "Some asymptotically efficient sequential procedures for ranking and slippage problems," J. Roy. Stat. Soc., Ser. B 28, 370-380.

Srivastava, M.S. and J. Ogilvie (1968). "The performance of some sequential procedures for a ranking problem," Ann. Math. Stat. 39, 1040-1047.

Srivastava, M.S. and A.K. Sen (1972). A Sequential Solution of Wilcoxon Types for a Slippage Problem, University of Toronto/University of Illinois at Chicago Circle (Center for Urban Studies, Chicago).

Srivastava, M.S. and V.S. Taneja (1972). Some Sequential Procedures for Ranking Multivariate Normal Populations, Department of Statistics, University of Toronto, Toronto.

Starr, N. (1966). "The performance of a sequential procedure for the fixed-width interval estimation of the mean," Ann. Math. Stat., 37, 36-50.

Statistical Theory and Method Abstracts (1959-1972). (Published until 1965 as Statistical Theory and Methods.) Oliver and Boyd, Edinburgh 1959-1972.

Taylor, R.J. and H.A. David (1962). "A multi-stage procedure for the selection of the best of several populations," J. Amer. Stat. Assoc. 57, 785-796.

Tocher, K.D. (1963). The Art of Simulation, The English Universities Press, London.

Tong, Y.L. (1969). "On partitioning a set of normal populations by their locations with respect to a control," Ann. Math. Stat. 40, 1300-1324.

Wetherill, G.B. (1966). Sequential Methods in Statistics, Methuen, London.

Zinger, A. (1961). "Detection of best and outlying normal populations with known variances," Biometrika 48, 457-461.

Chapter VI

MONTE CARLO EXPERIMENTATION WITH BECHHOFER AND BLUMENTHAL'S MULTIPLE RANKING PROCEDURE: A CASE STUDY[1)]

VI.1. INTRODUCTION AND SUMMARY

The purpose of this chapter is twofold:

(i) It is a case study demonstrating how various techniques discussed in previous chapters can be applied to the design and analysis of simulation and Monte Carlo experiments. We shall apply a 2_{IV}^{7-3} design, antithetic variates, sample size formulas, ANOVA and regression analysis, simultaneous inference, etc.

(ii) The subject of the case study is a Monte Carlo experiment with the multiple ranking procedure (abbreviated to MRP) devised by Bechhofer and Blumenthal (1962); the Monte Carlo experiment aims at investigating the robustness of this procedure.

In Section VI.2 we shall briefly describe the MRP and the purpose of the experiment, viz., the study of the sensitivity of the MRP to its underlying assumptions. In Section VI.3 we shall discuss the various factors that may influence this sensitivity, in detail. Nonnormality we shall specify by factor 1, the distribution can or cannot yield values smaller than a particular constant (so-called "truncation" factor) and by factor 2, the skewness and the tails of the distribution. Combining factors 1 and 2 we shall select four types of distributions (exponential, Erlang, weighted difference of two exponentials and sum of differences between exponentials). Heterogeneity of variances will be specified by three factors, factor 3 denoting whether the variance of the best population $(\sigma_{(k)}^2)$ is larger or smaller than the variance of the most competing inferior population (in the least favorable configuration), factor 4 measuring if these two variances do or do not differ much, factor 5 showing if the variances of the inferior populations (in the least favorable configuration) all are equal or all are different. Factor 6 specifies the number of populations (three or seven); factor 7 measures the distance $\delta = \delta^*$ between the best and the next best population in the least favorable configuration. The factor P^*, guaranteed minimum value of the probability of correct selection, is considered at 22 levels between 0.35 and 0.99 and yields vectors consisting of 22 dependent observations. All factors except P^* will be considered at their minimum number of levels, i.e., two levels. In Section VI.4 we shall derive a 2_{IV}^{7-3} design for the above 7 two-level factors; by definition the 16 factor combinations yield estimators of the main effects not biased by possible two-factor interactions and estimators of particular linear combinations of two-factor interactions. In Section VI.5 the number of replications will be so determined that the α error is smaller than 1% and the β error is smaller than 10% (for given alternatives); therefore eight hundred replications for each design point in the 2_{IV}^{7-3} design will be generated. In Section VI.6 two variance reducing techniques will be considered, viz., common random numbers and antithetic variates. Antithetics will be applied since they do not complicate the statistical analysis. In Section VI.7 the type of output of the experiment will be briefly described. In Section VI.8 we shall investigate the robustness of the MRP by testing if the estimated fractions are significantly lower than the corresponding P^*. Student's t-statistic will be applied with a familywise error rate. Three factor combinations give significantly low fractions of correct selection for any reasonable value of the error rates; eleven combinations give satisfactory fractions for any reasonable error rate. In Section VI.9 we shall consider several techniques for the detection of "important" factors, i.e., factors that cause the MRP to fail. The first technique involves contingency tables but is shown not to be applicable. The second technique consists of binomial tests of the probability that in the rejected factor combinations a factor is always at the same level and nevertheless is unimportant. These tests, however, are shown to have small power. The third technique is regression analysis. The usual assumptions of regression analysis and analysis of variance will be investigated and it will be shown why simple least squares are preferred over generalized least squares. We shall estimate the regression coefficients or "effects"; test lack of fit of the regression equations (for $P^* \geq 0.90$ good fit will be found); test if the regression coefficients are significantly different from zero (factor 1 gives a zero main effect; factor 5 may very well also have a zero main effect). In Sction VI.10 conclusions and directions for future research will be given. Finally we shall give appendices, exercises and references.

VI.2. PURPOSE OF THE EXPERIMENT

At the time the experiment with Bechhofer and Blumenthal's MRP was performed, it was planned to experiment with no less than nine procedures; see Kleijnen (1969a, pp. 1-2). Bechhofer and Blumenthal's procedure happened to be the first one that we investigated, partly because it looked as a most promising procedure at that time. During experimentation we discovered that a thorough investigation of all these procedures would require excessive amounts of computer time. The generation of the data for the 2^{7-4} design used to study the Bechhofer-Blumenthal procedure took about one hour on the IBM/360, model 75 computer. Therefore at this moment results are available only for the Bechhofer-Blumenthal procedure reported on in this chapter.

As we saw in the previous chapter the Bechhofer-Blumenthal MRP determines how many observations should be taken from each of the k populations $(k \geq 2)$ in order to guarantee that the population with the largest mean is selected with prescribed probability at least P^*. The probability of correct selection (abbreviated to CS) should be at least P^* provided the best population mean $\mu_{(k)}$ is at least δ^* better than the next best mean $\mu_{(k-1)}$ (so-called indifference zone approach), i.e.,

$$P(CS) \geq P^* \quad \text{if} \quad \mu_{(k)} - \mu_{(k-1)} \geq \delta^* \tag{1}$$

where P^* and δ^* are specified by the experimenter. The Bechhofer-Blumenthal procedure is sequential, i.e., in each stage s ($s = 1, 2, \ldots$) one observation is taken from each of the k populations and after each stage m ($m \geq 2$) the value of the stopping statistic $\underline{z}_m$, defined in V.C.3, is calculated; as soon as $\underline{z}_m \leq (1 - P^*)/P^*$ sampling is terminated and as the best population is selected the one yielding the highest sample mean. The procedure assumes that all observations $\underline{x}_{is}$ ($i = 1, \ldots, k$) ($s = 1, 2, \ldots$) are <u>normally</u> distributed independent observations with a <u>common variance</u> σ_x^2 and means μ_i, i.e.,

$$\underline{x}_{is} : \text{NID}(\mu_i, \sigma_x^2) \tag{2}$$

We want to investigate if the procedure is sensitive to nonnormality and heterogeneity of variance. (If more procedures guarantee the probability requirement in (1) we hoped to determine in a next experiment which procedure requires the smallest number of observations from the k populations.) The procedure may be insensitive to small deviations from normality and homogeneous variances. Therefore we shall consider small and large deviations from these two assumptions. Moreover the procedure may be robust (and require an acceptable number of observations) only for particular values of P^*, δ^*, and k. So next we shall consider several conceivably important factors in more detail.

VI.3. FACTORS IN THE EXPERIMENT

(i) <u>Nonnormality</u>. What we would like to investigate are those deviations from normality that might be expected in models of business and economic systems. Unfortunately it seems impossible to specify which distributions we might expect to hold for intricate simulation models. As far as the literature is concerned Bechhofer et al. (1968, pp. 266-267, 348, 363) used the uniform distribution in the Monte Carlo experiment with their MRP. The uniform distribution, however, seems unrealistic for simulation models. Neave and Granger (1968) utilized a particular bimodal nonsymmetric distribution generated by weighing two normal distributions and they tested the robustness of various two-sample tests for differences in means. Donaldson (1966) applied exponential and lognormal distributions for the investigation of the sensitivity of the F-test when applied to maintenance data. Andrews et al. (1972, pp. 56-57, 60) used, say, $\underline{x}/\underline{y}$ (where $\underline{x}$ is a standard normal variate, $\underline{y}$ is positive and independent of $\underline{x}$) to generate several types of symmetric, long-tailed distributions; also see Rogers and Tukey(1972). Additional references to Monte Carlo studies on the effect of nonnormality for various statistical procedures can be found in Scheffé (1964, pp. 345-351), Teichroew (1965, p. 35) and our Chapter I. Clearly, there is only one way to satisfy the assumption of normality but there are infinitely many ways to deviate from the assumption. We shall show how we made a decision on the specification of nonnormality.

One important characteristic of the distribution might be the "truncation" of the distribution, where here we use the word "truncation" if the distribution cannot yield values smaller than a particular constant. Gamma distributions cannot yield values smaller than zero. This characteristic is attractive in some models, e.g., queuing models where waiting times are studied that cannot be negative. A consequence of truncation is that even a small sample from the best population does not yield a sample mean smaller than the (lower) truncation point. It is conceivable that this increases P(CS). In other models where profit is studied the variable is not truncated and may range from $-\infty$ to $+\infty$. Therefore we decided to use as one factor in the Monte Carlo experiment, truncated versus non-truncated distributions.

A second important characteristic consists of the skewness and the tails of the distributions. We shall consider very skew distributions with heavy tails versus symmetric or nearly symmetric distributions having tails more compatible with the normal distribution.

An approximative measure for the heaviness of the tails is the kurtosis, say γ, defined as

$$\gamma = \sigma^{-4} E[(\underline{x} - \mu)^4] - 3 \tag{3}$$

If $\gamma > 0$ then the tails tend to be heavier than in a normal distribution; see e.g. Scheffé (1964, p. 332). (Kendall and Stuart (1963, p. 86) pointed out that the kurtosis is only an approximative measure for the tails.)

Combining the above two factors we decided to use the distributions summarized in Table 1 which we shall now discuss in more detail. The exponential and Erlang distributions are special cases of the gamma distribution. The gamma distribution has density function

$$\begin{aligned} f(x) &= \frac{b^p}{\Gamma(p)} x^{p-1} e^{-bx} && \text{for } x > 0 \\ &= 0 && \text{for } x \leq 0 \end{aligned} \tag{4}$$

with the kth moment given by

$$m_k = \frac{p(p+1) \cdots (p+k-1)}{b^k} \tag{5}$$

so that

$$E(\underline{x}) = \frac{p}{b} \quad \text{and} \quad \text{var}(\underline{x}) = \frac{p}{b^2} \tag{6}$$

TABLE 1

Types of Distributions Used to Specify Nonnormality

	Truncated	Nontruncated
Skew and heavy tails	Exponential	Weighted difference of two exponentials: $\underline{x}_1 - 0.1\,\underline{x}_2$
(Nearly) symmetric and less heavy tails	Erlang with parameter $p = 4$	Sum of differences between exponentials: $(\underline{x}_1 - \underline{x}_2) + (\underline{x}_3 - \underline{x}_4)$

see Fisz (1967, pp. 152-153). It is simple to show that its kurtosis is $6/p$. The exponential distribution is defined by setting $p = 1$ in (4). This distribution is extremely skew and has kurtosis 6. The Erlang distribution is the family of distributions obtained by restricting p in (4) to integer values. As p increases the Erlang distribution becomes more normal as is demonstrated in Mood and Graybill (1963, p. 127). We decided to take $p = 4$ which yields a density function as in Fig. 1. This distribution is nearly symmetric and its kurtosis is 1.5, i.e., compared with the exponential distribution its kurtosis is much smaller and is more compatible with the normal distribution. Its computer generation is straightforward since we simply add four independent exponential variates with parameter b. The third type of distribution function in Table 1 is a skew distribution with heavy tails as the exponential distribution has, but without truncation. We devised the following variate:

$$\underline{x} = \underline{x}_1 - 0.1\, \underline{x}_2 \qquad (7)$$

where $\underline{x}_1$ and $\underline{x}_2$ are independent exponential variates with a common parameter b. The density of $\underline{x}$ can be shown to be

$$\begin{aligned} f(x) &= 10/11b \exp(10\, bx) && \text{for } x \leq 0 \\ &= 10/11b \exp(-bx) && \text{for } x > 0 \end{aligned} \qquad (8)$$

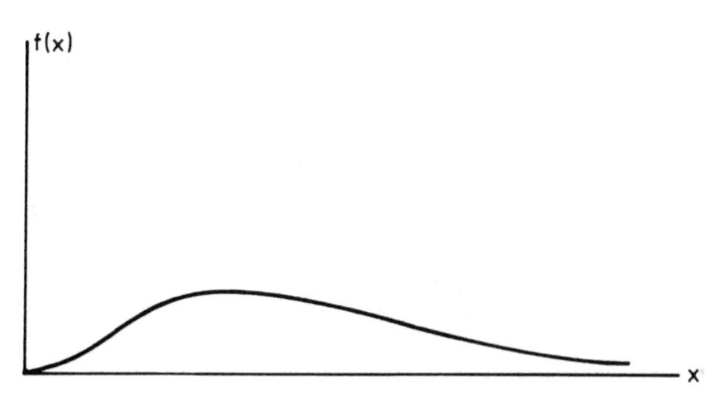

Fig. 1. The Erlang distribution with parameter $p = 4$.

which corresponds with Fig. 2. If $\underline{x}$ is positive its density is 10/11 of the exponential variate with parameter b; only 1/11 of its density volume is left of $\underline{x} = 0$. Obviously $\underline{x}$ has a very skew distribution and its kurtosis can be shown to be close to that of the exponential variate (i.e., $\gamma \approx 6$), but its values can range from $-\infty$ to $+\infty$. The fourth type of distribution in Table 1 is a symmetric distribution with less heavy tails and no truncation. We devised the following variable:

$$\underline{x} = (\underline{x}_1 - \underline{x}_2) + (\underline{x}_3 - \underline{x}_4) = (\underline{x}_1 + \underline{x}_3) - (\underline{x}_2 + \underline{x}_4) \qquad (9)$$

where the $\underline{x}_i$ $(i = 1, \ldots, 4)$ are independent exponential variables with common parameter b. It is easy to see that the density function of $\underline{x}$ in (9) is symmetric. We derived that its kurtosis is 1.5, i.e., exactly the value of the Erlang distribution with $p = 4$. Observe that the mean of any of the four types of distributions can be easily controlled by adding a constant. The constant does not influence the variance, skewness or kurtosis. Note that in our experiment all k populations in one particular factor combination have the same type of distribution (e.g., Erlang); it might be interesting to investigate cases where some populations have exponential distributions and some have Erlang distributions. In our approach, however, we can better identify the effect of truncation.

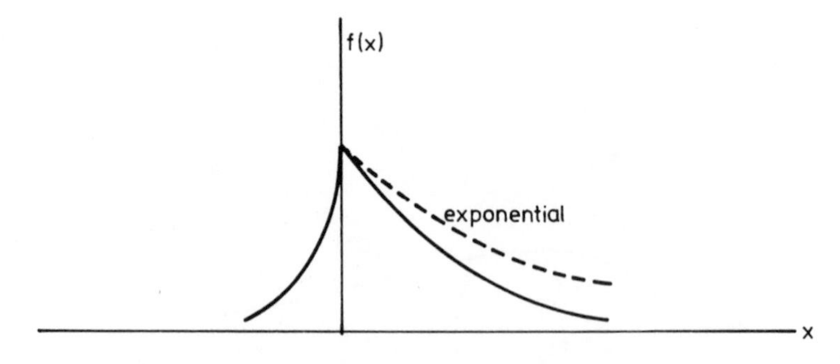

Fig. 2. The density function of $\underline{x}_1 - 0.1\,\underline{x}_2$ where $\underline{x}_1$ and $\underline{x}_2$ are exponentials.

(ii) Heterogeneity of variance. Let us consider the influence of the variances on the probability that an observation from some inferior population is higher than an observation from the best population. If we use normal approximations for the k distributions this probability is given by (10)

$$P(\underline{x}_{(j)} > \underline{x}_{(k)}) = P\left(\underline{z} > \frac{\mu_{(k)} - \mu_{(j)}}{[\sigma^2_{(k)} + \sigma^2_{(j)}]^{1/2}}\right) \qquad (10)$$

where the index (j) corresponds with the j'th best population and $\underline{z}$ is the standard normal variable. So the probability of a wrong selection increases if $\mu_{(k)} - \mu_{(j)}$ decreases or the variances $\sigma^2_{(k)}$ and $\sigma^2_{(j)}$ increase. Generalizing to any distribution type, the probability of a wrong selection increases if the population means are not far apart or if the observations show large fluctuations. We shall return to the configuration of the means later on; we point out that we shall restrict attention to the least favorable configuration (LFC) of the means, defined in V.C.2 as the configuration

$$\mu_{(1)} = \cdots = \mu_{(k-1)} = \mu_{(k)} - \delta^* \qquad (11)$$

From (11) it follows that the most competing inferior population is the one with the highest variance, $\max_{1 \le j \le k-1} \sigma^2_{(j)}$; also see Fig. 3.

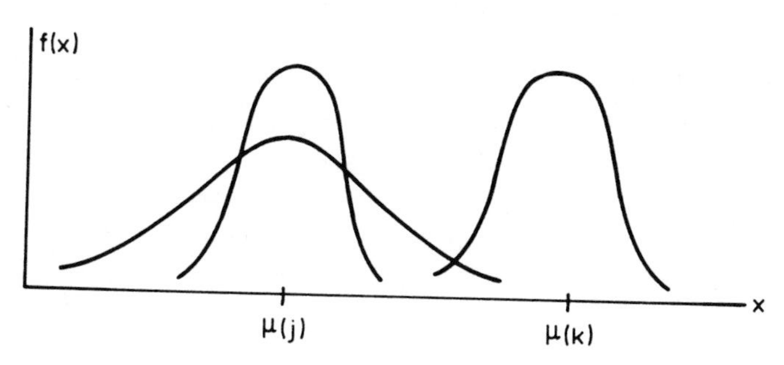

Fig. 3. The effect of the variances of the inferior populations.

We decided to compare the variance of the best population $(\sigma^2_{(k)})$ with the variance of the most competing inferior population $(\max \sigma^2_{(j)})$. One aspect is whether $\sigma^2_{(k)}$ is larger or smaller than $\max \sigma^2_{(j)}$; the other aspect is whether both variances differ much or not. We used the values specified in Table 2.[2)]

Another aspect is whether the variances of the other $(k - 2)$ inferior populations are the same or different. Therefore we studied two configurations of the variances of the inferior populations, viz., $\sigma^2_{(1)} = \cdots = \sigma^2_{(k-2)} = \sigma^2_{(k-1)}$ versus $\sigma^2_{(1)} < \cdots < \sigma^2_{(k-2)} < \sigma^2_{(k-1)}$. In the latter case we have the standard deviations increase linearly as in (12)

$$\sigma_{(j)} = \sigma_{(1)} + \frac{\{\sigma_{(k-1)} - \sigma_{(1)}\}}{k - 2} (j - 1) \qquad (j = 1, \ldots, k-1) \tag{12}$$

and take

$$\sigma_{(1)} = 0.25\sigma_{(k-1)} \tag{13}$$

We observe that as far as the literature is concerned, Donaldson (1966, p. 15) took $\sigma^2_i = \mu^2_i$ $(i = 1, \ldots, k)$ (but in our experiment all inferior populations have the same mean) and Box (1954a) and (1954b), reported on in Scheffé (1964, pp. 354-358), used $\max \sigma^2_i / \min \sigma^2_i$ $(i = 1, \ldots, k)$ equal to 3 or 7, all variances being different or all except one being the same.

TABLE 2

Standard Deviation of the Best Population $(\sigma_{(k)})$ Divided by Standard Deviation of Inferior Population Most Competing in the LFC $(\max \sigma_{(j)})$

	$\sigma_{(k)}/\max \sigma_{(j)} < 1$	$\sigma_{(k)}/\max \sigma_{(j)} > 1$
$\sigma_{(k)}/\max \sigma_{(j)} \approx 1$	4/5	5/4
$\sigma_{(k)}/\max \sigma_{(j)} \neq 1$	1/5	5

Summarizing, the heterogeneity of the variances is quantified by

- the variance of the best population being smaller or larger than that of the most competing inferior population,
- these two variances being nearly equal or much different,
- the configuration of the variances of the inferior populations (all equal or all different).

(iii) The number of populations. The number of populations (k) is a conceivably important factor. For if k is small even random selection of a population has a high probability $(1/k)$ of resulting in a correction selection. Bechhofer et al. (1968, p. 349) have k range from 2 to 25 in their Monte Carlo experiments. High values of k require much computer time since the number of observations per stage is proportional to k (take one observation from each population) and the number of stages increases with k as there are more competing populations. Therefore we decided to consider a maximum of seven populations. As minimum value for k we take three (not two since such a situation would not require multiple ranking procedures).

(iv) The configuration of the means and the factor δ^*. We decided to study only the LFC of the means, defined in (11) above. For we conjecture that for $\mu_{(k)} - \mu_{(k-1)} > \delta^*$ the procedure gives a $P(CS)$ not smaller than for $\mu_{(k)} - \mu_{(k-1)} = \delta^*$ and/or $\mu_{(j')} < \mu_{(k-1)}$ $(j' = 1, \ldots, k-2)$. If $\mu_{(k)} - \mu_{(k-1)} < \delta^*$ the procedure does not pretend to give a CS. (In future experiments other configurations, e.g., equal means, may be used to investigate the effect on the sample size.)

We can study the LFC for various values of $\delta \equiv \mu_{(k)} - \mu_{(k-1)} = \delta^*$. The population means may lie close together so that the probability is high that observations from the best population are worse than observations from an inferior population. Whether $\delta = \delta^*$ is considered to be small or large depends on the spread of the populations; compare Fig. 4. So we have to relate δ to the spread.

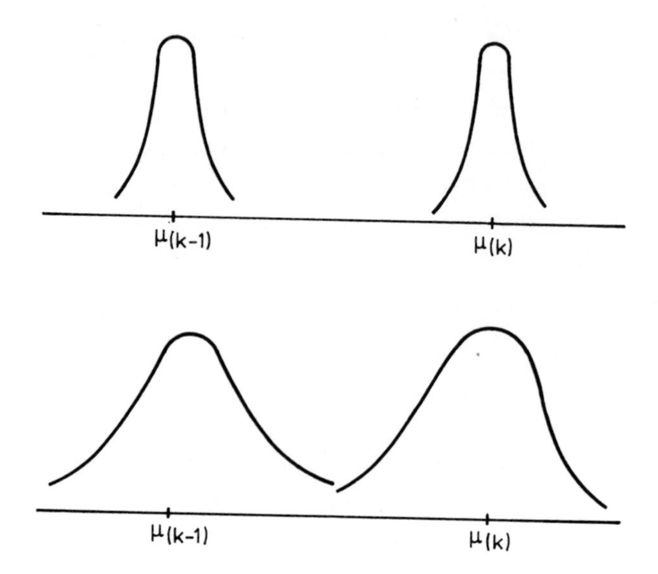

Fig. 4. The distance $\delta = \mu_{(k)} - \mu_{(k-1)}$ and the spread of the distributions.

Bechhofer et al. (1968, pp. 349, 365) studied $\sigma_i^2 = 1$ and $\delta = 0.15$ or 0.20. Since they used mainly normal distributions with $\delta = 0.20$ (and $\sigma_i^2 = 1$) we applied (10) above to derive that as much as 44% of the observations from an inferior population are better than observations from the superior population:

$$P(\underline{x}_{(j)} > \underline{x}_{(k)}) = P(\underline{z} > 0.20/\sqrt{2}) = 0.44 \tag{14}$$

Therefore we decided to apply as low value of $\delta = \delta^*$

$$\delta = \delta^* = 0.20 \frac{\{\sigma_{(k)} + \sigma_{(k-1)}\}}{2} \tag{15}$$

As high value of δ we use

$$\delta = \delta^* = 1.25 \frac{\{\sigma_{(k)} + \sigma_{(k-1)}\}}{2} \tag{16}$$

which corresponds with 19% instead of 44% in (14). The percentages 44 and 19 are only rough guidelines (they are based on normality). Anyhow, the high value of the "relative" distance, i.e., δ related to the average standard deviation $\{\sigma_{(k)} + \sigma_{(k-1)}\}/2$, is about six times as high as the low value.

(v) The guaranteed probability P^*. In the probability requirement in (1) P^* is another factor besides $\delta = \delta^*$. Bechhofer et al. (1968, pp. 349, 365) have P^* vary as shown in Table 3. In our experiment we take as lowest value $P^* = 0.35$ (which is higher than $1/k$). As highest value we take $P^* = 0.99$; higher values would require very many observations so that excessive computer time is needed. As Bechhofer et al. did we consider P^* at more than two levels. For fewer observations are needed in the Bechhofer-Blumenthal procedure as P^* decreases. (The stopping statistic $\underline{z}_m$ itself does not depend on P^* but sampling is terminated as soon as $\underline{z}_m < (1 - P^*)/P^*$.) So suppose we have decided to test the procedure with $P^* = 0.99$ and we start generating observations. Then any level of P^* smaller than 0.99 can be tested at the same time since such a level requires only part of the observations that must be generated for $P^* = 0.99$. Though considering more levels of P^* smaller than 0.99 does not require additional observations some extra calculations must be executed. We are willing to pay for the additional information through a slight increase in the amount of calculations. A more important consequence is that the responses for various P^* levels are dependent, for these responses partly use the same observations. The closer the P^* levels are, the more they use common observations and consequently the more their responses are correlated. (This correlation may be assumed positive.) This means that our experiment yields multivariate responses.

Except for the factor P^* the factors in the experiment are taken at their minimum number of levels, i.e., two levels--in order to limit the number of factor-level combinations. The seven two-level factors are summarized now.

TABLE 3

Values of P^* in the Monte Carlo Experiments in Bechhofer et al. (1968)

Main experiments	Additional experiments
0.35(0.05)0.85	0.40(0.05)0.85
0.89(0.01)0.99	0.991(0.001)0.999

1. Nonnormality: nontruncated versus truncated distributions.
2. Nonnormality: symmetric distributions with nearly normal tails versus skew distributions with heavy tails.
3. Heterogeneity of variances: $\sigma_{(k)}/\max \sigma_{(j)}$ lower than 1 or higher.
4. Heterogeneity of variances: $\sigma_{(k)}/\max \sigma_{(j)}$ approximately 1 or much different from 1.
5. Heterogeneity of variances: all variances $\sigma_{(j)}$ equal or all different.
6. Number of populations high ($k = 7$) or low ($k = 3$).
7. Distance between the best mean and the inferior means: $\delta = \delta^*$ equal to $0.20\{\sigma_{(k)} + \sigma_{(k-1)}\}/2$ versus $1.25\{\sigma_{(k)} + \sigma_{(k-1)}\}/2$.

Let us next consider how these seven factors taken together specify the populations. The factors 1 and 2 specify the type of distribution used in the Monte Carlo experiment. These distribution types are shown in Table 1 and have one free parameter, viz., the exponential parameter b. The factors, 3, 4, and 5 (together with factor 6) fix the relative variances through Table 2 and the Equations (12) and (13). Once we specify, say,

$$\max \sigma_{(j)} = 1 \tag{17}$$

all variances are fixed. Specification of the variances results in specific values for the exponential parameter b_i ($i = 1, \ldots, k$); cf. (6). These values of b_i (together with the fixed values of the parameter p) yield specific values for the means of the distributions in Table 1. These means will conflict with the configuration specified by factor 7. For, if we take, say

$$\mu_{(j)} = 0 \quad (j = 1, \ldots, k-1) \tag{18}$$

then factor 7 specifies

$$\mu_{(k)} = \delta^* \tag{19}$$

where δ^* follows from (15) and (16). The conflict between the means following from p and b_i and the means in (18) and (19) can be easily solved by adding suitable constants a_i $(i = 1, \ldots, k)$ to each observation from population i. These constants do not influence the variances, skewness, and tails of the distributions but simply shift the distributions away from the origin.

VI.4. THE EXPERIMENTAL DESIGN

Even if we consider the seven factors of Section VI.3 at only two levels we have as much as $2^7 = 128$ factor combinations. As we explained in Chapter IV we might assume that those seven factors have no important interactions and estimate their main effects from a 2_{III}^{7-4} design, i.e., from only eight combinations. We do not like to assume a priori that all two-factor interactions are zero. Therefore we decided to use a 2_{IV}^{7-3} design. By definition these sixteen combinations yield estimators of the main effects not biased by possible two-factor interactions and estimators of particular linear combinations of two-factor interactions. The latter class of estimators enables us to test if two-factor interactions are important.

Several 2_{IV}^{7-3} designs are possible, each having a different confounding or alias pattern. From the way a 2_{IV}^{7-3} design is constructed (see (20) below) it follows that all two-factor interactions involving only four different factors (e.g., factors 1, 2, 3, and 4) can be kept unconfounded. (So two-factor interactions involving five or more factors cannot be kept unaliased.) We conjectured that the first four factors listed near the end of Section VI.3 might be the most important ones and therefore we decided that the two-factor interactions involving only these four factors, should be unconfounded with each other. This alias structure for the two-factor interactions can be realized if we take as generators for the design:

$$\vec{5} = \overrightarrow{123} \qquad \vec{6} = \overrightarrow{124} \qquad \vec{7} = \overrightarrow{234} \tag{20}$$

These generators result in the defining relation (21).

$$\vec{I} = \overrightarrow{1235} = \overrightarrow{1246} = \overrightarrow{2347} = \overrightarrow{3456} = \overrightarrow{1457} = \overrightarrow{1367} = \overrightarrow{2567} \tag{21}$$

This defining relation shows that indeed no main effects are confounded with two-factor interactions. (They are aliased with three-factor interactions.) The two-factor interactions are confounded as in (22).

$$\begin{aligned}
\overrightarrow{12} &= \overrightarrow{35} = \overrightarrow{46} \\
\overrightarrow{13} &= \overrightarrow{25} = \overrightarrow{67} \\
\overrightarrow{14} &= \overrightarrow{26} = \overrightarrow{57} \\
\overrightarrow{23} &= \overrightarrow{15} = \overrightarrow{47} \\
\overrightarrow{24} &= \overrightarrow{16} = \overrightarrow{37} \\
\overrightarrow{34} &= \overrightarrow{27} = \overrightarrow{56} \\
\overrightarrow{17} &= \overrightarrow{45} = \overrightarrow{36}
\end{aligned} \tag{22}$$

The confounding pattern in (22) shows that indeed no two-factor interactions involving only the factors 1, 2, 3 and 4 are confounded with each other. Hence, if only the first four factors have important two-factor interactions then it follows from the first six equations in (22) that unbiased estimators of their two-factor interactions are possible. If, however, the factors 5, 6, and 7 also have two-factor interactions then bias results. If we have bad luck the aliased two-factor interactions may even compensate each other so that erroneously we conclude that no two-factor interactions exist.

The generators in (20) yield the design shown in Table 4. To specify the design in the original factors we associated the factors labeled 1 through 7 with the factors of Section VI.3 in the order in which they were listed, i.e., factor 1 is "truncated," factor 2 is "symmetry," etc. We further associated the plus and minus level of the factors in Table 4 with the two levels of the factors in Section VI.3 in a random way. This random association was performed using a table of random numbers and comparing these numbers with 1/2. This procedure resulted in the association shown in Table 5. Combination of Tables 4 and 5 gives the design in the original factors summarized in Table 6, where $\underline{x}_i$ $(i = 1, \ldots, 4)$ denotes the independent exponential distribution with parameter $b_i = b$. As an example consider combination 1 in Table 6. The factors 1 and 2 are at their + level in Table 4. Hence it follows from Table 5 that we should use a truncated, skew distribution with heavy tails. In Table 1 we see that such a distribution is the exponential distribution, say $\underline{x}_1$. Factor 6 is at the + level in Table 4 so that $k = 3$. The factors 3 and 4 are at their + levels in Table 4. Combining with Tables 5 and 2 yields $\sigma_{(3)}/\max \sigma_{(j)} = 5$. Since in (17) we set $\max \sigma_{(j)} = 1$ we know that $\sigma_{(3)} = 5$. Since factor 5 is at its + level we use (13) to find $\sigma_{(1)} = 0.25$. Finally, factor 7 is at its + level so that δ $(= \delta^*)$ equals

$$1.25\{\sigma_{(k)} + \sigma_{(k-1)}\}/2 .$$

VI.5. THE NUMBER OF REPLICATIONS

As we know Monte Carlo experimentation with the Bechhofer-Blumenthal procedure proceeds as follows. We have the computer generate stochastic variables, say $\underline{x}_{is}$ $(i = 1, \ldots, k)(s = 1,2,\ldots)$. To these variables we apply the MRP. The MRP determines when to stop sampling. Then the MRP selects as best population the one with the largest sample mean. In a Monte Carlo experiment we actually know which population has the highest population mean. So we can check if the MRP gave a correct selection. If the procedure yielded a CS

TABLE 4

A 2^{7-3}_{IV} Design

Combination	$\vec{1}$	$\vec{2}$	$\vec{3}$	$\vec{4}$	$5 = \vec{123}$	$6 = \vec{124}$	$7 = \vec{234}$
1	+	+	+	+	+	+	+
2	-	+	+	+	-	-	+
3	+	-	+	+	-	-	-
4	-	-	+	+	+	+	-
5	+	+	-	+	-	+	-
6	-	+	-	+	+	-	-
7	+	-	-	+	+	-	+
8	-	-	-	+	-	+	+
9	+	+	+	-	+	-	-
10	-	+	+	-	-	+	-
11	+	-	+	-	-	+	+
12	-	-	+	-	+	-	+
13	+	+	-	-	-	-	+
14	-	+	-	-	+	+	+
15	+	-	-	-	+	+	-
16	-	-	-	-	-	-	-

TABLE 5

Random Association between - and + Levels and the Two Levels of the Original Seven Factors

Factor	- Level	+ Level
1. Truncation	Nontruncated	Truncated
2. Symmetry and tails	Symmetric and nearly normal	Skew and Heavy Tails
3. $\sigma_{(k)} < \max \sigma_{(j)}$	$\sigma_{(k)} < \max \sigma_{(j)}$	$\sigma_{(k)} > \max \sigma_{(j)}$
4. $\sigma_{(k)} \approx \max \sigma_{(j)}$	$\sigma_{(k)} \approx \max \sigma_{(j)}$	$\sigma_{(k)} \neq \max \sigma_{(j)}$
5. $\sigma_{(1)} = \cdots = \sigma_{(k-1)}$	$\sigma_{(1)} = \cdots = \sigma_{(k-1)}$	$\sigma_{(1)} < \cdots < \sigma_{(k-1)}$
6. Number of populations (k)	7	3
7. Distance $\delta = \delta^*$	$0.20\{\sigma_{(k)} + \sigma_{(k-1)}\}/2$	$1.25\{\sigma_{(k)} + \sigma_{(k-1)}\}/2$

TABLE 6

The 2^{7-3}_{IV} Design in the Original Factors ($\underline{x}$ is Independent Exponential)

Combination	Distribution	k	$\sigma_{(k)}$	$\sigma_{(k-1)}$	$\sigma_{(k-2)}$		...		$\sigma_{(1)}$	$2\delta/\{\sigma_{(k)} + \sigma_{(k-1)}\}$
1	$\underline{x}_1$	3	5	1	1/4					1.25
2	$\underline{x}_1 - 0.1\underline{x}_2$	7	5	1	1	1	1	1	1	1.25
3	$\underline{x}_1 + \underline{x}_2 + \underline{x}_3 + \underline{x}_4$	7	5	1	1	1	1	1	1	0.20
4	$\underline{x}_1 - \underline{x}_2 + \underline{x}_3 - \underline{x}_4$	3	5	1	1/4					0.20
5	$\underline{x}_1$	3	1/5	1	1					0.20
6	$\underline{x}_1 - 0.1\underline{x}_2$	7	1/5	1	0.85	0.70	0.55	0.40	0.25	0.20
7	$\underline{x}_1 + \underline{x}_2 + \underline{x}_3 + \underline{x}_4$	7	1/5	1	0.85	0.70	0.55	0.40	0.25	1.25
8	$\underline{x}_1 - \underline{x}_2 + \underline{x}_3 - \underline{x}_4$	3	1/5	1	1					1.25
9	$\underline{x}_1$	7	5/4	1	0.85	0.70	0.55	0.40	0.25	0.20
10	$\underline{x}_1 - 0.1\underline{x}_2$	3	5/4	1	1					0.20
11	$\underline{x}_1 + \underline{x}_2 + \underline{x}_3 + \underline{x}_4$	3	5/4	1	1					1.25
12	$\underline{x}_1 - \underline{x}_2 + \underline{x}_3 - \underline{x}_4$	7	5/4	1	0.85	0.70	0.55	0.40	0.25	1.25
13	$\underline{x}_1$	7	4/5	1	1	1	1	1	1	1.25
14	$\underline{x}_1 - 0.1\underline{x}_2$	3	4/5	1	1/4					1.25
15	$\underline{x}_1 + \underline{x}_2 + \underline{x}_3 + \underline{x}_4$	3	4/5	1	1/4					0.20
16	$\underline{x}_1 - \underline{x}_2 + \underline{x}_3 - \underline{x}_4$	7	4/5	1	1	1	1	1	1	0.20

we score a 1, otherwise a zero. We repeat this application of the MRP to computer generated variates many times, say n times. After these n replications we estimate the probability of CS by the fraction of correct selections in the n replications. If the MRP does guarantee the probability requirement in (1) then this fraction should not be significantly smaller than P^*. These computations can be repeated for each of the sixteen factor combinations. We shall now introduce some more notation and decide on n, the number of replications per factor combination.

For simplicity of presentation we shall restrict attention to one factor combination in this section. Define

$$\begin{aligned} \underline{v}_r &= 0 \text{ if the MRP gives a wrong selection in replication } r \\ &= 1 \text{ if the MRP gives a correct selection in replication } r \ (r = 1, \ldots, n) \end{aligned} \tag{23}$$

where $\underline{v}_r$ has expected value, say μ_v, i.e.,

$$E(\underline{v}_r) = P(\underline{v}_r = 1) = \mu_v \tag{24}$$

If the MRP guarantees (1) then

$$\mu_v \geq P^* \tag{25}$$

We estimate μ_v from n replications by

$$\hat{\underline{\mu}}_v = \sum_{r=1}^{n} \underline{v}_r/n \tag{26}$$

Assuming independent replications (we shall return to this assumption) the $\underline{v}_r$ are binomial variables and

$$\text{var}(\hat{\underline{\mu}}_v) = \mu_v(1 - \mu_v)/n \tag{27}$$

We want to take so many replications that the probability of erroneously rejecting the MRP is small. For, once we reject the MRP we shall not consider it anymore in future experiments. In other words, our null-hypothesis is that the MRP works, i.e.,

$$H_0: \mu_v = {}_0\mu_v \geq P^* \tag{28}$$

where the lower index 0 corresponds with the null-hypothesis, and we want a small type I error or α-error. (See Chapter V for error types.) H_0 is rejected if the estimator $\hat{\underline{\mu}}_v$ is below a lower bound, say v_L. Hence

$$\begin{aligned} \alpha &= P(\hat{\underline{\mu}}_v < v_L | H_0) \\ &= P\left(\frac{\hat{\underline{\mu}}_v - {}_0\mu_v}{\{\text{var}(\hat{\underline{\mu}}_v)\}^{1/2}} < \frac{v_L - {}_0\mu_v}{\{\text{var}(\hat{\underline{\mu}}_v)\}^{1/2}} \middle| H_0 \right) \\ &\approx P\left(\underline{z} < \frac{v_L - {}_0\mu_v}{\{{}_0\mu_v(1 - {}_0\mu_v)/n\}^{1/2}} \middle| H_0 \right) \end{aligned} \tag{29}$$

where we approximate the binomial variable by the standard normal variable $\underline{z}$. Define z^α as the upper α point, i.e.,

$$P(\underline{z} > z^\alpha) = \alpha \tag{30}$$

or because of the symmetry of the normal distribution

$$P(\underline{z} < -z^\alpha) = \alpha \tag{31}$$

Hence, (29) and (31) give

$$n = (z^\alpha)^2 \; {}_0\mu_v(1 - {}_0\mu_v)/({}_0\mu_v - v_L)^2 \tag{32}$$

Under H_0, μ_v may exceed P^*, so that

$${}_0\mu_v - v_L \geq P^* - v_L \; (> 0) \qquad \text{for } {}_0\mu_v \geq P^* \tag{33}$$

and

$$ {}_0\mu_v(1 - {}_0\mu_v) \leq P^*(1 - P^*) \quad \text{for} \quad {}_0\mu_v \geq P^* \geq 0.5 \tag{34} $$

where in the latter relation we assume that the interesting levels of P^* are not smaller than 50%. Hence the type I error does not exceed α if in (32) we replace ${}_0\mu_v$ by P^*, i.e.,

$$ n = (z^\alpha)^2 P^*(1 - P^*)/(P^* - v_L)^2 \tag{35} $$

From (35) it follows that taking a low value for v_L results in a small number of replications. However, such a small v_L would mean that H_0 is seldomly rejected, even if H_0 is false. For consider the _type II error or β-error_. Defining the alternative hypothesis H_1 as in (36)

$$ H_1 : \mu_v = {}_1\mu_v < P^* \tag{36} $$

we have

$$ \beta = P(\hat{\underline{\mu}}_v > v_L | H_1) \approx P\left(\underline{z} > \frac{v_L - {}_1\mu_v}{\{{}_1\mu_v(1 - {}_1\mu_v)/n\}^{1/2}} \middle| H_1\right) \tag{37} $$

Analogous to (30) we define

$$ P(\underline{z} > z^\beta) = \beta \tag{38} $$

From (37) and (38) it follows that

$$ n = (z^\beta)^2 \, {}_1\mu_v(1 - {}_1\mu_v)/(v_L - {}_1\mu_v)^2 \tag{39} $$

If we want to control both the α- and β-error we have to satisfy both (35) and (39). These two equations yield

$$ v_L = (P^* \pm c \, {}_1\mu_v)/(1 \pm c) \tag{40} $$

where the symbol c denotes

$$c = \{(z^{\alpha})^2 P^*(1 - P^*)/(z^{\beta})^2 \ {}_1\mu_v(1 - {}_1\mu_v)\}^{1/2} \qquad (41)$$

The solution in (40) corresponding with the minus sign gives a value for v_L that does not satisfy

$${}_1\mu_v < v_L < P^* \qquad (42)$$

which must be satisfied to realize α and β errors smaller than 50%. The remaining solution for v_L is

$$v_L = \frac{P^* + c\ {}_1\mu_v}{1 + c} = \frac{1}{1 + c} P^* + \frac{c}{1 + c}\ {}_1\mu_v \qquad (43)$$

and this solution does meet the condition (42) as it is a weighted average of P^* and ${}_1\mu_v$. Substitution of v_L as given by (43) into (35) or (39) yields the desired number of replications, provided we have specified the α and β errors (or equivalently z^{α} and z^{β}), the P^* level we are interested in and ${}_1\mu_v$, the alternative μ_v we want to detect (with probability at least $1 - \beta$). Obviously these four quantities (α, β, P^*, ${}_1\mu_v$) are subjective. We decided to accept only a 1% probability of erroneously rejecting the MRP, i.e., $\alpha = 0.01$ or $z^{\alpha} = 2.33$, and a 10% probability of erroneously accepting the MRP, i.e., $\beta = 0.1$ or $z^{\beta} = 1.26$. We further decided to study those combinations of P^* and ${}_1\mu_v$ ($< P^*$) shown in Table 7. This table reveals that we looked at combinations of P^* and ${}_1\mu_v$ where ${}_1\mu_v$ is closer to P^* as P^* increases. For it seems reasonable to detect a small (absolute) difference between P^* and ${}_1\mu_v$ like 0.004 if P^* is as high as 99.9% but not if P^* is, say, 60%. The selected values of α, β, P^*, and ${}_1\mu_v$ are subjective but anyhow they give some basis for selecting the number of replications. This number n will determine the reliability of the conclusions of our experiment. At the outset of the experiment we decided to replicate each factor combination 1000 times. Because the

TABLE 7

Number of Replications (n) for Various P^* (to be Guaranteed) and ${}_1\mu_V$ (Actual Probability) for $\alpha = 0.01$ and $\beta = 0.1$

$100\ P^*$	50	60	70	75	80	85	90	95	97.5	99	99.5	99.9
$100\ {}_1\mu_V$	42	55	65	70	75	80	85	92.5	96.0	97.5	98.5	99.5
$100(P^* - {}_1\mu_V)$	8	5	5	5	5	5	5	2.5	1.5	1.5	1.0	0.4
n	499	1251	1114	1007	873	714	528	1128	1658	816	1008	1651

experiment turned out to require extremely much computer time we then decided to reduce the replications 20% so that 800 replications were taken. We observe that in most experiments Bechhofer et al. (1968, pp. 349, 365) also used 800 replications, while P^* ranged from 0.35 to 0.99 and about 200 replications in some additional experiments where P^* was as high as 0.999.

A few remarks can be added to the above derivation of the number of replications; also see Kleijnen (1969a, pp. 22-23).

(i) We used a testing approach, fixing the α- and β-errors. This is better than an estimation approach where we are, say, 90% certain that the estimated P(CS) is within 10% of the true P(CS), for we want to check if the MRP works indeed. The estimation approach would require only 90 replications for $P^* = 0.75$ and 30 replications for $P^* = 0.90$. If we want to estimate the probability of a wrong selection then we need 807 replications for $P^* = 0.75$ and 2421 replications for $P^* = 0.90$. So defining the variable of interest as the probability of a correct or a wrong selection yields completely different results. (See Exercise 4.)

(ii) In our derivation we looked only at one factor combination and one particular P^* level. Actually the experiment uses sixteen combinations of the seven two-levels factors and many P^* levels. The null-hypothesis might, for instance, be specified as "$\mu_v \geq P^*$ for any combination of P^* and the seven two-level factors." The error rate α may now be defined in an experimentwise way; see Section V.B.2. We did not apply such an approach when deciding on the number of replications, but we shall return to the problem of experimentwise error rates when analyzing the results of the experiment.

(iii) We assumed that all replications are independent. Actually as we shall see in the next section the replications for one particular factor combination are pairwise correlated, the correlations being negative, i.e.,

$$\text{cov}(\underline{v}_r, \underline{v}_{r'}) < 0 \qquad \text{if } r = 2r'', \ r' = 2r'' - 1 \quad (r'' = 1, \ldots, n/2)$$

$$\text{cov}(\underline{v}_r, \underline{v}_{r'}) = 0 \qquad \text{otherwise} \tag{44}$$

These correlations decrease the variance of $\hat{\underline{\mu}}_v$ below the value following from (27) so that the type I and II errors will be smaller than derived above.

VI.6. VARIANCE REDUCTION TECHNIQUES

The reliability of the results of a Monte Carlo experiment can be increased through the application of variance reduction techniques. In Chapter III we discussed a number of such techniques. Unfortunately, most techniques are quite complicated. Two techniques are suitable for routine applications, viz., "common random numbers" and "antithetic variates."

Common random numbers imply that replication r in each factor combination uses the same sequence of random numbers. (Actually the lengths of the sequences of random numbers fluctuates among factor combinations so that we use as many common random numbers as possible.) It is intuitively clear that in this way the responses for various factor combinations are subjected to the same random fluctuations so that the factor effects can be better evaluated. In Appendix 1 a formal analysis is given. This analysis shows that the main effect of the two-level factor j, say α_j, is estimated by

$$\hat{\underline{\alpha}}_j = \sum_{i=1}^{16} x_{ij} \underline{y}_i / 8 \qquad (j = 1, \ldots, 7) \tag{45}$$

where

$$x_{ij} = -1 \quad \text{if factor } j \text{ is at its minus level in factor combination } i \ (i = 1, \ldots, 16)$$

$$= +1 \quad \text{if factor } j \text{ is at its plus level in combination } i \tag{46}$$

(x_{ij} is specified in Table 4) and $\underline{y}_i$ denotes the response, i.e., the estimated fraction $\hat{\underline{\mu}}_v$ in (26), in combination i. (We use the symbols x and $\underline{y}$ in (45) since this is customary in experimental design and regression analysis; the symbol x_{ij} should not be confused with $\underline{x}_{is}$, the variables to which the MRP is applied.) Then the variance of $\hat{\underline{\alpha}}_j$ is given by (47).

$$\mathrm{var}(\hat{\underline{\alpha}}_j) = \frac{1}{4} \sigma_y^2 (1 - \rho) \tag{47}$$

where ρ denotes the correlation between the responses $\underline{y}_i$ and $\underline{y}_{i'}$ $(i \neq i')$ and $\rho > 0$ because of common random numbers in the same replication r of combinations i and i'. In the derivation of (47) it is assumed that the variances and covariances of the $\underline{y}_i$ are constant over all combinations. (This assumption is not automatically used in the rest of this chapter.)

Antithetic variates imply that replications 1 and 2, 3 and 4, etc. or in general $r = 2r''$ and $r' = 2r'' - 1$ $(r'' = 1, \ldots, n/2)$, use the complements of their random numbers. So replication 1, e.g. uses the random numbers $r_1, r_2, \ldots$ and replication 2 uses $(1 - r_1)$, $(1 - r_2)$, ... (The number of random numbers per replication is stochastic as it follows from the termination rule of the sequential MRP; consequently the sequences of random numbers have no constant length.) The random numbers within one replication must be independent as independent observations are a basic assumption of the MRP and this assumption is not violated in our Monte Carlo experiment. This assumption holds when applying antithetics (or common random numbers). Antithetics create negative correlation between the responses of replications $r = 2r''$ and $r' = 2r'' - 1$. For suppose that in a replication many random numbers for the best population are small so that the stochastic variables $\underline{x}_{is}$ to which the MRP is applied are, say, high; compare exponential variables generated as in (48).

$$\underline{x} = - E(\underline{x}) \ln \underline{r} \tag{48}$$

High values from the best population tend to yield a correct selection, i.e., $\underline{y} = 1$. In the antithetic partner replication many high random numbers will be generated for the best population. (There is no synchronization problem since the MRP requires that one observation be taken from each population in each step.) Consequently many low observations will be sampled from the best population and this tends to yield a wrong selection, i.e., $\underline{y} = 0$. So the responses of the two partner replications may be expected to be negatively correlated. (Negative correlation was indeed detected in the experimental results.[3]) This negative correlation reduces the variance of the average response per factor combination.[4]

From the above it follows that we may choose among

(i) Common random numbers
(ii) Antithetic variates
(iii) Joint application of (i) and (ii)

As we explained in Chapter III the third alternative does not necessarily give the highest variance reduction. We decided not to use alternative (i) or (iii) since they complicate the statistical analysis of the experiment. For these two alternatives imply that the (average) responses of the sixteen factor combinations are dependent. Alternative (ii) gives no problems of this kind as we take the average response of two partner replications, i.e.,

$$\tilde{\underline{v}}_{ir''} = \{\underline{v}_{i(2r'')} + \underline{v}_{i(2r''-1)}\}/2$$

$$(r'' = 1, \ldots, 400) \quad (i = 1, \ldots, 16) \qquad (49)$$

This yields $n/2 = 400$ independent observations per factor combination. The average response per combination, i.e., the estimated fraction of correct selections is given by (50).

$$\underline{y}_i = \sum_{r=1}^{800} \underline{v}_{ir}/800 = \sum_{r''=1}^{400} \tilde{\underline{v}}_{ir''}/400 \qquad (50)$$

Since the $\tilde{\underline{v}}_{ir''}$ are independent an unbiased estimator of the standard deviation of $\underline{y}_i$ is

$$\underline{s}_i = \{\hat{\underline{var}}(\tilde{\underline{v}}_{ir''})/400\}^{1/2} \tag{51}$$

where

$$\hat{\underline{var}}(\underline{v}_{ir''}) = \sum_{r''=1}^{400} (\tilde{\underline{v}}_{ir''} - \underline{y}_i)^2/399 \tag{52}$$

As output of the Monte Carlo experiment we recorded $\underline{y}_i$, the <u>estimated fraction</u> for each factor combination i, and its <u>standard deviation</u>, $\underline{s}_i$. This output was obtained for 22 levels of P^*, viz., for 0.35(0.05)0.85, 0.89(0.01)0.99.

VI.7. THE OUTPUT OF THE EXPERIMENT

As mentioned at the end of the previous section the Bechhofer-Blumenthal MRP was applied 800 times to each of the 16 combinations of the 7 two-level factors (forming the 2_{IV}^{7-3} design in Table 4) for each of the 22 levels of P^*; see Table 8. For each of these sixteen combinations independent random numbers were used so that $\underline{y}_{i\ell}$ and $\underline{y}_{i'\ell'}$ are independent for $i \neq i'$ ($\ell \neq \ell'$ or $\ell = \ell'$) where $i, i' = 1, \ldots, 16$ and $\ell, \ell' = 1, \ldots, 22$. The 2_{IV}^{7-3} design permits estimation of the overall or grand mean (β_0), the seven main effects ($\beta_1, \ldots, \beta_7$) of the 7 two-level factors listed near the end of Section VI.3 and the 7 sums of two-factor interactions ($\beta_{12}, \ldots, \beta_{67}$) confounded in sets of size 3, shown in (22). As in Chapter IV we can denote the grand mean, the 7 main effects and the 7 sums of 3 two-factor interactions by γ_1 through γ_{15}. These effects can be estimated for each P^* level since responses can be obtained for all levels of P^* smaller than the highest level with negligible extra computer time. The 22 P^* levels, however, yield dependent observations within one combination of the 2_{IV}^{7-3} design, i.e., $\underline{y}_{i\ell}$ and $\underline{y}_{i\ell'}$ are dependent. Summarizing,

TABLE 8

The Design and Output of the Experiment

	P^* levels		
	1		22
Factor combination	(0.35)	...	(0.99)
1	$\underline{y}_{1,1}$	...	$\underline{y}_{1,22}$
⋮	⋮		⋮
16	$\underline{y}_{16,1}$	...	$\underline{y}_{16,22}$

$$\begin{aligned} \mathrm{cov}(\underline{y}_{i\ell}, \underline{y}_{i\ell'}) &= 0 \quad \text{for } i \neq i' \\ &\neq 0 \quad \text{for } i = i' \end{aligned} \tag{53}$$

The sign of the covariance in (53) for $i = i'$ is not known a priori.[5)] (In additional experiments not incorporated in the original experiment of this chapter, the covariances in (53) were estimated; they were found to be positive and to decrease as the P^* levels are farther apart.) Besides 16 independent vectors each consisting of 22 dependent estimated fractions and each originating from a different population, we obtained estimates of the standard deviation of these 16 × 22 fractions. The estimators of these standard deviations are based on (51) above. Besides the fraction of correct selections we obtained the average number of stages required by the MRP to reach a selection, and the estimated standard deviations of these average numbers of stages. These data, however, are not yet analyzed since the purpose of this first experiment is to study the robustness of the MRP, not its efficiency. For $P^* = 0.99$ and all sixteen combinations if the 2^{7-3} design two variants of the original MRP were studied, where in both variants the MRP was not applied until a pilot number of observations $(n_0 > 2)$ from each population was taken. Two rules for the determination of n_0 were used; for details see Kleijnen (1969a, pp. 10-13). We shall not discuss these two variants further.

For the sake of completeness we remark that we used the multiplicative <u>random number generator</u>

$$x_{i+1} = ax_i \pmod{m} \tag{54}$$

with

$$a = 7^5 \tag{55}$$

$$m = 2^{31} - 1 \tag{56}$$

This generator was extensively tested by Lewis et al. (1969) who also give an assembler subroutine suitable for the IBM SYSTEM/360.

VI.8. TESTING THE ROBUSTNESS OF THE MRP

The first step in the analysis of the experiment is to test if the MRP does guarantee the probability requirement in (1), i.e., if the fraction of correct selections is not significantly lower than P^*, for all P^* levels and all combinations in the 2^{7-3}_{IV} design. To the second step, discussed in the next section, we shall proceed only if the MRP does not work under all circumstances. In that step we shall investigate which factors are important and hence may explain why the MRP did not work. We observe that in most experimental designs attention is concentrated on the effects of the factors that are varied in the design, i.e., on the "<u>relative</u>" <u>responses</u> or the variations of the responses when changing the levels of the factors. In our experiment, however, we should first of all investigate the "<u>absolute</u>" <u>responses</u> and test if $E(\underline{y}) \geq P^*$.

If no antithetic variates were applied then each estimated fraction, $\underline{y}_{i\ell}$ $(i = 1, \ldots, 16)$ $(\ell = 1, \ldots, 22)$, would be binomially distributed. Application of antithetics means that the individual independent observations $\underline{\tilde{v}}_{i\ell r''}$ $(r'' = 1, \ldots, 400)$ are no binomial variables any more. (They can assume the values 0, 1/2, and 1.) As Scheffé (1964, p. 335) concluded for large sample sizes the t-test is independent of the form of the population. Here we have $n = 400$ so that we decided to use the <u>t-statistic</u> for testing if the fraction of correct selections is at least P^*. Or

$$\underline{t}_{v_{i\ell}} = \frac{\underline{y}_{i\ell} - P^*_\ell}{\underline{s}_{i\ell}} \quad (i = 1, \ldots, 16) \quad (\ell = 1, \ldots, 22) \tag{57}$$

where $v_{i\ell}$ is the degrees of freedom of the t-statistic, i.e.,

$$v_{i\ell} = n - 1 = 399 \tag{58}$$

since each $\underline{s}_{i\ell}$ is estimated from four hundred independent observations. As we saw in Section V.B.2, when using a type I _error rate_ with value α then even if the hypothesis

$$H_0: E(\underline{y}_{i\ell}) \geq P^*_\ell \quad \text{for all } i \text{ and all } \ell \tag{59}$$

holds the expected number of false rejections of the MRP is $\alpha \times 16 \times 22$. So if we use a 5% level of significance in (57) then we may expect that more than 18 times we find a significantly low estimated fraction even if H_0 in (59) holds. Instead of fixing the so-called _per comparison_ error rate at, say, 5% in (57) we would prefer to fix the _experimentwise_ error rate. A disadvantage of the experimentwise error rate is that the individual tests like (57) have small probabilities of detecting deviations from the null-hypothesis. This low power of the individual tests may be remedied by not fixing the experimentwise error rate but the _familywise_ error rates, i.e., the probability of all statements being correct is fixed per family of statements; the family does not comprise all statements of the experiment but comprises a set of closely related statements. As we shall see we use the familywise approach in this chapter.

Denote the error rates used in the individual t-tests by α_C ("per comparison rate"); we use the same value for α_C in all tests. Let α_E denote the experimentwise error rate and α_F the familywise rate, where each family consists of the twenty-two statements for a particular combination i. Then we have

$$1 - \alpha_E = P[\underline{y}_{i\ell} \geq P^*_\ell - t_v^{\alpha_C} \underline{s}_{i\ell} \text{ for all } i \text{ and } \ell | H_0]$$
$$= P[\underline{y}_{1\ell} \geq P^*_\ell - t_v^{\alpha_C} \underline{s}_{1\ell} \text{ for all } \ell | H_0]$$
$$\times \cdots \times P[\underline{y}_{16\ell} \geq P^*_\ell - t_v^{\alpha_C} \underline{s}_{16\ell} \text{ for all } \ell | H_0]$$
$$= (1 - \alpha_F)^{16} \tag{60}$$

where the last equality but one holds since all 16 factor combinations give independent responses [cf. (53)], and the last equality holds since we fix all familywise error rates to the same value, α_F. From (60) results the relationship between the experimentwise and familywise error rates:

$$\alpha_F = 1 - (1 - \alpha_E)^{1/16} \tag{61}$$

From the Bonferroni inequality, explained in Section V.B.3, follows the relationship between the familywise and the per comparison error rates:

$$\alpha_F \leq m\, \alpha_C \tag{62}$$

where m is the number of statements per family. In our experiment we have $m = 22$ but we may also decide to improve the power of the individual tests by limiting attention to, say, the ten highest levels of P^*, viz., $P^* = 0.90(0.01)0.99$, so that m becomes 10. Equations (61) and (62) yield Tables 9 and 10, where we fixed α_E and α_F, respectively, to some reasonable values. Remember that these "joint" error rates may be taken much higher than the traditional values of, say, 5%. We observe that our derivation which employs the independence of the 16 factor combinations yields a higher α_C (and consequently higher power) for given α_E than using a Bonferroni inequality like (63).

TABLE 9

Familywise and Per Comparison Error Rates for Fixed Experimentwise Error Rates

α_E	α_F	α_C	
		m = 22	m = 10
0.20	0.013850	0.00062955	0.0013850
0.10	0.006563	0.00029832	0.0006563
0.05	0.003201	0.00014550	0.0003201

$$\alpha_E \leq 16m\ \alpha_C \tag{63}$$

For example, $\alpha_E = 0.20$, m = 10 yielded $\alpha_C = 0.001385$ in Table 9 but (63) results in $\alpha_C = 0.001250$.

To calculate the <u>critical points</u> corresponding with the various per comparison error rates we may use

$$t_v^{\alpha_C} = z^{\alpha_C} \tag{64}$$

since (58) showed that the number of degrees of freedom for each individual test is 399, so that we can legitimately approximate the t-statistic by the standard normal variable <u>z</u>; z^α was defined in (30) as the upper α point of the standard normal distribution.

TABLE 10

Per Comparison Error Rates for Fixed Familywise Error Rates

α_F	α_C	
	m = 22	m = 10
0.20	0.0090909	0.0200
0.10	0.0045455	0.0100
0.05	0.0022727	0.0050

There are many tables for z; we used Smirnov (1965) to derive Table 11. Smirnov tabulated the cumulative distribution $\Phi(z)$ for various z.[6] If $\Phi(z)$ does not exactly agree with $(1 - \alpha_C)$ we take the next largest value of $\Phi(z)$ so that the experimentwise or familywise error rates do not exceed α_E or α_F. (This approximation could be refined if the calculated t-value would turn out to be close to its critical point.)

As we pointed out in V.B.2 the selection of the type of error rate to be fixed in the experiment is a rather controversial matter. Since a per comparison error rate results in many false significance declarations we reject this error rate. The choice between experimentwise or familywise error rates is more difficult. To preserve the power of the tests we would recommend a familywise approach, a family being defined here as all statements pertaining to one particular combination in the 2_{IV}^{7-3} design. Besides the type of error rate we have to fix its value. Determination of these values is rather arbitrary. (A decision theoretic approach based on loss functions would also be arbitrary since it is hard to specify the losses in our experiment.) It is customary in statistics to use 5%, though 10% or 1% are also used. In simultaneous inference making we may follow e.g. Dunn (1964, p. 248) and apply a higher value of 20%. To enable the reader to choose his own value for α_E or α_F we give results for various values of α_E and α_F. We ourselves prefer $\alpha_F = 0.20$ protecting the power of the individual tests.

TABLE 11

Critical Points z^{α_C} for Various α_E and α_F

	α_C			α_C	
α_E	m = 22	m = 10	α_F	m = 22	m = 10
0.20	3.226	2.994	0.20	2.362	2.054
0.10	3.435	3.214	0.10	2.610	2.327
0.05	3.625	3.415	0.05	2.838	2.576

Having obtained the critical points we compare these points with the values of the t-statistics in (57).[7] From (57) and (58) it follows that a fraction is significantly low if the t-value is (algebraically) smaller than $-z^{\alpha_C}$, with z^{α_C} given in Table 11. To save space we do not list all 16 × 22 t-values but show in Table 12 the smallest t-value for each factor combination and the corresponding P^* level. Tables 11 and 12 give the following results:

TABLE 12

Smallest t-Value and Corresponding P^* Value for All Sixteen Combinations in the 2^{7-3}_{IV} Design

Factor combination	m = 22		m = 10	
	t_{min}	P^*	t_{min}	P^*
1	4.1666	0.99	4.1666	0.99
2	-4.6875	0.98	-4.6875	0.98
3	1.3214	0.99	1.3214	0.99
4	14.8686	0.50	29.7692	0.96
5	2.9855	0.75	3.7763	0.92
6	-5.2620	0.35	-3.7261	0.97
7	0.5434	0.98	0.5434	0.98
8	4.4871	0.97	4.4871	0.97
9	-6.1965	0.93	-6.1965	0.93
10	-2.4492	0.80	-2.3209	0.96
11	0.3636	0.99	0.3636	0.99
12	-1.4318	0.99	-1.4318	0.99
13	-0.7169	0.98	-0.7169	0.98
14	6.6923	0.99	6.6923	0.99
15	-2.4038	0.99	-2.4038	0.99
16	-1.9503	0.35	0.3636	0.99

(i) The factor combinations 2, 6, and 9 (specified in Table 6) give significantly low fractions at any of the three values of α_E or α_F given in Table 11.

(ii) Combination 15 is rejected at $m = 10$ and $\alpha_F = 0.10$ (and therefore also at $\alpha_F = 0.20$) or at $m = 22$ and $\alpha_F = 0.20$.

(iii) Combination 10 is rejected at $\alpha_F = 0.20$ and $m = 22$, and at $\alpha_F = 0.20$ and $m = 10$.

(iv) All other combinations are not rejected at any of the values of α_E, α_F, and m in Table 11.

Because of (i) the conclusion is that _the MRP does not work under all circumstances_. Because of (iv) we further conclude that nevertheless _the MRP guarantees the probability requirement in many situations_.

Once we know that the MRP does not work always we proceed to the next step and investigate which factors may have caused the MRP to break down. This will be examined in the next section.

VI.9. DETECTION OF IMPORTANT FACTORS

Several techniques may be considered for the detection of "_important_" factors, i.e., factors that cause the MRP to fail. One technique would involve the construction of a 2 × 2 _contingency table_ for each effect. For example, suppose that in Section 8 we reject only the combinations 2, 6, and 9. From Table 4 it follows that in the three rejected combinations factor 2 is always at its plus level. Hence for factor 2 Table 13 results. Using this table we may hope to determine if the attribute "level of factor 2" is associated with the attribute "rejection"; for a general discussion of "association" we refer to Kendall and Stuart (1961, pp. 536-591). A basic assumption for testing independence or lack of association is that all individuals (here all factor combinations) belonging to one particular cell (among the four cells) have the same probability; see Kendall and Stuart (1961, p. 547). But in the above table,

TABLE 13

Factor Levels versus Rejection and Nonrejection for Factor 2

Level of factor 2	Factor combinations		Total
	Rejected	Nonrejected	
+	3	5	8
-	0	8	8
Total	3	13	16

combinations 2 and 6 falling in the same cell do not have equal probabilities since these two combinations have different levels of the other factors (factor 3 is at its + level in combination 2 and at its minus level in combination 6; see Table 4). Therefore we do not analyze our results by contingency tables.

A second technique consists of conditional tests, viz., given the condition that three combinations are rejected as in Table 13, what is the probability that in all three combinations factor 2 is at the same level (in Table 13 at the plus level) under the null hypothesis of no factor 2 effect? Now we have

$$\begin{aligned} &P(\text{factor 2 is at } + \text{ level in rejected combination} \mid H_0) \\ &= P(\text{factor 2 is at } - \text{ level in rejected combination} \mid H_0) \\ &= 1/2 \end{aligned} \tag{65}$$

for the design is orthogonal.[8] Hence the probability of three equal signs follows from the binomial distribution with probability of "success" equal to 1/2. Or

$$\begin{aligned} P(3 \text{ equal signs} \mid H_0) &= P(3 \text{ pluses} \mid H_0) + P(3 \text{ minuses} \mid H_0) \\ &= (\tfrac{1}{2})^3 + (\tfrac{1}{2})^3 = 0.25 \end{aligned} \tag{66}$$

So even under the hypothesis of no important factor 2 the probability of factor 2 being at the same level in all three rejected combinations is still 25%. Consequently the null hypothesis would be rejected only at a (per comparison) error rate $\alpha > 25\%$ which is an unusually high value. Therefore we conclude that for three rejected factor combinations (viz., 2, 6, and 9) <u>no</u> <u>sensitive</u> <u>tests</u> can be based on (65) and (66). So if we look at Table 14 we know that it is very well possible that the effects $\vec{2}$, $\vec{6}$, and $\vec{26}$ are unimportant even while they have equal signs in all three rejected combinations. (The two-factor interaction $\vec{26}$ is confounded with $\vec{14}$ and $\vec{57}$ but it is reasonable to assume that if only the factors 2 and 6 are important then $\vec{14}$ and $\vec{57}$ are nonexistent.) At the end of Section 8 we concluded that at higher error rates combination 15 and at still higher error rates also combination 10 would be rejected. Consequently Table 14 can be extended as in Table 15. From (65) it follows that

$$P(4 \text{ equal signs in } 4 \text{ rejected combinations} \mid H_0) = (\tfrac{1}{2})^4\, 2 = 0.125 \quad (67)$$

$$P(3 \text{ equal signs in } 4 \text{ rejected combinations} \mid H_0) = \frac{4!}{3!\,1!}(\tfrac{1}{2})^3 (\tfrac{1}{2})^1\, 2 = 0.50 \quad (68)$$

$$P(4 \text{ equal signs in } 5 \text{ rejected combinations} \mid H_0) = \frac{5!}{4!\,1!}(\tfrac{1}{2})^4 (\tfrac{1}{2})^1\, 2 = 0.3125 \quad (69)$$

TABLE 14

Levels of the Main Effects and Two-Factor Interactions in the Rejected Three Combinations

Rejected combination	$\vec{1}$	$\vec{2}$	$\vec{3}$	$\vec{4}$	$\vec{5}$	$\vec{6}$	$\vec{7}$	$\vec{12}=\vec{35}=\vec{46}$	$\vec{13}=\vec{25}=\vec{67}$	$\vec{14}=\vec{26}=\vec{57}$	$\vec{15}=\vec{23}=\vec{47}$	$\vec{16}=\vec{24}=\vec{37}$	$\vec{34}=\vec{27}=\vec{56}$	$\vec{17}=\vec{45}=\vec{36}$
2	-	+	+	+	-	-	+	-	-	-	+	+	+	-
6	-	+	-	+	+	-	-	-	+	-	-	+	-	+
9	+	+	+	-	+	-	-	+	+	-	+	-	-	-

TABLE 15

Levels of Main Effects and Two-Factor Interactions in the Rejected Five Combinations

Rejected combination	$\vec{1}$	$\vec{2}$	$\vec{3}$	$\vec{4}$	$\vec{5}$	$\vec{6}$	$\vec{7}$	$\overrightarrow{12}=\overrightarrow{35}=\overrightarrow{46}$	$\overrightarrow{13}=\overrightarrow{25}=\overrightarrow{67}$	$\overrightarrow{14}=\overrightarrow{26}=\overrightarrow{57}$	$\overrightarrow{15}=\overrightarrow{23}=\overrightarrow{47}$	$\overrightarrow{16}=\overrightarrow{24}=\overrightarrow{37}$	$\overrightarrow{34}=\overrightarrow{27}=\overrightarrow{56}$	$\overrightarrow{17}=\overrightarrow{45}=\overrightarrow{36}$
2	-	+	+	+	-	-	+	-	-	-	+	+	+	-
6	-	+	-	+	+	-	-	-	+	-	-	+	-	+
9	+	+	+	-	+	-	-	+	+	-	+	-	-	-
15	+	-	-	-	+	+	-	-	-	-	+	+	+	-
10	-	+	+	-	-	+	-	-	-	+	+	-	-	+

From (67) through (69) we see that Table 15 does not lead to rejection of the hypothesis of no important effects at reasonable α error rates. The <u>most</u> <u>significant</u> result is given by the two-factor interaction $\overrightarrow{26}$, confounded with $\overrightarrow{14}$ and $\overrightarrow{57}$ and this may be interpreted as evidence that the <u>factors</u> <u>2</u> <u>and</u> <u>6</u> have the largest contribution to the rejection of the MRP for certain factor combinations. Remember that factor 2 denotes nonnormality as specified by the skewness and heaviness of the tails of the distributions, and factor 6 denotes the number of populations (k). The second technique is not powerful since it does not use the accepted factor combinations and in the rejected combinations it does not incorporate the magnitude of the deviations between the estimated fractions and the required probability level P^*.

The third technique is the well-known <u>analysis</u> <u>of</u> <u>variance</u> (ANOVA), or more generally <u>regression</u> <u>analysis</u>, of the 2_{IV}^{7-3} experiment explained in Chapter IV. These analyses use all sixteen factor combinations taking into account the actual values of the estimated fractions of correct selection. Moreover, the estimated standard deviations of these fractions are used as we shall see. In Section VI.4 we distinguished the grand mean β_0, the seven main effects β_1

through β_7, and the 21 two-factor interactions β_{12} through β_{67} confounded in seven sets of three two-factor interactions shown in (22). Denote these 15 coefficients by γ_1 through γ_{15}, where

$$\gamma_1 = \beta_0 \tag{70}$$

$$\gamma_2 = \beta_1, \ldots, \gamma_8 = \beta_7 \tag{71}$$

$$\gamma_9 = \beta_{12} + \beta_{35} + \beta_{46}, \ldots, \gamma_{15} = \beta_{17} + \beta_{45} + \beta_{36} \tag{72}$$

With γ_1 through γ_{15} correspond fifteen independent or explanatory variables, say x_1 through x_{15}. The variables x_j $(j = 1, \ldots, 15)$ can assume only the values -1 or +1. For x_2 through x_8 these values are specified in Table 4; for x_9 through x_{15} they follow through multiplication of the appropriate columns in Table 4 (e.g. for x_9 multiply column $\vec{1}$ with $\vec{2}$, or $\vec{3}$ with $\vec{5}$, etc.); x_1 is identical to +1. The 2_{IV}^{7-3} design implies that all independent variables are <u>orthogonal</u>, i.e., (73) holds

$$\sum_{i=1}^{16} x_{ij}\, x_{ij'} = 0 \qquad \text{if} \quad j \neq j' \quad (j, j' = 1, \ldots, 15) \tag{73}$$

Obviously

$$\sum_{i=1}^{16} x_{ij}^2 = 16 \qquad (j = 1, \ldots, 15) \tag{74}$$

Each combination of the independent variables x_j yielded 22 correlated "responses" also called dependent or explained variables, say $\underline{y}_\ell$ $(\ell = 1, \ldots, 22)$ corresponding with the 22 P^* levels. We hypothesize that each dependent variable can be expressed as a linear combination of the 15 independent variables x_j plus a disturbance, noise or error term $\underline{e}$. So a <u>regression equation</u> of the form

$$\underline{y}_i = \sum_{j=1}^{15} \gamma_j x_{ij} + \underline{e}_i \qquad (i = 1, \ldots, 16) \tag{75}$$

is hypothesized for each of the 22 dependent variables, or

$$\underline{y}_{i\ell} = \sum_{j=1}^{15} \gamma_{j\ell} x_{ij} + \underline{e}_{i\ell} \qquad (i = 1, \ldots, 16)\ (\ell = 1, \ldots, 22) \tag{76}$$

At first sight it may seem nonsensical to fit a regression equation with 15 parameters $(\gamma_1 \cdots \gamma_{15})$ to only sixteen observations $(\underline{y}_1 \cdots \underline{y}_{16})$ as in (75). However, we have to remember that each $\underline{y}_{i\ell}$ is actually the <u>average</u> of four hundred independent observations shown in (50). Consequently we fit the regression equation with 15 parameters to $16 \times 400 = 6400$ independent observations. The computer program was so written that as output we have available only the averages $\underline{y}_{i\ell}$ and their standard deviations $\underline{s}_{i\ell}$ and not the individual observations. In ANOVA or regression analysis, however, we do not need these individual replications to estimate the effects. They are required only for the calculation of the standard errors when testing the effects, and our estimates of $\underline{s}_{i\ell}$ are indeed based on the individual replications as (52) shows. (For the investigation of the true form of the response function it is advantageous to have many experimental points rather than duplicates at a few points; see lack of fit tests later on.)

A complication is that ANOVA is based on particular <u>assumptions</u>, viz., normally distributed independent homoscedastic error terms $\underline{e}$. So $\underline{e}_i$ in (75) should satisfy

$$\underline{e}_i : \text{NID}(0, \sigma^2) \tag{77}$$

Moreover, ANOVA (as opposed to MANOVA, multivariate analysis of analysis) assumes univariate responses. In our experiment the assumption of <u>normality</u> seems noncrucial since $\underline{y}_i$, or more generally $\underline{y}_{i\ell}$, is based on many observations, viz., 400 independent observations. The <u>independence</u> assumption holds for the $\underline{y}_i$ (but not for the $\underline{y}_{i\ell}$, i.e., only for one particular ℓ value the observations are independent). The common variance assumption will be violated for the $\underline{y}_i$ and the $\underline{y}_{i\ell}$ as we can show as follows. If no antithetic variates were applied then the binomial relation in (78) would hold:

$$\mathrm{var}(\underline{y}) = c(1 - c)/800 \tag{78}$$

where c is the probability of "success", i.e., P(CS). If the factors in the 2^{7-3}_{IV} experiment influence P(CS) then c will vary over the 16 factor combinations. (Obviously c varies over the 22 levels of P^* since c will increase as P^* increases.) This reasoning should be adapted for the antithetic variates. Nevertheless it is a priori clear that the variances are not equal. The output of the experiment gave estimates that indeed strongly suggest heterogeneity of variances. For $P^* = 0.99$ (or $\ell = 22$) the estimated standard deviations $s_{i(22)}$ range between 0.0000 and 0.0074, and for $P^* = 0.35$ (or $\ell = 1$) between 0.0105 and 0.0160. Finally, we note that the experiment yields multivariate responses; see e.g. (53).

Because the assumptions of equal variances and univariate responses are violated we considered the possibility of applying other techniques than simple least squares per regression equation. In Appendix 2 we present the following three alternatives in detail.

(i) Ordinary least squares (OLS) per regression equation, resulting in the traditional ANOVA

(ii) Generalized least squares (GLS) per equation which can take into account heteroscedasticity

(iii) Generalized least squares for the whole set of equations which can cope with both heteroscedasticity and dependence among the error terms in the 22 regression equations.

We prefer estimators of the regression coefficients $\gamma_{i\ell}$ that are efficient and can be tested for significance. The estimators are efficient if they are "best linear unbiased estimators" (BLUE) where "best" refers to the minimum variance property. GLS estimators are BLUE but require that we know the covariance matrix of the error terms (Σ_ℓ and Σ in (2.8) and (2.17) in Appendix 2). Unfortunately we do not know the covariance matrix. We can estimate the elements in this matrix. (Its diagonal elements, viz., the variances, are estimated by $\underline{s}^2_{i\ell}$; GLS for the whole set would also require estimation of covariances; these covarainces are not estimated in the original experiment but were estimated in an additional experiment.) Replacing the covariance matrix in the GLS by an estimated matrix leaves the estimators of γ_{ij} unbiased but these estimators are no longer BLUE; we do not know if they still have a smaller variance than the OLS estimators; compare the literature.[9)] We do know that the OLS estimators have the advantage of requiring only simple calculations since the orthogonality of the matrix of independent variables implies that no matrix inversion is needed. Matrix inversion by computer may give serious rounding errors. The OLS estimators remain unbiased even if the covariance matrix is not of the form $\sigma^2 \vec{I}$. Testing the significance of the estimated regression coefficients can be performed for OLS estimators by using the well-known ANOVA F-tests per equation. These F-tests are known to be quite insensitive to heterogeneity of variances (and nonnormality) especially for large equal numbers of observations per experimental point; compare Scheffé (1964, pp. 331-369). We can cope with the presence of more than one equation by applying a Bonferroni inequality. (An asymptotic test of the significance of the GLS estimators is also possible using results from multivariate analysis; see Zellner (1962, p. 355) who based his test on Roy (1957). The resulting formulas are much more complicated. Moreover, in the additional experiment where covariances were estimated the matrix of estimated covariances and variances, and even submatrices for only three P^* levels, turned out to be singular so that the test could not be applied.)

In Appendix 2 it is derived that OLS per equation yield the following formula for the estimated regression coefficients:

$$\hat{\underline{\gamma}}_{j\ell} = \sum_{i=1}^{16} x_{ij}\, \underline{y}_{i\ell}/16 \qquad (j = 1, \ldots, 15)\ (\ell = 1, \ldots, 22) \tag{79}$$

This formula agrees with the well-known formula for the estimation of the effects in a 2^{k-p} design. (Depending on the definition of these effects, the regression coefficients are exactly equal to the effects or are half these effects; see (45) above or Chapter IV. Here we define the effects such that (70) through (72) holds.) From (79) the estimated standard deviations of $\hat{\underline{\gamma}}_{j\ell}$ follow:

$$\underline{s}(\hat{\underline{\gamma}}_{j\ell}) = \left\{ \sum_{i=1}^{16} \underline{s}_{i\ell}^2/16^2 \right\}^{1/2} \tag{80}$$

Observe that contrary to the familiar formula for 2^{k-p} designs, (80) allows from unequal variances; all regression coefficients have the same standard deviation for one particular level ℓ of P^*. Formulas (79) and (80) yield Table 16.

When studying Table 16 we can understand the size of some effects on hindsight. Besides the factors that are really the subject of study in our experiment (viz., the factors 1 through 5 specifying nonnormality and unequality of variances) there are two more factors in the 2^{7-3} design (viz., the number of populations k being factor 6 and the distance $\delta = \delta^*$ being factor 7) and there is the multilevel factor P^*. Table 16 together with Table 5 shows that a "small" number of populations has a favorable effect and this makes sense since with only three populations even a random choice procedure would give $P(CS) = 1/3$. Further, with large distance $\delta = \delta^*$ any procedure will give a correct selection most times. Finally as P^* increases the factor effects become smaller. Two explanations may be given. First, as P^* increases more observations are required by the MRP so that the sample means $\bar{\underline{x}}_i$ converge to μ_i and the seven two-level factors may become less important. Second, a deviation between the estimated fraction $\underline{y}_{i\ell}$ and the minimum probability P_ℓ^* of, say, 0.05 is less important at $P^* = 0.35$ than at $P^* = 0.99$.

TABLE 16

The Effects in the 2_{IV}^{7-3} Design for Twenty-two Levels of P^*

		Main effects (× 1000)							Two-factor interactions (× 1000)							
									12= 35= 46	13= 25= 67	14= 26= 57	23= 15= 47	24= 16= 37	34= 27= 56	17= 45= 36	Standard deviation
P^*	β_0	1	2	3	4	5	6	7								$s(\hat{\underline{\beta}})$ (× 1000)
0.35	60.0	-1	-22	53	8	3	54	175	-8	13	20	-21	-4	28	-30	3.4
0.40	62.6	-2	-21	49	16	10	46	156	-2	15	24	-26	-2	27	-26	3.4
0.45	65.3	2	-24	46	21	8	41	138	3	18	26	-29	-2	29	-26	3.5
0.50	68.3	1	-24	40	21	8	38	124	6	17	26	-27	1	31	-23	3.5
0.55	72.7	-2	-30	36	24	6	36	111	13	17	25	-30	-3	33	-19	3.4
0.60	75.0	-4	-27	31	24	5	32	95	14	21	21	-29	-3	30	-14	3.4
0.65	78.4	-6	-28	30	25	5	31	79	15	21	17	-31	-5	32	-9	3.2
0.70	81.4	-8	-26	27	20	2	30	64	15	17	19	-27	-7	29	-3	3.1
0.75	84.2	-6	-26	21	17	4	26	51	14	17	15	-24	-6	28	-3	2.9
0.80	87.2	-3	-24	15	18	-0	23	38	15	15	15	-20	-5	26	-1	2.7
0.85	90.1	-2	-24	9	14	-1	23	25	12	10	17	-19	-2	21	3	2.5
0.89	91.3	-3	-19	4	12	-2	20	18	10	8	15	-15	-2	17	6	2.2

0.90	92.9	-3	-18	3	11	-2	20	15	9	7	15	-16	-2	15	6	2.1
0.91	93.5	-3	-17	0	11	-2	20	14	7	5	15	-14	-1	14	6	2.1
0.92	93.9	-3	-17	0	10	-2	19	12	8	5	15	-13	-1	12	6	2.0
0.93	94.1	-2	-15	-1	10	-2	18	11	8	4	15	-13	-2	10	6	1.9
0.94	95.0	-1	-13	-2	10	-3	16	9	9	4	14	-12	-1	9	6	1.8
0.95	95.7	-0	-12	-3	8	-2	15	8	8	3	12	-11	-1	8	6	1.7
0.96	96.3	1	-11	-4	5	-3	13	7	7	2	11	-10	-1	7	5	1.6
0.97	97.1	0	-10	-3	5	-3	12	6	5	1	11	-9	-2	6	4	1.4
0.98	97.8	-1	-8	-4	4	-2	10	4	4	2	9	-7	-2	4	4	1.3
0.99	98.6	0	-4	-3	3	-3	7	3	2	-0	6	-4	-2	3	3	1.0

The regression coefficients explain absolute differences, $\underline{y} - P^*$, not relative differences, e.g. $(\underline{y} - P^*)/(1 - P^*)$. Consequently a particular value of an effect has more influence at a high P^* level than at a low P^* level. (On hindsight it seems better to measure the response not as $\underline{y}_{i\ell}$ but as $(\underline{y}_{i\ell} - P^*_{i\ell})/(1 - P^*_{i\ell})$. An alternative approach, suggested to us by Professor J. Tukey, might be based on the "logit" transformation; see Finney (1952) and Kendall and Buckland (1971).)

Our next step could be testing the significance of the estimated effects. But first we test if the fitted regression equations are adequate approximations, i.e., do equations with interactions between no more than two factors give good fit? To test <u>lack of fit</u> we proceed as follows. (Also see Section IV.6.) It is well-known, see Johnston (1963, p. 112), that the residual mean squares $\underline{S}_R$ have expected value σ^2 provided the fitted regression equation does not show specification error (and the error terms satisfy the usual assumptions (77)). So we determine

$$_{\ell}\underline{S}_R = \sum_{i=1}^{16} (\underline{y}_{i\ell} - \hat{\underline{y}}_{i\ell})^2/(16 - 15) \qquad (\ell = 1, \ldots, 22) \qquad (81)$$

where

$$\hat{\underline{y}}_{i\ell} = \sum_{j=1}^{15} \hat{\underline{\gamma}}_{j\ell}\, x_{ij} \qquad (82)$$

and $_{\ell}\underline{S}^2_R/\sigma^2$ is a χ^2-variable with one degree of freedom. A second estimator of σ^2 in (77) can be based on the four hundred independent replications of each factor combination as we specified in (51). Since (77) implies that all factor combinations i have the same experimental error σ^2, we can take the average or pooled estimator based on duplication, say $\underline{s}^2_D$,

$$_{\ell}\underline{s}^2_D = \sum_{i=1}^{16} \underline{s}^2_{i\ell}/16 \qquad (\ell = 1, \ldots, 22) \qquad (83)$$

Under the assumptions (77) (and the stricter assumption that not only the $\underline{y}_{i\ell}$ but also the individual observations $\underline{\tilde{v}}_{i\ell r}$ are normally distributed) ${}_{\ell}\underline{s}_{D}^{2}\, v/\sigma^{2}$ is a χ^{2}-variable with $v = 16 \times 399$ degrees of freedom and is independent of ${}_{\ell}\underline{S}_{R}$.[10)] Hence the ratio of ${}_{\ell}\underline{S}_{R}$ and ${}_{\ell}\underline{s}_{D}^{2}$ is F-distributed under the above assumptions. Or

$$ {}_{\ell}\underline{F}_{1,v} = \frac{{}_{\ell}\underline{S}_{R}}{{}_{\ell}\underline{s}_{D}^{2}} \qquad (\ell = 1, \ldots, 22) \tag{84} $$

with $v = 16 \times 399 = 6384$ so that we can put $v = \infty$. High values of the statistics (84) indicate lack of fit.

Next consider the influence of the usual <u>assumptions</u> (77) on this lack of fit test. We know that the assumption of independence is satisfied. The $\underline{y}_{i\ell}$ (and consequently the $\hat{\underline{y}}_{i\ell}$) in ${}_{\ell}\underline{S}_{R}$ are approximately normal since each of them is the average of 400 independent observations. The observations on which the $\underline{s}_{i\ell}^{2}$ are based, however, are not normal. Our discussion of (78) further showed that the observations at different combinations i $(i = 1, \ldots, 16)$ have different variances. So we need to know, how sensitive the F-statistics in (84) are to nonnormality and heterogeneity of variances. We know from Scheffé (1964) that the F-statistic is robust when applied to tests on means and is nonrobust when applied to tests on variances. A lack of fit test concerns the null-hypothesis

$$ H_0\!: E(\underline{y}_{i\ell}) = \sum_{j=1}^{15} \gamma_{i\ell}\, x_{ij} \qquad (i = 1, \ldots, 16)\ (\ell = 1, \ldots, 22) \tag{85} $$

Consequently it is a test on means, not on variances! Therefore we did perform the tests of (84).

We have 22 regression equations and our null hypothesis is that all these equations give good fit. Table 10 above gave various per comparison error rates for fixed <u>familywise error rates</u> of 20, 10, and 5%, and 22 statements in the family (or 10 statements, if we restrict attention to the 10 highest levels of P^{*}). In the tables for $\underline{F}_{v_1,v_2}$ are given the significance levels for standard values of α like 1% but not for odd values like 0.90909%. From Dixon and Massey (1957, p. 402) Table 17 follows, where the row for $F_{7,\infty}$ will be used later on. (Dixon and Massey also list values for α higher than 25%.) In the last column of Table 18 we show the smallest value of α for which the calculated value of the F-statistic for goodness of fit in (84) exceeds the tabulated value $F^{\alpha}_{1,\infty}$. This value of α we call the "critical" level α_L. So in general (86) is implied.

$$F_{v_1,v_2} > F^{\alpha_L}_{v_1,v_2} \tag{86}$$

For instance, for $P^* = 0.40$ we calculated $F_{1,\infty} = 9.63$ and this value 9.63 exceeds $F^{\alpha}_{1,\infty}$ for $\alpha = 0.005$ but not for $\alpha = 0.001$ so that $\alpha_L = 0.005$. Consequently a per comparison error rate α_C not smaller than the critical level α_L leads to rejection of the null hypothesis, for

$$F^{\alpha_L}_{v_1,v_2} \geq F^{\alpha_C}_{v_1,v_2} \quad \text{if} \quad \alpha_L \leq \alpha_C \tag{87}$$

From Tables 10 and 18 it follows that we have to reject the hypothesis of all 22 regressions giving good fit at any of the three familywise error rates (α_F is 20, 10, or 5%). If, however, we restrict our study to the highest ten levels of P^* (i.e., $P^* = 0.90(0.01)0.99$) then we conclude that the hypothesis of good fit would be rejected only at a per comparison rate higher than 0.10. For Table 18 shows

TABLE 17

Significance Level $F^{\alpha}_{1,\infty}$ and $F^{\alpha}_{7,\infty}$ Where

$P(\underline{F}_{v_1,v_2} \geq F^{\alpha}_{v_1,v_2}) = \alpha$

α	0.25	0.10	0.05	0.025	0.01	0.005	0.001	0.0005
$F^{\alpha}_{1,\infty}$	1.32	2.71	3.84	5.02	6.63	7.88	10.8	12.1
$F^{\alpha}_{7,\infty}$	1.29	1.72	2.01	2.29	2.64	2.90	3.47	3.72

that $F_{1,\infty}$ becomes significant at 0.25 but not at 0.10. From the last column in Table 10 we see that for the 10 highest P^* levels we are far from rejecting the hypothesis of good fit. So we conclude that <u>for $P^* \geq 0.90$ the fitted regression equations are good approximations</u> but for $P^* < 0.90$ the fitted regressions give no good fit. Consequently for $P^* < 0.90$ it does not make much sense investigating the estimated regression coefficients $\hat{\underline{\gamma}}_{j\ell}$. (Because of the lack of fit we conjecture that for $P^* \leq 0.90$ three-factor interactions exist so that main effects are confounded with these interactions; see (21).) In the sequel we shall concentrate on tests for the significance of the effects when $P^* \geq 0.90$.

We want to test the <u>significance</u> of the estimated regression coefficients, i.e.,

$$H_0: E(\hat{\underline{\gamma}}_{j\ell'}) = 0 \qquad (j = 1, \ldots, 15)\ (\ell' = 13, \ldots, 22) \tag{88}$$

Per equation ℓ' we can test (88) if we suppose that the usual assumptions in (77) hold. For in that case we can apply an F-test per effect or, since the numerator degrees of freedom is 1, a t-test:

$${}_{\ell'}\underline{F}_{1,v} = {}_{\ell'}\underline{t}_v^2 = {}_{\ell'}\underline{SS}_j / {}_{\ell'}\underline{s}_D^2 \tag{89}$$

where $\underline{SS}_j$ is the sum of squares for effect j, i.e.,

$${}_{\ell}\underline{SS}_j = 16(\hat{\underline{\gamma}}_{j\ell'})^2 \tag{90}$$

and ${}_{\ell'}\underline{s}_D^2$ defined in (83) is the estimator of σ^2 in (77) and has $v = 16 \times 399 = 6384$ degrees of freedom; see Section IV.3 or Johnston (1963, p. 135). The tests are tests on means, not on variances, for the effects $\hat{\underline{\gamma}}_{j\ell}$ in (88) are differences between the mean response at the plus level of variable x_j and the minus level of x_j; cf. (79). Consequently the F-tests based on (89) are known to be quite insensitive to heterogeneity of variances and nonnormality. We further tested all two-factor interactions <u>jointly</u>, i.e.,

TABLE 18. Critical Levels (α_L) in Tests of

		Main effects						
P^*	β_0	1	2	3	4	5	6	7
0.35	0.0005	0.25+	0.0005	0.0005	0.025	0.0005	0.0005	0.0005
0.40	0.0005	0.25+	0.0005	0.0005	0.0005	0.01	0.0005	0.0005
0.45	0.0005	0.25+	0.0005	0.0005	0.0005	0.025	0.0005	0.0005
0.50	0.0005	0.25+	0.0005	0.0005	0.0005	0.05	0.0005	0.0005
0.55	0.0005	0.25+	0.0005	0.0005	0.0005	0.25	0.0005	0.0005
0.60	0.0005	0.25+	0.0005	0.0005	0.0005	0.25	0.0005	0.0005
0.65	0.0005	0.10	0.0005	0.0005	0.0005	0.25	0.0005	0.0005
0.70	0.0005	0.025	0.0005	0.0005	0.0005	0.25+	0.0005	0.0005
0.75	0.0005	0.05	0.0005	0.0005	0.0005	0.25	0.0005	0.0005
0.80	0.0005	0.25	0.0005	0.0005	0.0005	0.25+	0.0005	0.0005
0.85	0.0005	0.25+	0.0005	0.0005	0.0005	0.25+	0.0005	0.0005
0.89	0.0005	0.25	0.0005	0.05	0.0005	0.25+	0.0005	0.0005
0.90	0.0005	0.25	0.0005	0.25	0.0005	0.25+	0.0005	0.0005
0.91	0.0005	0.25	0.0005	0.25+	0.0005	0.25+	0.0005	0.0005
0.92	0.0005	0.25	0.0005	0.25+	0.0005	0.25+	0.0005	0.0005
0.93	0.0005	0.25+	0.0005	0.25+	0.0005	0.25+	0.0005	0.0005
0.94	0.0005	0.25+	0.0005	0.25	0.0005	0.25	0.0005	0.0005
0.95	0.0005	0.25+	0.0005	0.05	0.0005	0.25	0.0005	0.0005
0.96	0.0005	0.25+	0.0005	0.025	0.001	0.25	0.0005	0.0005
0.97	0.0005	0.25+	0.0005	0.025	0.0005	0.10	0.0005	0.0005
0.98	0.0005	0.25+	0.0005	0.005	0.005	0.25	0.0005	0.005
0.99	0.0005	0.25+	0.0005	0.001	0.005	0.005	0.0005	0.001

a) 0.25+ means $\alpha_L > 0.25$.

Significance for Effects and Lack of fit.[a)]

Two-factor interactions							Joint inter- actions	Lack of fit
12=35 =46	13=25 =67	14=26 =57	23=15 =47	24=16 =37	34=27 =56	17=45 =36		
0.025	0.0005	0.0005	0.0005	0.25	0.0005	0.0005	0.0005	0.25
0.25+	0.0005	0.0005	0.0005	0.25+	0.0005	0.0005	0.0005	0.005
0.25+	0.0005	0.0005	0.0005	0.25+	0.0005	0.0005	0.0005	0.0005
0.10	0.0005	0.0005	0.0005	0.25+	0.0005	0.0005	0.0005	0.0005
0.0005	0.0005	0.0005	0.0005	0.25+	0.0005	0.0005	0.0005	0.0005
0.0005	0.0005	0.0005	0.0005	0.25+	0.0005	0.0005	0.0005	0.0005
0.0005	0.0005	0.0005	0.0005	0.25	0.0005	0.01	0.0005	0.001
0.0005	0.0005	0.0005	0.0005	0.025	0.0005	0.25+	0.0005	0.0005
0.0005	0.0005	0.0005	0.0005	0.10	0.0005	0.25+	0.0005	0.005
0.0005	0.0005	0.0005	0.0005	0.05	0.0005	0.25+	0.0005	0.0005
0.0005	0.0005	0.0005	0.0005	0.25+	0.0005	0.25	0.0005	0.025
0.0005	0.0005	0.0005	0.0005	0.25+	0.0005	0.01	0.0005	0.10
0.0005	0.005	0.0005	0.0005	0.25+	0.0005	0.01	0.0005	0.25+
0.0005	0.025	0.0005	0.0005	0.25+	0.0005	0.005	0.0005	0.25+
0.0005	0.01	0.0005	0.0005	0.25+	0.0005	0.005	0.0005	0.25+
0.0005	0.05	0.0005	0.0005	0.25+	0.0005	0.005	0.0005	0.25
0.0005	0.05	0.0005	0.0005	0.25+	0.0005	0.001	0.0005	0.25
0.0005	0.10	0.0005	0.0005	0.25+	0.0005	0.0005	0.0005	0.25+
0.0005	0.25	0.0005	0.0005	0.25+	0.0005	0.001	0.0005	0.25+
0.0005	0.25+	0.0005	0.0005	0.25+	0.0005	0.005	0.0005	0.25+
0.005	0.10	0.0005	0.0005	0.25	0.005	0.001	0.0005	0.25+
0.025	0.25+	0.0005	0.0005	0.05	0.005	0.001	0.0005	0.25+

$$H_0: E(\hat{\underline{\gamma}}_{j'\ell'}) = 0 \qquad (j' = 9, \ldots, 15)\ (\ell' = 13, \ldots, 22) \qquad (91)$$

So (91) hypothesizes that first order regression equations give adequate approximations. This hypothesis can be tested per equation by pooling the sums of squares for the two-factor interactions as in (92)

$$_{\ell'}\underline{F}_{7,v} = \frac{\Sigma_{j'=9}^{15}\ {}_{\ell'}\underline{SS}_{j'}}{7} \frac{1}{_{\ell'}\underline{s}_D^2} \qquad (92)$$

see Section IV.2. The critical level of $\underline{F}_{7,v}$ can be found using the last row of Table 17. The critical levels (α_L) for the individual tests (89) and the joint test (92) are shown in Table 18. Table 18 can be interpreted as follows.

As we mentioned above we restrict our conclusions to the ten regression equations corresponding with $P^* = 0.90(0.01)0.99$. (For the sake of completeness Table 18 also gives the critical levels of (89) and (92) for $P^* < 0.90$.) In the last column of Table 10 the per comparison error rates were shown that correspond with familywise error rates of 0.20, 0.10, and 0.05 and--as in (88)--10 statements per family. From Table 18 we conclude that even at these small per comparison error rates all effects are significantly different from zero, except for the main effect of factor 1 ("truncation") and the confounded two-factor interactions $\vec{16}$, $\vec{24}$, and $\vec{37}$. In agreement with the conditional tests based on (65) the main effects 2 ("skewness"), 6 (populations k) and their interaction give the most significant results (together with the confounded interactions $\vec{23} + \vec{15} + \vec{47}$ and of course the general mean). But also very significant are the factors 7 $(\delta = \delta^*)$ and $4(\sigma(k)/\max \sigma_{(j)} \neq 1)$ (and the interactions $\vec{34} + \vec{27} + \vec{56}$ and $\vec{12} + \vec{35} + \vec{46}$). If we inferred from Table 18 that the factors 1, 3, and 5 are nonexistent (implies $\alpha_F < 0.01$) then we could pool the sums of squares for the main effects of the factors 1, 3, and 5 with the residual sums of squares. This would yield a lack of fit F-statistic with value 4.94 for $P^* = 0.99$ (at $P^* = 0.99$ the significance of the main effects 1, 3 and 5 is highest).

$F_{4,\infty} = 4.94$ is significant at $\alpha_C = 0.001$. So elimination of 1, 3 and 5 would give bad fit!. Elimination of only factors 1 and 5 (α_F = 0.01) gives lack of fit $F_{3,\infty} = 2.86$ at $P^* = 0.99$ and this F is significant only at $\alpha_C = 0.05$ so that we may very well conclude that factors 1 and 5 have no influence.

We have to realize that an effect may be significantly different from zero but nevertheless unimportant. For, if a regression coefficient is different from zero but the overall mean β_0 is high (compare "overprotection") then it is quite well possible that the expected fraction of correct selections, $E(\underline{y})$ in (75), is still not smaller than the required probability P^*. We conjecture that such a phenomenon indeed exists in our experiment for, since all factors except one (or two) give significant effects we would expect that nearly all 16 factor combinations would yield estimated fractions significantly below the corresponding P^*. Actually we saw in Section 8 that the MRP worked most of the times. So it seems incorrect to conclude from the regression coefficients that all factors except factor 1 (and 5) are "important". We conjecture that primarily factors 2 and 6 make the MRP break down. The estimated regression coefficients are yet useful since they can guide us in future research of the MRP, as we shall see in Section 10, but first we briefly discuss the multi-level factor P^*.

The factor P^* exercises its influence in a special way. As P^* increases the fraction of correct selections, $\underline{y}$, is expected to increase too, even if the MRP does not guarantee probability requirement (1). For as P^* increases the MRP will require more observations so that $P(CS) = E(\underline{y})$ increases. So it does not make sense to measure the effect of P^* as, say,

$$E\left(\frac{\Sigma_{i=1}^{16} \underline{y}_{i\ell}}{16}\right) - E(\bar{\underline{y}}) = E(\bar{\underline{y}}_{\ell}) - E(\bar{\underline{y}}) \quad (\ell = 1, \ldots, 22) \quad (93)$$

where $\bar{\underline{y}}$ is the average of the $\bar{\underline{y}}_{\ell}$. The factor P^* is important if at some levels of P^* the MRP works and at other levels it does not work, i.e., at some levels ℓ

$$E(\underline{y}_{i\ell}) \geq P_{\ell}^{*} \tag{94}$$

holds and at some levels (94) does not hold. Applying the t-tests as in Section 8 shows that restricting attention to the high levels of P^*, viz., $P^* = 0.90(0.01)0.99$, leads to rejection of (94) and so do the low levels of P^*; compare Table 12. Since the most relevant levels of P^* are the high values we may restrict attention to these high values in future experiments. (A more detailed analysis of the influence of P^* could be made by expressing $\underline{y}_{i\ell}$ or $(\underline{y}_{i\ell}-P^*)/(1-P^*)$ as a regression equation in the quantitative variable P_{ℓ}^*; such an equation can be fitted for each of the sixteen combinations in the 2^{7-3} design. If P^* does not interact with the seven two-level factors then all regression equations have the same coefficients except for their grand mean β_0. In the analysis of such equations we have to allow for the dependence among the observations per equation.)

VI.10. CONCLUSIONS AND FUTURE RESEARCH

From Section VI.8 we know that the MRP does not guarantee the probability requirement in (1) always so that some factors must be important. Hence we like to know for <u>which</u> factor combinations the procedure breaks down. The analysis in Section VI.9 could not reveal exactly which factors are really important and which are not important. For the regression analysis (or ANOVA) showed that at the usual error rate values all effects are significantly different from zero (except for β_1 and $\beta_{16} + \beta_{24} + \beta_{37}$). But as we explained some of these effects may nevertheless be unimportant. The binomial tests on rejected combinations suggested that the factors 2 and 6 and their interaction are most important, but definitive conclusions were not possible. Therefore in a next experiment we may investigate all 2^7 factor combinations, collect the rejected combinations and analyze them with binomial tests as in Section VI.9, hoping that the increased number of combinations will improve the power of the tests so that definitive conclusions will be possible. Remember, however, that generating data for 2^{7-3} combinations took one hour on the IBM 360/75 so that the additional $2^7 - 2^4$ experiment would require seven more hours of computer time and this seems a prohibitive amount of computer time. Therefore we may predict the fraction of correct selections at each of the $2^7 - 2^4$ combinations. These predictions can be based on the regression equations of Section VI.9; compare (82) above. Unfortunately as the analysis in Section VI.9 showed, the two-factor interactions are significant so that the 2^{7-3}_{IV} experiment does not yield unique estimators of the individual interactions. For $P^* < 0.90$ the fitted regression equations give no good fit. So in a possible next stage of experimentation it is advisable to study only $P^* \geq 0.90$. Moreover we should eliminate factor 1 since this factor was insignificant at any reasonable error rate. Additional elimination of factor 5 is correct at an α_F as low as 1% and may be contemplated in order to save computer time in the next stage. Additional elimination of a third factor, viz., factor 3, is not justified since the fitted regressions give no good fit in that case. If we eliminate less than three factors then some two-factor interactions remain confounded so that additional experimentation is indeed necessary (unless we simply assume that particular two-factor interactions are zero but this seems a very arbitrary procedure). If we want our conclusions to be valid for all values of k (number of populations) and all $\delta = \delta^*$ (distance) then we may perform experiments only at the unfavorable levels of these two factors, i.e., k is "high" (say $k = 7$) and $\delta = \delta^*$ is "small" (say $0.20\{\sigma_{(k)} + \sigma_{(k-1)}\}/2$) so that they are no factors anymore in the experimental design. For the choice of the design in the next stage we refer to the discussion of resolution V designs and Rechtschaffner's saturated designs in Section IV.5 and to sequentialization in Section IV.6. We can test if a predicted fraction deviates from P^* significantly using a t-test with

$$\hat{\underline{\text{var}}}(\hat{y}) = \hat{\underline{\text{var}}}\left(\sum_j \hat{\underline{\gamma}}_j x_{ij}\right)$$

$$= \sum_j \hat{\underline{\text{var}}}(\hat{\underline{\gamma}}_j) \qquad (95)$$

where the last equality holds if the estimated regression coefficients are orthogonal (and x_{ij} is ± 1).

Besides the probability of CS we may investigate the required sample sizes as a function of k, δ^* and P^* and investigate other configurations of the means (e.g. EMC and MFC). Other ranking procedures can be studied in the same way in order to find a robust and efficient procedure for simulation experiments.

APPENDIX VI.1. COMMON RANDOM NUMBERS AND THE ESTIMATION OF FACTOR EFFECTS

The main effect of factor j, say α_j, is estimated by

$$\underline{\hat{\alpha}}_j = 2 \sum_{i=1}^{N} x_{ij}\, \underline{y}_i / N \qquad (j = 1, 2, \ldots, J) \tag{1.1}$$

where x_{ij} is $+1$ or -1 if factor j is at its high level or low level respectively in factor combination i, and $\underline{y}_i$ is the response of combination i $(i = 1, \ldots, N)$; see Chapter IV. From (1.1) it follows that

$$\operatorname{var}(\underline{\hat{\alpha}}_j) = \frac{4}{N^2} \operatorname{var}\Big(\sum_i x_{ij}\, \underline{y}_i\Big)$$

$$= \frac{4}{N^2} \Big\{ \sum_i x_{ij}^2 \operatorname{var}(\underline{y}_i) + \sum_{i \neq i'}\sum x_{ij}\, x_{i'j} \operatorname{cov}(\underline{y}_i, \underline{y}_{i'}) \Big\} \tag{1.2}$$

In the design $x_{ij} = +1$ in half the combinations and $x_{ij} = -1$ in the other half. Hence

$$x_{ij}^2 = +1 \qquad (\text{all } i \text{ and } j) \tag{1.3}$$

and

$$\sum_{i \neq i'}\sum x_{ij} x_{i'j} = \{(N/2)(-1) + (N/2-1)(+1)\} N = -N \tag{1.4}$$

Assume that the covariances between the responses of the combinations are the same, say γ, i.e.,

$$\text{cov}(\underline{y}_i, \underline{y}_{i'}) = \gamma \qquad (i \neq i') \tag{1.5}$$

and the variances are constant, i.e.,

$$\text{var}(\underline{y}_i) = \sigma_y^2 \qquad (i = 1, \ldots, N) \tag{1.6}$$

Substitution of (1.3) through (1.6) into (1.2) gives

$$\text{var}(\hat{\underline{\alpha}}_j) = \frac{4}{N} \sigma_y^2 (1 - \rho) \tag{1.7}$$

where ρ is the constant correlation coefficient for the response $\underline{y}_i$ and $\underline{y}_{i'}$ $(i \neq i')$. To study the sign of ρ we consider

$$\text{cov}(\underline{y}_i, \underline{y}_{i'}) = \text{cov}\left(\frac{\Sigma_{r=1}^n \underline{y}_{ir}}{n}, \frac{\Sigma_{r=1}^n \underline{y}_{i'r}}{n}\right) \tag{1.8}$$

where $\underline{y}_{ir}$ is the response in replication r of combination i (i.e., $\underline{y}_{ir}$ is 0 or 1 corresponding with a failure or success of the MRP in replication r of combination i). So

$$\text{cov}(\underline{y}_i, \underline{y}_{i'}) = \frac{1}{n^2} \sum_r \sum_{r'} \text{cov}(\underline{y}_{ir}, \underline{y}_{i'r'}) \tag{1.9}$$

where

$$\begin{aligned} \text{cov}(y_{ir}, y_{i'r'}) &= 0 \qquad \text{for } r \neq r' \\ &> 0 \qquad \text{for } r = r' \end{aligned} \tag{1.10}$$

so that $\text{cov}(\underline{y}_i, \underline{y}_{i'})$ and therefore ρ is positive.

APPENDIX VI.2. SIMPLE AND GENERALIZED LEAST SQUARES

Because the assumptions of equal variances (per regression equation) and a single regression equation are violated in our experiment, we consider the following three types of estimators.

(i) Simple least squares per regression equation, i.e., we neglect the heteroscedasticity and the existence of more than one equation. The best linear unbiased estimator (BLUE) is then the simple or ordinary least squares (OLS) estimator. The regression equations of (76) are in matrix notation

$$\vec{\underline{Y}}_\ell = \vec{X}_\ell \vec{\Gamma}_\ell + \vec{\underline{E}}_\ell \qquad (\ell = 1, \ldots, 22) \tag{2.1}$$

$\vec{\underline{Y}}_\ell$ being the vector of observations on the dependent variables, i.e.,

$$\vec{\underline{Y}}'_\ell = (\underline{y}_{1\ell}, \ldots, \underline{y}_{16\ell}) \tag{2.2}$$

$\vec{X}_\ell$ being the matrix of values of the independent variables

$$\vec{X}_\ell = \{x_{ij}\} \qquad (i = 1, \ldots, 16)\ (j = 1, \ldots, 15) \tag{2.3}$$

consisting of orthogonal columns as (73) shows (actually $\vec{X}_\ell$ does not vary with ℓ); $\vec{\Gamma}_\ell$ being the vector of regression coefficients

$$\vec{\Gamma}_\ell = (\gamma_{1\ell}, \ldots, \gamma_{15\ell}) \tag{2.4}$$

specified in (70) through (72); $\vec{\underline{E}}_\ell$ being the vector of error terms

$$\vec{\underline{E}}'_\ell = (\underline{e}_{1\ell}, \ldots, \underline{e}_{16\ell}) \tag{2.5}$$

From (2.1) the OLS estimators of Γ_ℓ follow:

$$\hat{\vec{\underline{\Gamma}}}_\ell = (\vec{X}'_\ell \vec{X}_\ell)^{-1} \vec{X}'_\ell \vec{\underline{Y}}_\ell \tag{2.6}$$

Because of (73) and (74) (2.6) reduces to

$$\hat{\underline{\gamma}}_{j\ell} = \sum_{i=1}^{16} x_{ij}\, \underline{y}_{i\ell}/16 \tag{2.7}$$

(ii) Generalized least squares per regression equation, i.e., we estimate the regression coefficients per equation taking into account that the covariance matrix of the error terms (of one particular equation) is not a matrix $_{\ell}\sigma^2 \vec{I}$. In our experiment this covariance matrix, say $\vec{\Sigma}_{\ell}$, is a diagonal matrix as specified in (2.8) and (2.9)

$$\vec{\Sigma}_{\ell} = \{_{\ell}\sigma_{ii'}\} \qquad (i, i' = 1, \ldots, 16) \tag{2.8}$$

where

$$\begin{aligned} _{\ell}\sigma_{ii'} &= 0 \qquad \text{for } i \neq i' \\ &= {_{\ell}\sigma_i^2} \qquad \text{for } i = i' \end{aligned} \tag{2.9}$$

From Johnston (1963, p. 183) it follows that the BLUE of Γ_{ℓ} is now:

$$\underline{\hat{\vec{\Gamma}}}^{*}_{\ell} = (\vec{X}'_{\ell}\ \vec{\Sigma}^{-1}_{\ell}\ \vec{X}_{\ell})^{-1}\ \vec{X}'_{\ell}\ \vec{\Sigma}^{-1}_{\ell}\ \underline{\vec{Y}}_{\ell} \tag{2.10}$$

Since we do not know $\vec{\Sigma}_{\ell}$ we can replace $\vec{\Sigma}_{\ell}$ in (2.10) by its estimator, say $\underline{\hat{\vec{\Sigma}}}_{\ell}$ which is a diagonal matrix with on its diagonal the elements

$$_{\ell}\underline{\hat{\sigma}}_i = \underline{s}^2_{i\ell} \tag{2.11}$$

where the $\underline{s}^2_{i\ell}$ are estimators defined as in (51). Substitution of $\underline{\hat{\vec{\Sigma}}}_{\ell}$ for $\vec{\Sigma}_{\ell}$ in (2.10) yields estimators that are not necessarily BLUE.

(iii) Generalized least squares for the whole set of regression equations, i.e., we take into account heteroscedasticity and the existence of more than one regression equation. Following Zellner (1962) we can write the various regression equations in the single formula (2.12), that corresponds with formula (2.1) for a one-equation problem.

$$\vec{\underline{Y}} = \vec{X}\,\vec{\Gamma} + \vec{\underline{E}} \tag{2.12}$$

$\vec{\underline{Y}}$ being a vector of 22 subvectors or 22 × 16 elements, namely

$$\vec{\underline{Y}}' = (\vec{\underline{Y}}_1, \ldots, \vec{\underline{Y}}_{22}) \tag{2.13}$$

$\vec{X}$ being a matrix "diagonal" in its submatrices $\vec{X}_\ell$:

$$\vec{X} \begin{bmatrix} \vec{X}_1 & \vec{0} & \ldots & \vec{0} \\ \vec{0} & \vec{X}_2 & \ldots & \vec{0} \\ \vec{0} & \vec{0} & \ldots & \vec{X}_{22} \end{bmatrix} \tag{2.14}$$

where $\vec{0}$ is a 16 × 15 matrix of zero's (the submatrices $\vec{X}_\ell$ are actually identical); $\vec{\Gamma}$ and $\vec{E}$ are defined analogous to $\vec{\underline{Y}}$, i.e.,

$$\vec{\Gamma}' = (\vec{\Gamma}_1, \ldots, \vec{\Gamma}_{22}) \tag{2.15}$$

$$\vec{\underline{E}}' = (\vec{\underline{E}}_1, \ldots, \vec{\underline{E}}_{22}) \tag{2.16}$$

The covariance matrix, say $\vec{\Sigma}$, of the error terms in (2.16) is a symmetric matrix specified in (2.17)

$$\vec{\Sigma} = \begin{bmatrix} \vec{\Sigma}_1 & \vec{\Sigma}_{1,2} & \cdots & \vec{\Sigma}_{1,22} \\ \vec{\Sigma}_{2,1} & \vec{\Sigma}_2 & \cdots & \vec{\Sigma}_{2,22} \\ \vec{\Sigma}_{22,1} & \vec{\Sigma}_{22,2} & \cdots & \vec{\Sigma}_{22} \end{bmatrix} \tag{2.17}$$

where the matrices $\vec{\Sigma}_\ell$ on the "diagonal" are the 16 × 16 diagonal covariance matrices of the individual regression equations specified in (2.8) and (2.9), and the "off-diagonal" matrices $\vec{\Sigma}_{\ell\ell'}$ $(\ell \neq \ell')$ are 16 × 16 diagonal matrices of covariances because of (53). Since (2.12) is completely analogous to (2.1) but now showing the covariance matrix (2.17) the generalized least squares estimators for the set of regression equations are

$$\hat{\vec{\underline{\Gamma}}} = (\vec{X}' \vec{\Sigma}^{-1} \vec{X})^{-1} \vec{X}' \vec{\Sigma}^{-1} \vec{\underline{Y}} \tag{2.18}$$

Since $\vec{\Sigma}$ is unknown we have to estimate the variances and covariances in it. Its main diagonal elements are estimated by (2.11). Its covariances were not obtained in the (original) experiment but can be estimated from the covariances between the individual observations that also yielded estimates of the variances. For let $\underline{v}_{i\ell r'}$ denote replication r' ($r' = 1, \ldots, 2R$) of factor combination i ($i = 1, \ldots, 16$) for level ℓ ($\ell = 1, \ldots, 22$) of P^* and let $\tilde{\underline{v}}_{i\ell r}$ denote the average of two antithetic replications ($r = 1, \ldots, R$). We then have

$$\begin{aligned} \text{cov}(\underline{y}_{i\ell}, \underline{y}_{i\ell'}) &= \text{cov}\left(\frac{\Sigma_r^R \tilde{\underline{v}}_{i\ell r}}{R}, \frac{\Sigma_r^R \tilde{\underline{v}}_{i\ell' r}}{R} \right) \\ &= \frac{1}{R^2} \sum_{r=1}^{R} \text{cov}(\tilde{\underline{v}}_{i\ell r}, \tilde{\underline{v}}_{i\ell' r}) \\ &= \text{cov}(\tilde{\underline{v}}_{i\ell}, \tilde{\underline{v}}_{i\ell'})/R \qquad \ell \neq \ell' \end{aligned} \tag{2.19}$$

where we can estimate $\text{cov}(\tilde{\underline{v}}_{i\ell}, \tilde{\underline{v}}_{i\ell'})$ by

$$\begin{aligned} \hat{\underline{\text{cov}}}(\tilde{\underline{v}}_{i\ell}, \tilde{\underline{v}}_{i\ell'}) &= \sum_{r=1}^{R} (\tilde{\underline{v}}_{i\ell r} - \tilde{\underline{v}}_{i\ell})(\tilde{\underline{v}}_{i\ell' r} - \bar{\underline{v}}_{i\ell'})/(R - 1) \\ &= \left\{ \sum_{r=1}^{R} \tilde{\underline{v}}_{i\ell r} \tilde{\underline{v}}_{i\ell' r} - R\, \bar{\underline{v}}_{i\ell} \bar{\underline{v}}_{i\ell'} \right\} /(R - 1) \end{aligned} \tag{2.20}$$

with $\underline{v}_{i\ell} = \underline{y}_{i\ell}$. (The last equality in (2.20) may be convenient for computer calculations.) We observe that Zellner (1962, p. 350) assumed that within each diagonal submatrix of the covariance matrix $\vec{\Sigma}$ in (2.17), the diagonal elements are the same. (This implies homoscedasticity per equation.) He estimates the diagonal elements in each submatrix from the deviations between the actual values of the dependent variables and the values predicted by the OLS regression equations. We, however, estimate these elements from replications for the same factor combination.

EXERCISES

1. Prove that the variable defined in (7) has mean 0.9/b and variance $1.01/b^2$; derive its density function.

2. Show that in a 2_{IV}^{7-3} design all two-factor interactions involving only four different factors can be kept unconfounded.

3. Derive factor combination 6 in Table 6.

4. Derive a formula for the number of replications if we want to estimate P(CS) within 10% with 90% certainty.

5. What is wrong with the following proof of the unbiasedness of the GLS estimators with estimated covariance matrix?

$$E[(\vec{X}'\underline{\hat{\vec{\Sigma}}}^{-1}\vec{X})^{-1}\vec{X}'\underline{\hat{\vec{\Sigma}}}^{-1}(\vec{X}\,\vec{B} + \underline{\vec{E}})$$

$$= E[\vec{I}\,\vec{B} + (\vec{X}'\underline{\hat{\vec{\Sigma}}}^{-1}\vec{X})^{-1}\,\vec{X}'\underline{\hat{\vec{\Sigma}}}^{-1}(\vec{X}\,\vec{X}')(\vec{X}\,\vec{X}')^{-1}\underline{\vec{E}}]$$

$$= E[\vec{B} + \vec{I}\,\vec{X}'(\vec{X}\,\vec{X}')^{-1}\underline{\vec{E}}] = \vec{B}$$

6. Show that ANOVA and regression analysis do not use the individual replications except for the calculation of the standard deviation.

7. Assume that in (53) $\text{cov}(\underline{y}_{i\ell}, y_{i\ell'}) = \sigma_{\ell\ell'}$ (independent of i). Prove that in OLS $\text{cov}(\underline{\hat{\gamma}}_{j\ell}, \underline{\hat{\gamma}}_{j\ell'}) = \sigma_{\ell\ell'}/16$ and $\text{cov}(\underline{\hat{\gamma}}_{j\ell}, \underline{\hat{\gamma}}_{j'\ell'}) = 0$ for $j \neq j'$.

8. How can computer rounding errors be reduced in a simple way when estimating the regression coefficients $\gamma_{j\ell}$? (Hint: do not use the original output y.)

9. Show that in (84) ${}_{\ell}\underline{s}_D^2 = \sigma_y^2\,\underline{\chi}_v^2/v$ with $v = 16 \times 399$ holds and that ${}_{\ell}\underline{s}_D^2$ is independent of ${}_{\ell}\underline{S}_R$.

10. Show that if three factors were found to have no effect in the 2_{IV}^{7-3} experiment then all interactions could be estimated without additional experimentation. How can $\vec{17} + \vec{45} + \vec{36}$ in Table 18 be interpreted if we eliminate the factors 1, 3, and 5?

11. Why does $\hat{\beta}_0 - P^*$ underestimate the overprotection of the MRP?

NOTES

1. This chapter is based on two preliminary reports of Kleijnen (1969a, 1969b); a summary was presented in Kleijnen (1972). Professor D. Burdick (Duke University) gave helpful comments on the design of the experiment, and Professor G. Shorack (University of Washington, temporarily Mathematisch Centrum Amsterdam) commented on the analysis. Computer time for the data generation was made available by Professor T. Naylor (Duke University; IBM 360/75); these data were analyzed at the Katholieke Hogeschool Tilburg (IBM 1620). Additional experimentation was made possible by Professor H. Lombaers (Technische Hogeschool Delft; IBM 360/65). The Fortran program for the data generation was written by Mr. B. Fitzgerald (Duke University).

2. $\sigma_{(k)}/\max \sigma_{(j)} > 1$ means that the best population also has the highest standard deviation. If the experimenter would know that $\sigma_i = a\mu_i$ he could use a logarithmic transformation to obtain a common variance; see Table 4 in Chapter IV.

3. If the replications would be independent we could estimate the variance of their average from the binomial formula $\hat{\underline{p}}(1 - \hat{\underline{p}})/n$. The actual estimated variances are lower.

4. Note that computer time can also be saved not by a statistical technique like antithetics but by a more efficient numerical technique. For instance, instead of using a logarithmic transformation as in (48) we may apply a linear approximation to generate approximately exponentially distributed variables.

5. $\text{cov}(\underline{y}_{i\ell}, \underline{y}_{i\ell'}) = \Sigma_r^{800} \Sigma_{r'}^{800} \text{cov}(\underline{v}_{i\ell r}, \underline{v}_{i\ell' r'})/800^2 = (c_1+c_2)/800$ where c_1 denotes the negative covariance between replications $2r''$ and $2r'' - 1$ $(r'' = 1, \ldots, 400)$ of levels ℓ and ℓ', respectively, and c_2 denotes the positive covariance between replication r of levels ℓ and ℓ', respectively. Compare the discussion of the possible conflict between antithetic and common randon numbers in Chapter III.

6. Viz., z = 0.000(0.001)2.500(0.002)3.400(0.005)4.00(0.01)6.00.

7. The t-values were obtained by feeding in the (card) output of the 2_{IV}^{7-3} design for 22 P^* levels shown in Table 8, as input for an analysis computer program.

8. Suppose that the only important factor is factor 1 so that all combinations with factor 1 at its plus level are rejected. Then in the eight rejected combinations factor 2 is four times at its plus level and four times at its minus level.

9. Literature: Zellner (1962) and (1963), Zellner and Huang (1962), Kakwani (1967), Parks (1967), Kmenta and Gilbert (1968), Plasmans (1970), Plasmans and van Straelen (1970), and Rao and Mitra (1971, pp. 155-167).

10. Our presentation differs slightly from Section IV.6 where $\underline{S}_R$ is based on all observations including duplications.

REFERENCES

Andrews, D.F., P.J. Bickel, F.R. Hampel, P.J. Huber, W.H. Rogers, and J.W. Tukey (1972). Robust Estimates of Location. Princeton University Press, Princeton, N.J.

Bechhofer, R.E. and S. Blumenthal (1962). "A sequential multiple-decision procedure for selecting the best one of several normal populations with a common unknown variance, II: Monte Carlo sampling results and new computing formulae," Biometrics, 18, 52-67.

Bechhofer, R.E., J. Kiefer and M. Sobel (1968). Sequential Identification and Ranking Procedures, the University of Chicago Press, Chicago, Ill.

Box, G.E.P. (1954a). "Some theorems on quadratic forms applied in the study of analysis of variance problems: I. Effect of inequality of variance in the one-way classification," Ann. Math. Stat., 25, 290-302.

Box, G.E.P. (1954b). "Some theorems on quadratic forms applied in the study of analysis of variance problems: II. Effect of inequality of variance and of correlation of errors in the two-way classification," Ann. Math. Stat., 25, 484-498.

Dixon, W.J. and F.J. Massey (1957). Introduction to Statistical Analysis, 2nd ed., McGraw-Hill, New York.

Donaldson, T.S. (1966). Power of the F-Test for Nonnormal Distributions and Unequal Error Variances, RM-5072-PR, The Rand Corp., Santa Monica, Calif.

Dunn, O.J. (1964). "Multiple comparisons using rank sums," Technometrics, 6, 241-252.

Finney, D.J. (1952). Statistical Method in Biological Assay, Griffin, London.

Fisz, M. (1967). Probability Theory and Mathematical Statistics, Wiley, New York.

Johnston, J. (1963). Econometric Methods. McGraw-Hill, New York.

Kakwani, N.C. (1967). "The unbiasedness of Zellner's seemingly unrelated regression equations estimators," J. Amer. Stat. Assoc., 62, 141-142.

Kendall, M.G. and W.R. Buckland (1971). A Dictionary of Statistical Terms; 3rd ed., Oliver and Boyd, Edinburgh.

Kendall, M.G. and A. Stuart (1961). The Advanced Theory of Statistics, Vol. 2, Griffin, London.

Kendall, M.G. and A. Stuart (1963). The Advanced Theory of Statistics, Vol. 1, 2nd ed., Griffin, London.

Kleijnen, J.P.C. (1969a). The Use of Multiple Ranking Procedures to Analyze Simulations of Business and Economic Systems II: The Design of a Monte Carlo Experiment for the Evaluation of Multiple Ranking Procedures, Econometric System Simulation Program Working Paper, Duke University, Durham, N.C.

Kleijnen, J.P.C. (1969b). The Design and Analysis of a Monte Carlo Experiment with a Multiple Ranking Procedure, Katholieke Hogeschool, Tilburg (Netherlands).

Kleijnen, J.P.C. (1972). The Design and Analysis of Monte Carlo Experiments: A Case Study on the Robustness of a Multiple Ranking Procedure, in Working papers, Vol. 1, Symposium Computer simulation versus analytical solutions for business and economic models, Graduate School of Business Administration, Gothenburg (Sweden).

Kmenta, J. and R.F. Gilbert (1968). "Small sample properties of alternative estimators of seemingly unrelated regressions," J. Amer. Stat. Assoc., 63, 1180-1200.

Lewis, P.A.W., A.S. Goodman, and J.M. Miller (1969). "A pseudo-random number generator for the System/360," IBM Systems J., 8, 136-146.

Mood, A.M. and F.A. Graybill (1963). Introduction to the Theory of Statistics, 2nd ed., McGraw-Hill, New York.

Neave, H.R. and C.W.J. Granger (1968). "A Monte Carlo study comparing various two-sample tests for differences in mean," Technometrics, 10, 509-522.

Parks, R.W. (1967). "Efficient estimation of a system of regression equations when disturbances are both serially and contemporaneously correlated," J. Amer. Stat. Assoc., 62, 500-509.

Plasmans, J. (1970). The General Linear Seemingly Unrelated Regression Problem, I: Models and Inference. EIT 18, Tilburg Institute of Economics, Department of Econometrics, Tilburg (Netherlands).

Plasmans, J. and R. van Straelen (1970). The General Linear Seemingly Unrelated Regression Problem, II: Feasible Statistical Estimation and an Application, EIT 19, Tilburg Institute of Economics, Department of Econometrics, Tilburg (Netherlands).

Rao, C.R. and S.K. Mitra (1971). Generalized Inverse of Matrices and its Applications, Wiley, New York.

Rogers, W.H. and J.W. Tukey (1972). "Understanding some long-tailed symmetrical distributions," Stat. Neerl., 26, 211-226.

Roy, S.N. (1957). Some Aspects of Multivariate Analysis, Wiley, New York.

Scheffé, H. (1964). The Analysis of Variance, Wiley, New York.

Smirnov, N.V. (ed.) (1965). Tables of the Normal Probability Integral, the Normal Density, and its Normalized Derivatives, Pergamon Press, Oxford.

Teichroew, D. (1965). "A history of distribution sampling prior to the era of the computer and its relevance to simulation," J. Amer. Stat. Assoc., 60, 27-49.

Zellner, A. (1962). "An efficient method of estimating seemingly unrelated regressions and tests for aggregation bias," J. Amer. Stat. Assoc., 57, 348-368.

Zellner, A. (1963). "Estimators for seemingly unrelated regression equations: some exact finite sample results," J. Amer. Stat. Assoc., 58, 977-992.

Zellner, A. and D.S. Huang (1962). "Further properties of efficient estimators for seemingly unrelated regression equations," Int. Econ. Rev., 3, 300-313.

SOLUTIONS TO EXERCISES

Chapter I

1. After the modulo operation m integers remain between 0 and (m-1); so no more than m different numbers can result. In Fibonnaci the same two numbers must be reproduced.

2. Sample $\underline{x}$ from a normal distribution with $E(\underline{x}) = b$ and $var(\underline{x}) = c^2$. Substitute $\underline{x}$ into $g(\underline{x}) = c\sqrt{2\pi}/\ln x$ for $\underline{x} \geq a$ and $g(\underline{x}) = 0$ for $\underline{x} \leq a$.

3. Generate AT and ST; calculate DIF = WT + ST - AT; if DT > 0 then WT = DIF, or else IT = -DIF.

4. $P(\underline{x} < a) = 1 - P(\underline{x} \geq a) = 1 - P(\underline{x}_1 \geq a)\, P(\underline{x}_2 \geq a)$, etc. For $a = 83$, $p = 0.39$.

5. Move "SG2 = 1" from below connector 2 to above connector 2, and add "SG2 = 1" in the right hand side.

6. Faster on the average.

7. From $N(0, \hat{\sigma}^2)$ with $\underline{\hat{\sigma}}^2 = \Sigma_1^n (\underline{y}_i - \hat{\underline{y}}_i)^2/(n-2)$. Do not sample $\hat{\alpha}$ and $\hat{\beta}$ (sensitivity analysis may be useful).

Chapter II

1. $var[(\bar{\underline{x}}_1 + \bar{\underline{x}}_2)/2] = [var(\bar{\underline{x}}_1) + var(\bar{\underline{x}}_2) + 2\, cov(\bar{\underline{x}}_1, \bar{\underline{x}}_2)]/4$. Last term zero in two runs, positive in prolonged run.

2. Antithetic variates.

3. Common random numbers. Smaller $\sigma(\bar{\underline{x}}_1 - \bar{\underline{x}}_2)$. Use t-test for differences $\underline{d}_i = \underline{x}_{1i} - \underline{x}_{2i}$ $(i = 1, 2, 3, \ldots)$.

4. $(1 - \alpha)^2$. No, for $\hat{\underline{\mu}}_1 - \hat{\underline{\mu}}_2$ and $\hat{\underline{\mu}}_1 - \hat{\underline{\mu}}_3$ are dependent.

5. $a = \rho(\hat{\underline{\mu}}, \bar{\underline{y}})\, \sigma(\hat{\underline{\mu}})/\sigma(\bar{\underline{y}})$, where ρ denotes correlation coefficient.

6. Yes, formula applied only to estimate quantities at the end of the planning period [compare Naylor et al. (1967c)].

7.-10. Compare text of Chap. II.

Chapter III

1. $E\{[(\bar{y} - \eta) + (\eta - \mu)]^2\}$, etc.

2. $E(\bar{v}_k) = \Sigma\Sigma\, E(\bar{v}_k | n_k = i, n = j)\, P(n_k = i, n = j)$. Proceed as in Appendix III.2.

3. Procced as in Appendix III.4.

4. $E(\hat{\vec{A}}) = (\vec{Z}'\vec{Z})^{-1} \vec{Z}'\, E(\vec{X}) = (\vec{Z}'\vec{Z})^{-1} \vec{Z}'[\vec{Z}\,\vec{A} + E(\vec{U})] = \vec{A}$.

5. $J_1 = 2\hat{\theta} - \Sigma_1^M x_j/\Sigma_1^M y_j$, $J_2 = 2\hat{\theta} - \Sigma_{M+1}^n x_g/\Sigma_{M+1}^n y_g$ $(M = n/2)$,

   $\bar{J} = 2\hat{\theta} - (\Sigma\, x_j/\Sigma\, y_j + \Sigma\, x_g/\Sigma\, y_g)/2$.

6. Analogous to Appendix III.9.

7. Analogous to Appendix III.8.

8. $\int_{-\infty}^{\infty} \int_{-\infty}^{\infty} h_0(\vec{R})\, d\vec{R} = 1$

9. Symmetry: $F(y) = 1 - F(2\eta - y)$. Hence, $F(y^*) = 1 - r$ gives same results as $1 - F(2\eta - y^*) = 1 - r$ or $y^* = - F^{-1}(r) + 2\eta$.

10. $E[(y - \eta)(2\eta - y - \eta)] = \eta^2 - E(y^2) = \sigma_y^2$; $\rho = \sigma_y^2/\sigma_y\sigma_y = 1$

    Exponential: $\rho = \int_0^1 \ln(r)\ln(1 - r)\,dr - 1 = 1 - \pi^2/6 \approx - 0.645$.

    Correlation between x and $y = a + bx$ is 1 if $b > 0$; -1 if $b < 0$.

11. For $0 \le a \le c$, $P(r_2 \le a) = P(c - a \le r_1 \le c) = a$. For $c \le b \le 1$, $P(r_2 \le b) = b$.

12. $Cov(x,y) = E(cy) - E(c)\, E(y) = cE(y) - cE(y) = 0$.

13. Service time j may be used for customer j in FIFO and customer j' in LIFO ($j \neq j'$). Store service and arrival times of each customer for later use in LIFO.

14.-15. See text of Chap. III.

Chapter IV

1. Erlang family, yields, e.g. exponential for small parameter value.

2. $\underline{MSE} = \Sigma_{i=1}^{r_j} \Sigma_{j=1}^{J} (\underline{y}_{ij} - \underline{y}_{.j})^2 / \Sigma_j (r_j - 1)$. Substitute $r_j = r$.

3. Insignificant at 5%: main effect 1, interactions 13 = 25 = 67, 24 = 16 = 37. Good fit. See also Chap. VI.

4. (a) Substitute Eqs. (3.1) and (3.8) into Eq. (3.6). Rows 1 through (k+1) yield $(\vec{U}', -\vec{U}')\vec{\underline{y}} = 2\vec{U}'\vec{U}\,\vec{\underline{b}}_M$.
   (b) Apply: if $\vec{\underline{w}} = A\vec{\underline{v}}$ with covariance matrix Σ_v then $\Sigma_w = A\Sigma_v A'$. Here $A = 0.5(\vec{U}'\vec{U})^{-1}(\vec{U}', -\vec{U}')$.
   (c) Apply Eq. (177). Orthogonal design: $\text{var}(\underline{b}_1) = \sigma^2/N = \sigma^2/6 < \sigma^2/4$.
   (d) $\vec{\underline{b}}_M = \vec{A}\,\vec{y}$ [$\vec{A}$ defined in 4(b)] with $E(\vec{\underline{y}}) = \vec{X}\,\vec{\beta}$. Substitute Eq. (3.1) for $\vec{X}$ and find $\vec{A}\,\vec{X} = \vec{I}$.

5. Total number of important factors, say s, is kp. Total number of groups in stage i is $g_i s = p^{-1/(n+1)}\, kp$.

6.-8. See Chap. IV.

Chapter V.
Part V.A.

1. $\text{var}(\bar{\underline{x}}) = \Sigma_t \Sigma_{t'} \text{cov}(\underline{x}_t, \underline{x}_{t'})/N$ and use Eq. (3).

2. Substitute $\rho_s = \rho_{-s}$ and $\rho_0 = 1$.

3. $P\left(\frac{\underline{x}_1}{\underline{y}_2} < \frac{\mu_1}{\mu_2} < \frac{\underline{x}_2}{\underline{y}_1}\right) \geq 1 - 2\alpha$ [see Eq. (26)].

4. Use Bonferroni inequality as in Eq. (26). Alternative: Fit regression equation to all observations at λ_1 and λ_2, and test regression parameters; see Chap. IV.

5. Terminating systems.

6. 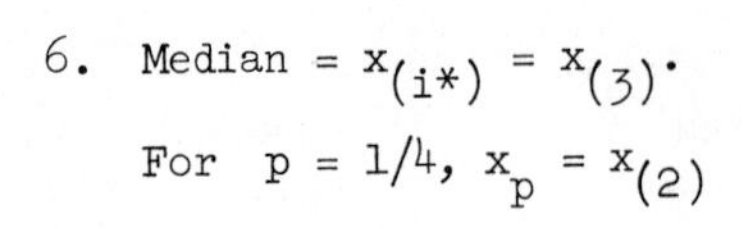
Median $= x_{(i*)} = x_{(3)}$.
For $p = 1/4$, $x_p = x_{(2)}$

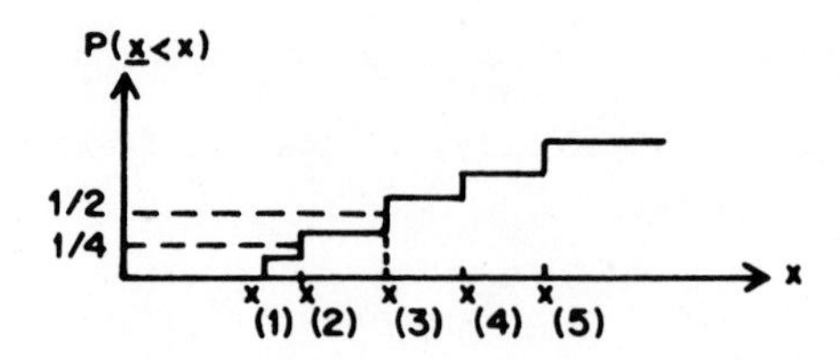

7. $n = 528$ (see Chap. VI, Table 7).

8. Solve $d[\mathrm{var}(\bar{\underline{x}}_1 - \bar{\underline{x}}_2)]/dn_1 = -\sigma_1^2/n_1^2 + \sigma_2^2/(n - n_1)^2 = 0$.

9. $(\sigma_1 - \sigma_2)^2 - 4\mathrm{cov}(\underline{x}_1, \underline{x}_2) < 0$ if $\sigma_1 = \sigma_2$.

10. Figure 2. Joint probability is $(0.5)^m$.

11. Substitute denominator of Eqs. (38), (39), and (104) into Eq. (102) and take squares.

12. Substitute Eq. (108) into $\Sigma_1^{n_1} (\underline{u}_i - \bar{\underline{u}})^2 = \Sigma(\underline{u}_i^2) - (\Sigma\, \underline{u}_i)^2/n_1$.

13. $\mu(c) = 8\, \Sigma_{D \leq c} DP(\underline{D} = D) - 2c.\ P(\underline{D} \leq c) + 6c.\ P(\underline{D} > c)$.

14. See text of Chap. V.A.

Chapter V.
Part V.B.

1. The subtraction term $P(\underline{S}_1 = 0 \wedge \underline{S}_2 = 0)$ is largest if $\underline{S}_1$ and $\underline{S}_2$ are positively correlated.

2. Put $\bar{\underline{x}}_i' = -\bar{\underline{x}}_i$ so that $\mu_i < \mu_0$ is equivalent to $\mu_i' > \mu_0'$.

3. $\underline{d}_{(2)} = -1$, $\underline{d}_{(5)} = 1$; hence $-1 \leq \mu_i - \mu_0 \leq 1$.

4. The variance of $\underline{d}_j$ is exactly known [see Eq. (42)].

5. a_i is exceeded by the $[1 - (1 - \alpha)^{1/k}]/2$ fraction of $\underline{t}$.

6. $P(\bar{\underline{x}}_i - \bar{\underline{x}}_0 < d_{k,v}^{\alpha}\, \underline{s}_v (\sqrt{2/n}) | \mu_i = \mu_0) = 1 - \alpha$ [see Eq. (17)]. Draw pictures for alternative hypotheses.

7. $\underline{s}_v^2 \, v/\sigma^2$ is χ^2-distributed; χ^2-distribution is a gamma distribution.

8. $\text{cov}(\hat{\underline{\alpha}}_1^A, \hat{\underline{\alpha}}_2^A) = \text{cov}(\underline{x}_{1.}, \underline{x}_{2.}) - \text{cov}(\underline{x}_{1.}, \underline{x}_{..}) - \text{cov}(\underline{x}_{..}, \underline{x}_2)$

$$+ \text{cov}(\underline{x}_{..}, \underline{x}_{..}) = \text{etc.}$$

9. $q = ab$ and $q = ab - 1$, respectively.

10. $\Sigma_{i=1}^a \alpha_{ib}^{AB} = (- \Sigma_{g=1}^{b-1} \alpha_{1g}^{AB}) + \cdots + (- \Sigma_{g=1}^{b-1} \alpha_{ag}^{AB})$, etc.

11. Eq. (67) yields $\max_j \hat{\underline{\alpha}}_j^2 \leq (m_{J,v}^{\alpha})^2 \, 4\hat{\underline{\sigma}}^2/N$; Eq. (99) in Chap. IV gives $\underline{SS}_j = N\hat{\underline{\alpha}}_j^2/4$.

12. Numerator of ρ is: $\text{cov}(\bar{\underline{x}}_i - \bar{\underline{x}}_0, \bar{\underline{x}}_{i'} - \bar{\underline{x}}_0) = \text{var}(\bar{\underline{x}}_0)$. Write out denominator.

13. (a) 25 Averages per level with variance $\sigma^2(1 + \rho)/2$. Degrees of freedom reduced from 5 (50 - 1) to 5 (25 - 1). Compare $S[\Sigma \, c_i^2/n_i)^{1/2} \, \hat{\underline{\sigma}}$.

14.-15. See text of Chap. V.B.

Chapter V.
Part V.C.

1. $n_1 = 12$, $n_2 = 17$, $n_3 = 25$.

2. $E(\underline{s}_0^2) = v^{-1} \, \Sigma \, a_i^{-1}(a_i n_0 - 1) \, \sigma_i^2 = v^{-1} \, \Sigma(a_i n_0 - 1) \, \sigma^2 = \sigma^2$.

3. $2s_0^2(h/\delta^*)^2 = 2(51{,}901) \, \{2.62/(\sqrt{2}.100)\}^2 = 36 > 50$. Enough runs.

4. $p_2 = 1 - P(\underline{x}_2 \leq \underline{x}_1) = 1 - \int_{-\infty}^{\infty} F_2(x) \, f_1(x) \, dx$ and $F_2(x) < F_1(x)$ for all x.

5. Use $F_i(x) = \Phi[(x - \mu_i)/\sigma]$ and $F_i(x) = 1 - \exp[-x/E(\underline{x})]$.

6. CS for factor A if $\underline{y}_j = \underline{x}_{(a).} - \underline{x}_{(j).} > 0$ for all $j = 1, \ldots, a-1$; for B if $\underline{z}_g = \underline{x}_{.(b)} - \underline{x}_{.(g)} > 0$ for $g = 1, \ldots, b-1$. Write out $\mathrm{cov}(\underline{y}_j, \underline{z}_g)$.

7. For same form $p_1 = p_2 = p$ (= 1 for exponentials). Different means if $p/b_1 \neq p/b_2$. Different b implies different forms.

8. Apply procedure to $\underline{x}_{ig} = -\ln \underline{s}^2_{ig}$ with $\delta^* = -\ln(0.9) \approx 0.11$.

9. $E(\bar{\underline{x}}^2) = E(\underline{x}^2)/m + [E(\underline{x})]^2 (m-1)/m$; $E(\underline{x}_i\bar{\underline{x}})$

$$E(\underline{x}_i\bar{\underline{x}}) = m^{-1}\{[E(\underline{x})]^2 (m-1) + E(\underline{x}^2)\}, \text{ etc.}$$

10. See text of Chap. V.C.

Chapter VI

1. $E(\underline{x}) = 1/b - (0.1)1/b$, $\mathrm{var}(\underline{x}) = 1/b^2 + (0.1)^2/b^2$. $f(x)$ in (8) can be derived applying, e.g. Eq. (2.4.19) in Fisz (1967).

2. See analogue of Eq. (20) involving $p = 3$ generators, identifying three factors with three three-factor interactions; three three-factor interactions require at least four different factors.

3. See Sec. VI.4.

4. $n = (z^{\alpha/2}/0.1)^2 (1 - P^*)/P^*$.

5. $(\vec{X}\vec{X}')^{-1}$ does not exist since $(\vec{X}\vec{X}')$ is singular.

6. See, e.g. the Scheffé (1964, p. 123) formulas for a design with three factors and M replications. Work out the Johnston (1963, p. 134) formulas for regression analysis.

7. $\vec{\Sigma}_\gamma = \vec{A}\vec{\Sigma}_y\vec{A}'$ where $\vec{A} = (\vec{X}'\vec{X})^{-1}\vec{X}'$ and $\vec{\Sigma}_y = \sigma_{\ell\ell'}\vec{I}$.

8. Calculate $\hat{\gamma}_{j\ell}$ from $(y_{i\ell} - P^*_\ell)$. Next $\hat{\gamma}_{0\ell}$ becomes $\hat{\gamma}_{0\ell} + P^*_\ell$.

9. $\underline{s}^2_{i\ell} = \chi^2_{399}\,\sigma^2_y/399$ with $\sigma^2_y = \sigma^2_{\underset{\sim}{v}}/400$ and independent of $\underline{y}_{i\ell}$.

10. See Eq. (20). $(\overrightarrow{17} + \overrightarrow{45} + \overrightarrow{36}) = (12\overrightarrow{34}) = (24\overrightarrow{67})$ = four-factor interaction among remaining factors.

11. Overprotection is $\Sigma_j\, \gamma_j\, x_{uj} - P^*$, where each x_j is at its favorable level, say x_{uj}, so that the basic assumptions of MRP are (nearly) satisfied.

AUTHOR INDEX

Underlined numbers give the page on which the complete reference is given.

C

T

SUBJECT INDEX

D

V

W

Y